Bulk Metallic Glasses

Michael Miller • Peter Liaw
Editors

Bulk Metallic Glasses

An Overview

 Springer

Edited by:

Michael Miller
Oak Ridge National Laboratory
Materials Science and Technology Division
Building 4500S, MS 6136
PO Box 2008
Oak Ridge, TN 37831-6136
USA

Peter Liaw
University of Tennessee
Dept. of Materials Science & Engineering
427B Dougherty Engineering Building
Knoxville, TN 37996-2200
USA

Library of Congress Control Number: 2007929110

ISBN 978-0-387-48920-9   e-ISBN 978-0-387-48921-6

Printed on acid-free paper.

9 8 7 6 5 4 3 2 1

springer.com

In memory of
Prof. Raymond A. Buchanan

CONTENTS

PREFACE

Natural glass has existed from the early days of the Earth and was formed from the rapid solidification of molten rock produced during volcanic eruptions, lightning strikes, and meteorite impacts. Phoenician merchants were aware of its existence in Syria from around 5000 BCE. Man-made glass objects from Egypt and Eastern Mesopotamia are thought to date back to around 3500 BCE.

In comparison, amorphous metals or metallic glasses are newcomers to the field of amorphous materials. Prior to the 1960s, some amorphous thin films were fabricated by metal deposition onto very cold substrates. In 1960 Klement, Willens, and Duwez at the California Institute of Technology reported the synthesis of an amorphous metal by rapidly quenching a Au-Si alloy from ~1,300°C to room temperature. A high cooling rate of ~10^6 K/s was required to bypass crystallization and this restricted the thickness of the sample to the micrometer range. In the 1960s, Chen and Turnbull developed amorphous alloys of Pd-Si-Ag, Pd-Si-Cu, and Pd-Si-Au. Chen also fabricated an amorphous Pd-Cu-Si alloy with a diameter of up to 1 mm that could be considered to be a bulk metallic glass. In 1974, Chen made systematic investigations on Pd-Si-, Pd-P-, and Pt-P-based alloys and obtained a critical casting diameter of 1-3 mm by quenching the melt, contained in a drawn fused quartz capillary, into water. In the early 1980s, Turnbull's group studied Pd-Ni-P alloys and they were able to produce glassy ingots of $Pd_{40}Ni_{40}P_{20}$ with diameters of 5 mm. In 1984, they extended the critical casting diameter to 10 mm by processing the Pd-Ni-P melt in a boron oxide flux. From the late 1980s, Inoue's group has discovered many new bulk metallic glasses in a variety of multicomponent alloy systems including the rare-earth-based systems that have cooling rates less than 100 K/s and thicknesses reaching several centimeters. These systems are discussed in detail in Chapter 1. In 1992, Johnson and Peker developed a pentary $Zr_{41.2}Cu_{12.5}Ni_{10}Ti_{13.8}Be_{22.5}$ metallic glass with a critical cooling rate of 1 K/s. This alloy became the first commercial bulk metallic glass and is known as Vitreloy 1. Over the last four decades, the critical casting thickness has been increased by more than three orders of magnitudes and amorphous components weighing several kilograms can be fabricated. To date, more than a thousand different bulk metallic glasses have been produced in Zr-, Fe-, Pd-, Ni-, Cu-, Mg-, and Ti-based systems.

Bulk metallic glasses are a new emerging field of materials with many desirable and unique properties, such as high strength, good hardness, good wear resistance, and high corrosion resistance that can be produced in near net shape components. These amorphous materials have many diverse applications from structural applications to microcomponents. Some unique applications of these novel materials such as pressure sensors, microgears for motors, magnetic cores for power supplies, and nano-dies for replicating next generation DVDs are documented in Chapter 1. An atomistic theory of local topological fluctuations is introduced in Chapter 2 to describe the atomistic movements in glasses and liquids. In this theory, topological fluctuations are represented by the atomic level stresses, and evolution of their distribution with temperature determines various thermal properties. This theory describes the glass transition, structural relaxation, glass formation and mechanical deformation, and the importance of Poisson's ratio. This theory promises to replace the free volume theory in elucidating the complex behaviors of metallic glasses. Atomistic simulations, including empirical potentials and *ab initio* calculations, are presented in Chapter 3. The application of classical nucleation theory and role of Molecular Dynamics simulations to bulk metallic glasses are discussed. In Chapter 4, glass formation, glass forming ability, and the underlying mechanisms and physical insights of these criteria are presented. The unique microstructures of these amorphous materials are discussed in Chapter 5. The state-of-the-art techniques (XRD, SEM, HREM, FEM, FIM, APT, SANS/SAXS, and PAS) that have been used to characterize the microstructures of these bulk metallic glasses from the as-produced to the crystallized material are described. The mechanical deformation of bulk metallic glasses including the topics of strength, plasticity (Poisson's ratio and solidity index), homogenous deformation, dynamic deformation, strain rate effects, strain hardening, and shear band nucleation and propagation, and the yielding criterion of metallic glasses are reviewed in Chapter 6. The structure of the shear band and the temperature associated with the propagation of a shear band are also reviewed. The fatigue, fracture and corrosion behaviors of these materials are also reviewed in Chapters 7 and 8.

This book is based on a short course that was organized by Peter Liaw and taught by several of the contributing authors at the Department of Materials Science and Engineering, The University of Tennessee in 2005 and 2006.

Michael K. Miller, Oak Ridge National Laboratory
Peter K. Liaw, The University of Tennessee

ACKNOWLEDGEMENTS

We would like to thank Kaye F. Russell and Dr. Gongyao Wang for their assistance in preparing this monograph. We would also like to thank Profs. A. L Greer, T. C. Huffnagel, C. A. Schuh, K. M. Flores and B. D. Wirth and Drs. M. P. Brady and G. S. Painter for their helpful suggestions on the manuscript.

We would also like to thank Greg Franklin and Caitlin Womersley of Springer for their assistance in producing this monograph.

Research at the Oak Ridge National Laboratory SHaRE User Facility was sponsored by Basic Energy Sciences, U.S. Department of Energy.

MKM acknowledges support from the SHaRE User Facility sponsored by Basic Energy Sciences, U. S. Department of Energy. PKL acknowledges support from the National Science Foundation (NSF) (1) the Combined Research and Curriculum Development (CRCD) Program (DGE-0203415), (2) Integrative Graduate Education and Research Training (IGERT) Program (DGE-9987548), and (3) the International Materials Institutes (IMI) Program (DMR-0231320) with Ms. M. Poats, Dr. C. V. Van Hartesveldt, and Dr. C. Huber, as the program directors, respectively.

.

CONTRIBUTING AUTHORS

Akihisa Inoue, Baolong Shen and Nobuyuki Nishiyama
*Institute for Materials Research, Tohoku University, Sendai 980-8577,
Japan*
*RIMCOF Tohoku University. Laboratory, R&D Institute of Metals
and Composites for Future Industries, Sendai 980-8577, Japan*

T. Egami
*Department of Materials Science and Engineering and Department
of Physics and Astronomy, The University of Tennessee, Knoxville, TN
37996, and Oak Ridge National Laboratory, Oak Ridge, TN 37831*

Rachel S. Aga and James R. Morris
*Materials Science and Technology Division, Oak Ridge National
Laboratory, Oak Ridge, TN 37831-6115*
*Department of Materials Science and Engineering, The University
of Tennessee, Knoxville, TN 37996-2200*

Z. P. Lu, Y. Liu and C. T. Liu
*Materials Science and Technology Division, Oak Ridge National
Laboratory, Oak Ridge, TN 37831-6115*
*Department of Materials Science and Engineering, The University
of Tennessee, Knoxville, TN 37996-2200*

M. K. Miller
*Materials Science and Technology Division, Oak Ridge National
Laboratory, Oak Ridge, TN 37831-6136*

T. G. Nieh
*Department of Materials Science and Engineering, The University
of Tennessee, Knoxville, TN 37996-2200*

Gongyao Wang and Peter K. Liaw
*Department of Materials Science and Engineering, The University
of Tennessee, Knoxville, TN 37996-2200*

**Brandice A. Green, Peter K. Liaw,
and Raymond A. Buchanan**
*Department of Materials Science and Engineering, The University
of Tennessee, Knoxville, TN 37996-2200*

Chapter 1

DEVELOPMENT AND APPLICATIONS OF LATE TRANSITION METAL BULK METALLIC GLASSES

Akihisa Inoue,[1] Baolong Shen,[1] and Nobuyuki Nishiyama[2]

[1]Institute for Materials Research, Tohoku University, Sendai 980-8577, Japan
[2]RIMCOF Tohoku University Laboratory, R&D Institute of Metals and Composites for Future Industries, Sendai 980-8577, Japan

1.1 INTRODUCTION

Bulk metallic glasses (BMGs) in metal–metal systems such as La-, Mg-, and Zr-based alloys were first prepared in the early 1990s by the stabilization of supercooled liquid.[1–4] Since then much effort has been devoted to the development of BMGs for both fundamental scientific research and for industrial applications. As a result, many unique and useful properties of BMGs have been found.[5–8] In particular, research at the Institute for Materials Research has been concentrated primarily on early transition metal (Zr-, Ti-, and Hf-based) systems, lanthanide metal (Ln-based) systems, simple metal (Mg- and Ca-based) systems, and noble metal (Pd- and Pt-based) systems.[5–8] Because of their excellent properties, BMGs are expected to emerge as a new type of industrial or engineering material. The development of late transition metal (LTM)-based BMGs is strongly encouraged due to material costs and the availability of raw material deposits. Therefore, an Fe-based BMG in the Fe–Al–Ga–P–C–B alloy system was successfully developed in 1995.[9] Also at that time, three empirical component rules for the stabilization of a supercooled metallic liquid were proposed.[5–7] These rules stated that (1) the multicomponent system should consist of three or more elements, (2) there should be a significant difference (greater than ~12%) in the atomic sizes of the main constituent elements, and (3) the elements should have negative heats of mixing. A variety of Fe-based,[10–14] Co-based,[15–17] Ni-based,[18–20] and Cu-based[21–25] BMGs have been synthesized in accordance with these rules and other topological and chemical criteria. As a result, various unique properties of LTM-based BMGs have been obtained. These properties have not been obtained in any crystalline alloys. Therefore, it should be possible to extend the range of applications. This chapter reviews recent results on the formation, properties, thermal stability, workability, and applications of LTM-based BMGs.

1.2   FEATURES OF ALLOY COMPONENTS
IN LTM-BASED BMGS

Since the first synthesis of LTM-based BMG containing more than 50% LTM in the Fe–(Al,Ga)–(P,C,B),[9] other systems including Co–Ga–(Cr,Mo)–(B,C,P),[15] Ni–Nb–(Zr,Ti,Hf)–(Co,Fe,Cu,Pd),[26] and Cu–(Zr,Hf)–Ti[21] BMGs were developed between 1996 and 2001. It is important to note that the research for LTM-based BMGs began just over a decade ago in 1995.

The typical BMG-forming systems containing more than 50% LTM as a main constituent element are summarized in Table 1.1. The systems can be classified into two different groups: metal–metalloid and metal–metal systems. The metal–metalloid group systems are primarily in Fe-, Co-, Ni-, Pd-, and Pt-based alloys, and metal–metal group systems are primarily in Ni- and Cu-based alloys. Only Ni-based alloys belong to both groups. For engineering applications, low-cost LTM-based BMGs with simplified composition that are easy to process should be preferred. To fit these requirements, metal–metal type BMGs such as Ni–Nb–(Ti,Zr,Hf), Cu–Ti–(Zr,Hf), and Cu–Al–(Zr,Hf) systems are more applicable than metal–metalloid types such as Fe–(Al,Ga)–metalloid, Fe–(Cr,Mo)–(C,B), Fe–(early transition metal)–B, Fe–Ln–B, and Fe–(B,Si)–Nb. Unfortunately, these metal–metal type BMGs generally exhibit lower glass-forming abilities than the metal–metalloid type. The critical diameter, D_{max}, for metal–metal type alloy systems is typically limited to 1.5 mm. To extend the range of applications, it will be necessary to optimize the composition and further to develop the metal–metal type BMGs.

The other classification is summarized in Table 1.2. All ternary or pseudo-ternary glass-forming systems can be divided into five groups. Group I consists of LTM, simple metal, and early transition metal as exemplified by Cu–Zr–Al and Cu–Hf–Al systems. Group II includes LTM, metalloid, and early transition metal or Ln such as Fe–(B,Si)–Nb, Fe–(Zr, Hf, Nb)–B, Fe–Ln–B, and Fe–(Cr,Mo)–(C,B) systems. Group III is composed of Fe, metalloid, and Al or Ga. Group IV is exemplified by Ni–Nb–Ti and Cu–(Zr,Hf)–Ti systems. Group V consists of Ni–Pd–P and Cu–Pt–P systems. As evident in Table 1.2, all the glass-forming systems classified into different groups belong to ternary or pseudoternary alloys, which are composed of three types of elements with different atomic radii. From this, it is concluded that the stabilization of supercooled liquid is dominated by the atomic size mismatch rather than the negative heat of mixing between the constituent elements.

Table 1.1. Typical bulk metallic glass systems in late transition metal (LTM) base containing more than 50 at.% LTM reported to date[49] (courtesy Japan Institute of Metals)

Base metal	Metal–metalloid	Metal–metal
Fe	Fe–(Al,Ga)–(P,C,B,Si)	Fe–Nd–Al
	Fe–Ga–(P,C,B,Si)	
	Fe–Ga–(Nb,Cr,Mo)–(P,C,B) (Los Alamos)	
	Fe–(Cr,Mo)–(B,C)	
	Fe–Ln–B, Fe–(Zr,Hf,Nb,Ta)–B,	
	Fe–(B,Si)–Nb	
Co	Co–Ga–(Cr,Mo)–(P,C,B)	Co–Nd–Al
	Co–(Zr,Hf,Nb,Ta)–B	Co–Sm–Al
	Co–Ln–B	
Ni	Ni–(Nb,Cr,Mo)–(P,B)	Ni–Nb–Ti, Ni–Nb–Zr, Ni–Nb–Hf
	Ni–(Ta,Cr,Mo)–(P,B)	Ni–Nb–Zr–Ti
	Ni–Zr–Ti–Sn–Si (Yonsei University)	Ni–Nb–Zr–Ti–M (M = Fe, Co, Cu)
	Ni–Pd–P	Ni–Nb–Hf–Ti
		Ni–Nb–Hf–Ti–M
		Ni–Nb–Sn (Cal Tech)
Cu	Cu–Pd–P	Cu–Zr–Ti, Cu–Hf–Ti
	Cu–Ni–Pd–P	Cu–Zr–Ti–Ni, Cu–Hf–Ti–Ni
		Cu–Zr–Ti–Y, Cu–Hf–Ti–Y
		Cu–Zr–Ti–Be, Cu–Hf–Ti–Be
		Cu–Zr–Al, Cu–Hf–Al
		Cu–Zr–Al–M, Cu–Hf–Al–M
		(M = Ni, Co, Pd, Ag)
		Cu–Zr–Ga, Cu–Hf–Ga
		Cu–Zr–Ga–M, Cu–Hf–Ga–M
		Cu–Zr–Al–Y (Cal Tech)
Pt	Pt–Cu–P	
	Pt–Cu–Co–P (Cal Tech)	
	Pt–Pd–Cu–P	
Pd	Pd–Cu–Ni–P	

Table 1.2. Features of three metallic components in ternary base systems where bulk metallic glasses are formed by the copper mold casting method[49] (courtesy Japan Institute of Metals) *ETM* early transition metal, *LTM* late transition metal, *Ln* lanthanide metal

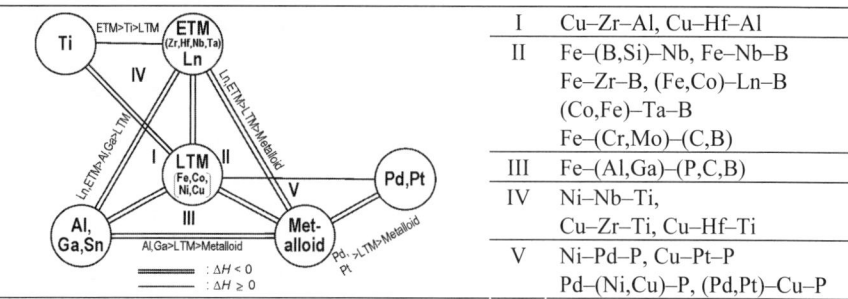

I	Cu–Zr–Al, Cu–Hf–Al
II	Fe–(B,Si)–Nb, Fe–Nb–B
	Fe–Zr–B, (Fe,Co)–Ln–B
	(Co,Fe)–Ta–B
	Fe–(Cr,Mo)–(C,B)
III	Fe–(Al,Ga)–(P,C,B)
IV	Ni–Nb–Ti,
	Cu–Zr–Ti, Cu–Hf–Ti
V	Ni–Pd–P, Cu–Pt–P
	Pd–(Ni,Cu)–P, (Pd,Pt)–Cu–P

The tendency of glass-forming ability (GFA), temperature interval of supercooled liquid region (ΔT_x), and reduced glass transition temperature $(T_g/T_l)^{27}$ for all LTM-based BMGs reported to date are summarized in Table 1.3. The highest GFA can be obtained for Pd- and Pt-based alloys,

Table 1.3. Features of glass-forming ability, temperature interval of supercooled liquid, and reduced glass transition temperature for late transition metal-based bulk metallic glasses (base metals > 50 at.%)[49] (courtesy Japan Institute of Metals)

Glass-Forming Ability

Alloy Type	Fe	Co	Ni	Cu	Pd	Pt
Metal-Metalloid	◎	○	◎	◎	◎*	◎*
Metal-Metal	-	-	○	◎	-	-

Base Metal > 50 at%
◎*; $d_{max} \geq 30$ mm,
◎; $d_{max} \geq 5$ mm;
○; $d_{max} \geq 3$ mm,
△; $d_{max} < 3$ mm

GFA: Pd Pt >> Cu > Ni > Fe > Co

Temperature Interval of Supercooled Liquid (ΔT_x)

Alloy Type	Fe	Co	Ni	Cu	Pd	Pt
Metal-Metalloid	◎	△	○	○	◎	◎
Metal-Metal	-	-	○	◎	-	-

Base Metal > 50 at%
◎; $\Delta T_x > 90$ K,
○; $\Delta T_x = 60 \sim 90$ K,
△; $\Delta T_x < 60$ K

ΔT_x : Cu > Fe > Ni > Pd > Pt > Co

Reduced Glass Transition Temperature (T_g/T_l)

Alloy Type	Fe	Co	Ni	Cu	Pd	Pt
Metal-Metalloid	○	○	○	○	◎*	◎
Metal-Metal	-	-	○	○	-	-

Base Metal ≥ 50 at%
◎*; $T_g/T_l \geq 0.70$,
◎ ; $T_g/T_l \geq 0.65$,
○ ; $T_g/T_l \geq 0.60$,
△ ; $T_g/T_l < 0.60$

followed by Cu-, Ni-, Fe-, and then Co-based alloys. Generally, the GFA of ternary alloys is enhanced with decreasing liquidus temperature. In fact, it can be seen that there is a relationship between GFA and ΔT_x or T_g/T_l.

The mechanical fracture strength, σ_f, and fracture elongation, ε_f, under compressive load for the LTM-based BMGs are summarized in Table 1.4. All the BMGs exhibit high σ_f, exceeding 1,000 MPa. In addition to their elastic elongations of 2%, Fe-, Ni-, Cu-, Pd-, and Pt-based metal–metalloid BMGs also have good ductility as evident from the achievement of plastic elongation. Conversely, Co-based metal–metalloid and metal–metal BMGs and Fe-based metal–metal BMGs exhibit no plastic elongation. The 2% elastic elongation property has been recognized for other BMGs such as Zr-, Mg-, and Ln (lanthanide metal)-based alloy systems. The elastic limit of a crystalline alloy is typically less than 0.65% due to the presence of dislocations. BMGs do not contain dislocations and ideally should exhibit an elastic limit of 2%. However, some BMGs do not achieve this 2% elastic limit due to cast defects. Therefore, the 2% elastic limit reflects the random atomic configuration of an ideal glass and is an essential factor for the achievement of high fracture strength.

The development of Fe- and Co-based BMG alloys has not been as rapid as the other systems. However, BMGs in (Fe,Co,Ni)–Nb–(B,Si) and Co–Fe–Ta–B alloy systems were developed over the last 3 years. Therefore, the formation and fundamental properties of these new Fe- and Co-based BMGs are described in Sect. 1.3.

Table 1.4. Features of static mechanical strength and compressive ductility for late transition metal-based bulk glassy alloys[49] (courtesy Japan Institute of Metals)

Static Mechanic Strength

Alloy Type	Fe	Co	Ni	Cu	Pd	Pt
Metal-Metalloid	◎	◎*	○	△	△	△
Metal-Metal	△	△	◎	○	-	-

Base Metal ≥ 50 at%
◎*; $\sigma_f \geq$ 5000 MPa
◎ ; $\sigma_f \geq$ 3000 MPa
○ ; $\sigma_f \geq$ 2000 MPa
△ ; $\sigma_f <$ 2000 MPa

Strength (σ_f): Co > Fe > Ni > Cu > Pd > Pt

Compressive Ductility

Alloy Type	Fe	Co	Ni	Cu	Pd	Pt
Metal-Metalloid	◎*	◎	◎*	◎*	◎*	◎*
Metal-Metal	△	△	◎*	◎*	-	-

◎*; ε_f = 0.02 + plastic strain
◎ ; ε_f = 0.02
○ ; ε_f = 0.017~0.02
△ ; $\varepsilon_f <$ 0.017

Ductilty: Cu ≫ Ni > Pd > Pt > Fe > Co

1.3   FORMATION AND FUNDAMENTAL PROPERTIES OF Fe- AND Co-BASED BMGS IN (Fe,Co,Ni)–(B,Si)–Nb SYSTEMS

Small Nb additions were found to increase the stability of supercooled liquid and enhance the GFA in amorphous (Fe,Co,Ni)–(B,Si) alloys.[14] The compositional dependence of T_g in (Fe,Co,Ni)–B–Si–4%Nb BMG alloys is shown in Fig. 1.1. The glass transition phenomenon can be observed over the entire composition range in $[(Fe_{1-x-y}Co_xNi_y)_{0.75}B_{0.2}Si_{0.05}]_{96}Nb_4$ alloys. The glass transition temperature, T_g, shows a significant change with Ni content and decreases almost linearly with increasing Ni content from 810 to 760 K. There is no distinct change in T_g with the Co:Fe concentration ratio.

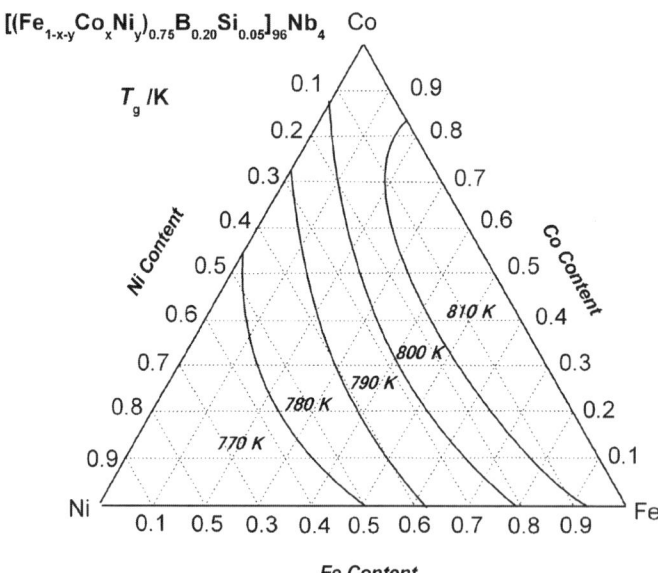

Fig. 1.1.   Compositional   dependence   of   glass   transition   temperature   (T_g)   for $[(Fe_{1-x-y}Co_xNi_y)_{0.75}B_{0.2}Si_{0.05}]_{96}Nb_4$ BMG alloys[49] (courtesy Japan Institute of Metals)

As shown in Fig. 1.2, the ΔT_x shows a maximum value of approximately 65 K in the range of 0.50–0.65Fe, 0.35–0.45Co, and 0–0.15Ni and keeps relatively large values of over 60 K in the Ni content range up to approximately 0.35Ni. In addition, the large T_g/T_1 values above 0.61 can also be obtained, leading to the formation of BMGs with diameters up to at least 5 mm by copper mold casting, as shown in Fig. 1.3. Considering that the conventional Fe–Co–Ni–B–Si system does not show glass transition phenomenon, it is important to note that, as it is a pseudoternary system, the

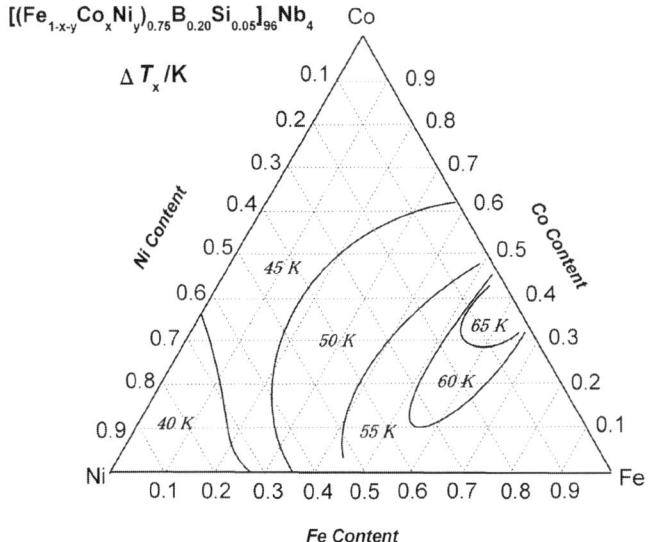

Fig. 1.2. Compositional dependence of supercooled liquid region (ΔT_x) for $[(Fe_{1-x-y}Co_xNi_y)_{0.75}B_{0.2}Si_{0.05}]_{96}Nb_4$ BMG alloys[49] (courtesy Japan Institute of Metals)

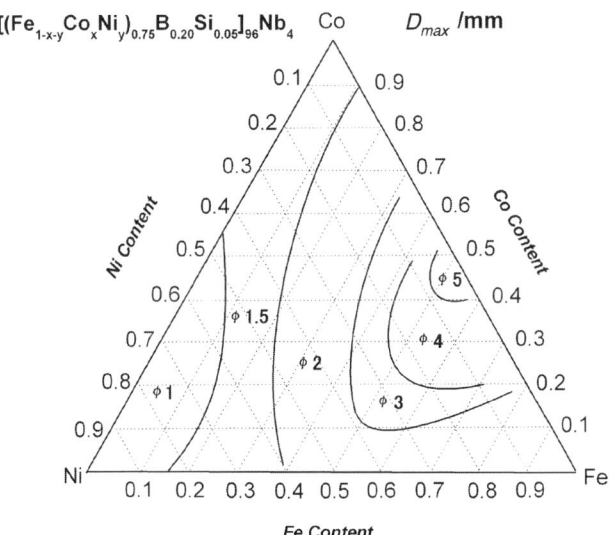

Fig. 1.3. Compositional dependence of maximum sample diameter (D_{max}) for $[(Fe_{1-x-y}Co_xNi_y)_{0.75}B_{0.2}Si_{0.05}]_{96}Nb_4$ BMG alloys[49] (courtesy Japan Institute of Metals)

addition of Nb satisfies the three empirical component rules for stabilization of supercooled liquid. In other words, the satisfaction of the rules leads to the formation of BMG. As an example, the outer shape and surface appearance of the cast Fe–Co-based BMG rods with diameters of up to 5 mm are shown in Fig. 1.4. The rods exhibit good metallic luster with a smooth surface, and no crystalline peaks are recognized in X-ray diffraction (XRD) patterns even in the 5-mm-diameter rods.

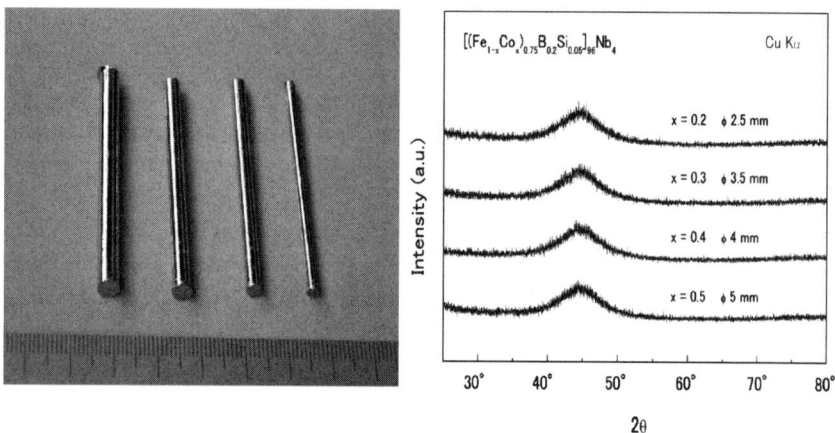

Fig. 1.4. Outer shape and X-ray diffraction patterns of $[(Fe_{1-x}Co_x)_{0.75}B_{0.2}Si_{0.05}]_{96}Nb_4$ BMG alloy rods[49] (courtesy Japan Institute of Metals)

The compositional dependence of compressive σ_f for the cast Fe–Co–Ni–B–Si–Nb alloy rods is shown in Fig. 1.5. High strengths exceeding 4,000 MPa can be obtained in the wide composition range of 0–1.0Co and 0–0.7Ni. Further increasing the Ni content decreases the strength to ~3,700 MPa. By fixing the B, Si, and Nb contents, the highest σ_f can be obtained for the Fe-based alloy, followed by the Co-based alloy and then the Ni-based alloy. In addition to the high strength, the Fe–Co–Ni–B–Si–Nb alloy rods also exhibit distinct plastic elongation up to about 0.5% before final fracture as revealed in Fig. 1.6. The alloy rod subjected to the plastic elongation up to 0.3% shows a distinct shear band along the maximum shear stress plane. Traces of viscous flow deformation were also observed on the shear band, indicating a significant temperature rise in the shear band.

Fe–Co-based BMG alloys also exhibit good soft magnetic properties. High saturation magnetizations, J_s, reaching 1.3 T were obtained in the Fe-rich composition range above 0.8 and low coercivities, H_c, of 1.0–2.5 A m^{-1} in the wide composition range of 0.25–1.0Fe and 0–0.6Ni. Thus, the appearance of room temperature ferromagnetic properties is dependent on the Ni and Fe contents. The decrease of H_c with increasing Co content has been recognized to originate from the reduction of saturation magnetostriction.[28]

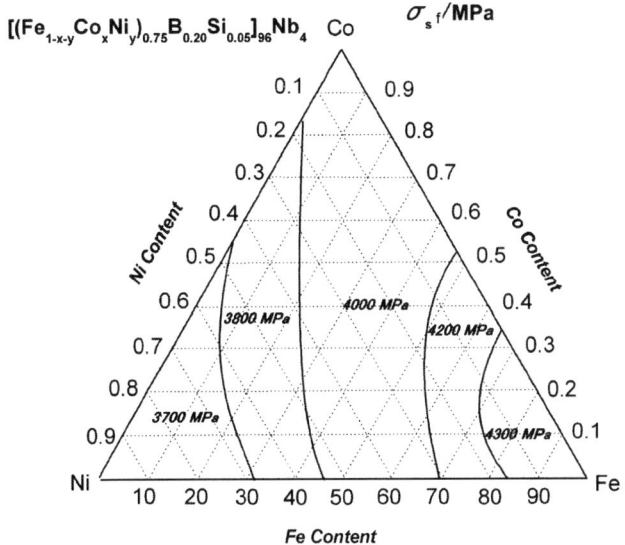

Fig. 1.5. Compositional dependence of compressive fracture strength of $[(Fe_{1-x-y}Co_xNi_y)_{0.75}$ $B_{0.2}Si_{0.05}]_{96}Nb_4$ BMG rods produced by copper mold casting[49] (courtesy Japan Institute of Metals)

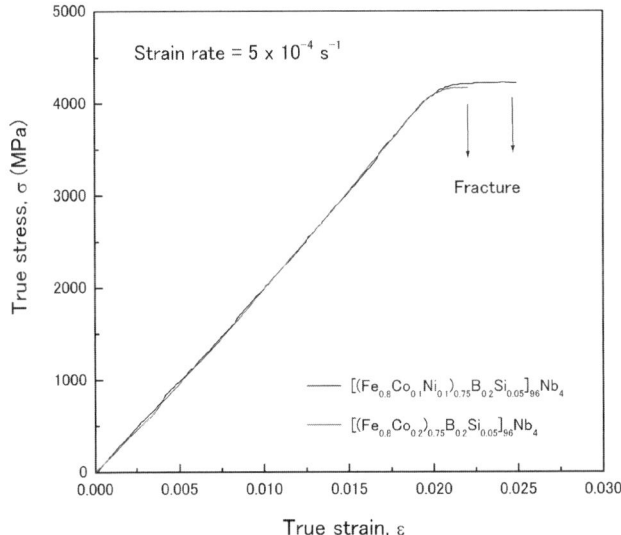

Fig. 1.6. True stress–strain curves of $[(Fe_{0.8}Co_{0.1}Ni_{0.1})_{0.75}B_{0.2}Si_{0.05}]_{96}Nb_4$ and $[(Fe_{0.8}Co_{0.2})_{0.75}$ $B_{0.2}Si_{0.05}]_{96}Nb_4$ BMG rods with a diameter of 2 mm[49] (courtesy Japan Institute of Metals)

The relationship between H_c and electrical resistivity, ρ, is shown in Fig. 1.7 for Fe-based BMGs in Fe–B–Si–Nb- and Fe–Ga–P–C–B-based systems, together with the data of amorphous and nanocrystalline alloys which require high cooling rates of over 10^5 K s^{-1} for preparation as well as $Co_{43}Fe_{20}Ta_{5.5}B_{31.5}$ BMG. The Fe- and Co-based BMG alloys have a better combination of lower H_c and higher ρ among all soft magnetic metallic alloys. The lower H_c is presumably due to the smaller magnetic anisotropy and lower internal stress, σ. The contribution of σ to H_c has been examined in more detail. It has previously been reported that the H_c is proportional to the ratio of saturation magnetostriction, λ_s, to J_s, i.e., $H_c \propto \Delta V \sqrt{\rho_d} \left(\lambda_s / J_s \right)$,[29] and hence the slope is related to the volume and density of internal defects consisting mainly of free volumes in the glassy structure. Good linear relationships between H_c and the ratio of λ_s to J_s for Fe-based BMGs and amorphous alloys are shown in Fig. 1.8. It is also evident that the slopes are clearly distinguished and are much smaller between the BMG alloys and the amorphous alloys. This difference indicates that the structure of the BMG alloys is distinguished from that of amorphous alloys and includes much lower volume and density of internal defects. The formation of a more homogenized disordered atomic configuration is concluded to be the origin for the lower H_c for the BMG alloys as compared with the amorphous alloys including crystalline nuclei and density fluctuations.

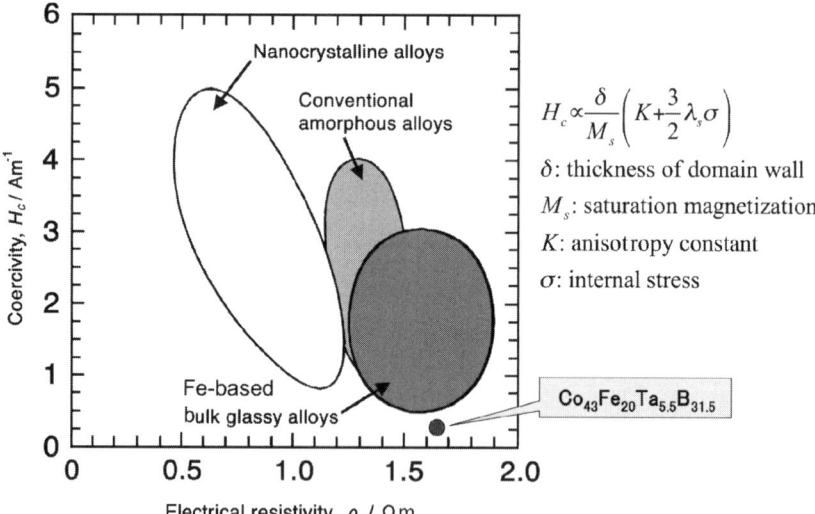

Fig. 1.7. Relationship between coercivity and electrical resistivity for Fe- and Co-based bulk metallic glasses. The data of conventional amorphous and nanocrystalline alloys are also shown for comparison[49] (courtesy Japan Institute of Metals)

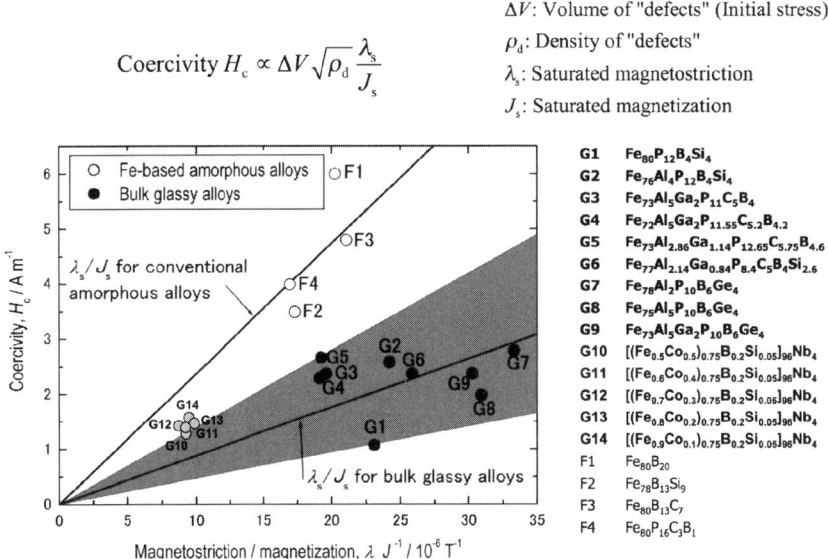

$$\text{Coercivity } H_c \propto \Delta V \sqrt{\rho_d} \, \frac{\lambda_s}{J_s}$$

ΔV: Volume of "defects" (Initial stress)
ρ_d: Density of "defects"
λ_s: Saturated magnetostriction
J_s: Saturated magnetization

G1	$Fe_{80}P_{12}B_4Si_4$
G2	$Fe_{76}Al_4P_{12}B_4Si_4$
G3	$Fe_{73}Al_5Ga_2P_{11}C_5B_4$
G4	$Fe_{72}Al_5Ga_2P_{11.55}C_{5.2}B_{4.2}$
G5	$Fe_{73}Al_{2.86}Ga_{1.14}P_{12.65}C_{5.75}B_{4.6}$
G6	$Fe_{77}Al_{2.14}Ga_{0.84}P_{8.4}C_5B_4Si_{2.6}$
G7	$Fe_{78}Al_2P_{10}B_6Ge_4$
G8	$Fe_{75}Al_5P_{10}B_6Ge_4$
G9	$Fe_{73}Al_5Ga_2P_{10}B_6Ge_4$
G10	$[(Fe_{0.5}Co_{0.5})_{0.75}B_{0.2}Si_{0.05}]_{96}Nb_4$
G11	$[(Fe_{0.6}Co_{0.4})_{0.75}B_{0.2}Si_{0.05}]_{96}Nb_4$
G12	$[(Fe_{0.7}Co_{0.3})_{0.75}B_{0.2}Si_{0.05}]_{96}Nb_4$
G13	$[(Fe_{0.8}Co_{0.2})_{0.75}B_{0.2}Si_{0.05}]_{96}Nb_4$
G14	$[(Fe_{0.9}Co_{0.1})_{0.75}B_{0.2}Si_{0.05}]_{96}Nb_4$
F1	$Fe_{80}B_{20}$
F2	$Fe_{78}B_{13}Si_9$
F3	$Fe_{80}B_{13}C_7$
F4	$Fe_{80}P_{16}C_3B_1$

Fig. 1.8. Relationship between coercivity and the ratio of saturation magnetostriction to saturation magnetization for Fe-based bulk metallic glasses. The data of amorphous type alloys are also shown for comparison[49] (courtesy Japan Institute of Metals)

Co–Fe–Ta–B-based BMGs exhibit a large ΔT_x above 70 K before crystallization. The large ΔT_x value leads to the formation of BMG rods with diameters up to at least 2 mm.[16] In addition, it has been reported that the Co–Fe–Ta–B BMG rods exhibit exceptionally high-yield strength of ~5,200 MPa at room temperature as well as high elevated temperature strength of over 2,000 MPa in the wide temperature range up to 585°C, as shown in Fig. 1.9.[16] Co-based BMGs exhibit not only an ultrahigh strength but also excellent soft magnetic properties. For instance, a Co-based BMG with a ring shape form of 1 mm in thickness, 10 mm outer, and 5 mm inner diameters exhibits an extremely high maximum permeability reaching 500,000 and low H_c of 0.26 A m^{-1}. As these excellent soft magnetic properties are attributed to originate from a highly homogeneous magnetic domain structure in the cast ring, a soft magnetic amorphous thin film was fabricated by a sputtering technique that exhibited a unique soft magnetic property through the control of the structure-sensitive magnetic domain. A Co–Fe–Ta–B glassy alloy film with a thickness of 2.6 μm was deposited at 298 K and had a fine perpendicular-type domain structure with a spacing of ~1.7 μm. The domain structure changed to an in-plane type for a film deposited at 473 K and was accompanied by a significant change in the magnetic properties, as shown in Fig. 1.10.[30] The success of synthesizing these thin films with a fine perpendicular-type domain structure even at a large thickness of 2.6 μm is promising for future development of new types of perpendicular-type data storage media, because the previous magnetic thin film thickness with the perpendicular domain structure is limited to less than several hundreds nanometers.[31]

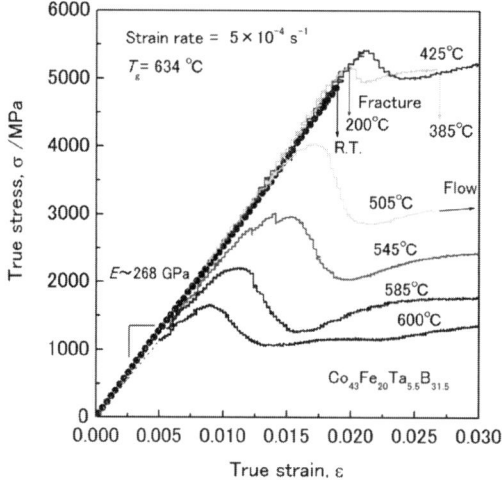

Fig. 1.9. Compressive true stress–true strain curves of $Co_{43}Fe_{20}Ta_{5.5}B_{31.5}$ BMG rod deformed at various temperatures between room temperature and 873 K[49] (courtesy Japan Institute of Metals)

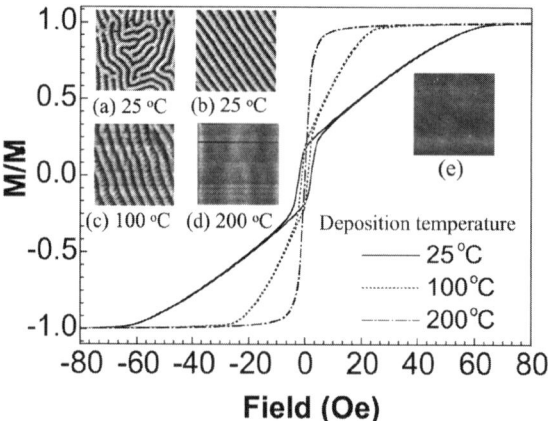

Fig. 1.10. Spin reorientation from perpendicular to in-plane in Co–Fe–Ta–B thin films with increasing deposition temperature. Hysteretic loops at 300 K for Co–Fe–Ta–B thin films deposited at different temperatures. *Insets* show the MFM images (area 20 × 20 mm²) for (*a*) film deposited at 25°C – in virgin state, (*b*) at 25°C – in remanence state, (*c*) at 100°C – in remanence state, (*d*) at 200°C – in remanence state, and (*e*) typical topography of Co–Fe–Ta–B thin films[49] (courtesy Japan Institute of Metals)

Fe- and Co-based BMGs belonging to the metal–metalloid type have a unique network-like atomic configuration in which distorted trigonal prisms of Fe or Co and B are connected with each other in edge- or face-shared configuration modes through glue atoms of Ln, Zr, Hf, Nb, or Ta.[6,7] The networked short-range ordered atomic configuration can effectively suppress the progress of crystallization due to the difficulty of long-range rearrangement of the constituent elements, leading to the stabilization of supercooled liquid. All Fe- and Co-based BMG alloys in metal–metalloid alloy systems have a unique primary crystallization phase of fcc-$(Fe,Co)_{23}B_6$ that has a large lattice parameter of ~1.2 nm and a complicated structure with 96 atoms in the unit cell.[14,16] This phase is different from the primary crystalline phase consisting of the mixture of α-Fe, Fe_2B, Fe_3B, and Fe_3Si equilibrium phases for Fe-based amorphous type alloys which require high cooling rates for amorphous phase formation.[32]

1.4   Ni- AND Cu-BASED BMGS

Recent progress in Ni- and Cu-based BMGs is discussed in this section. The stress–strain curves of Ni-based bulk glassy alloys under tensile and compressive loads are shown in Fig. 1.11. The σ_f under tensile load is as high as 2,700 MPa for the metal–metal type alloys,[19] whereas metal–metalloid type alloys exhibit a high compressive σ_f of 1,800 MPa as well as large compressive plastic elongation of 7.5%.[33] The high σ_f of the Ni-based metal–metal type BMG is believed to be the highest for all the BMGs. Much higher strength has been obtained under a compressive deformation mode. Therefore, metal–metal type Ni-based BMGs are appropriate for structural materials which require simultaneously high strength and good ductility.

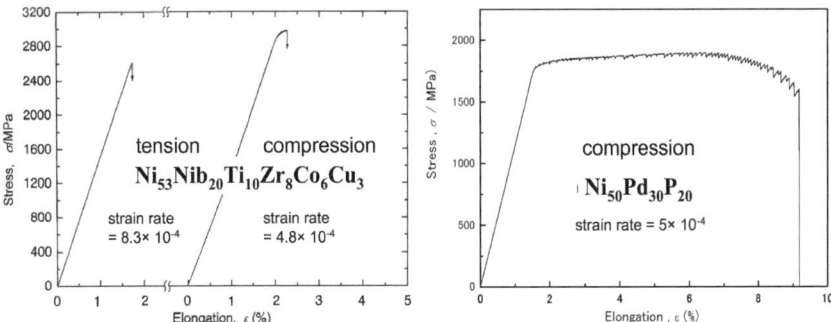

Fig. 1.11. Stress–strain curves of Ni-based bulk glassy alloys belonging to metal–metal and metal–metalloid types under tensile and compressive deformation model[49] (courtesy Japan Institute of Metals)

Ni-based BMGs containing Nb or Ta as a solute element exhibit high corrosion resistance in extremely severe circumferential condition which is required for fuel cell applications, i.e., in pH 2 H_2SO_4 at 353 K and in pH 2 H_2SO_4 containing 500 ppm NaCl or NaF at 353 K. The addition of Ta causes an increase in anodic potential and a decrease in corrosive current density, as shown in Fig. 1.12 and results in a much higher corrosion resistance compared to SUS316L.

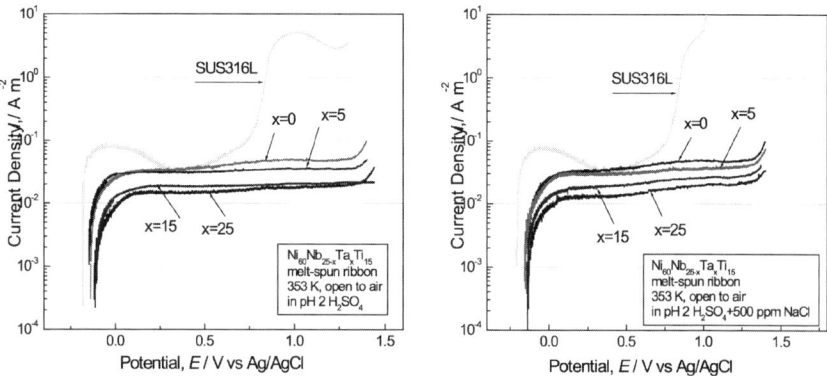

Fig. 1.12. Anodic polarization curves of $Ni_{60}Nb_{25-x}Ta_xTi_{15}$ bulk metallic glass[49] (courtesy Japan Institute of Metals)

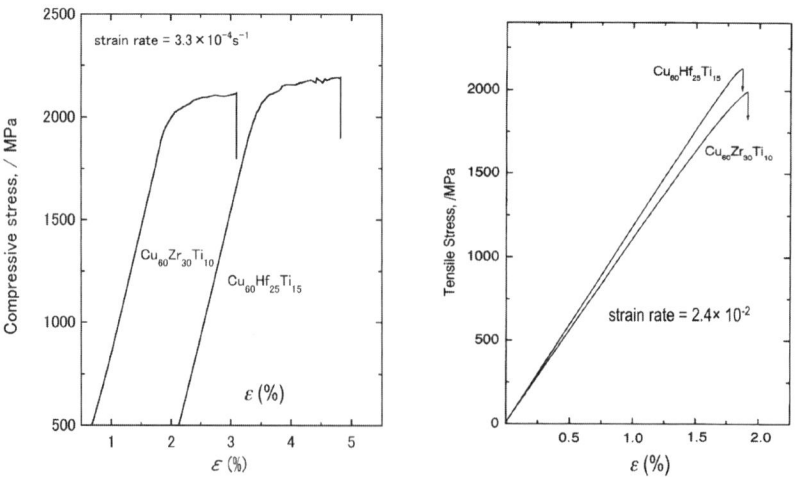

Fig. 1.13. Stress–strain curves of Cu-based bulk metallic glasses under tensile and compressive deformation modes[49] (courtesy Japan Institute of Metals)

Tensile and compressive stress–strain curves of Cu-based BMGs in Cu–Zr–Ti and Cu–Hf–Ti systems are shown in Fig. 1.13.[21] The Cu-based BMG rods exhibit high σ_f of 2,000–2,100 MPa and have plastic elongations of ~1.5% under compressive loads. It has previously been reported that the tensile σ_f of the Cu–Zr–Ti alloy increases further to ~2,500 MPa for higher multicomponent Cu-based alloys with Be or Y additions.[22,24] A large ΔT_x of more than 100 K is also obtained for the Cu–Hf–Al-based alloys containing 5 at.% Ag or Pd, and the largest ΔT_x reaches 110 K[25] which is the largest value for all LTM-based BMGs.

The fatigue strength of a Cu–Zr–Hf BMG rod with high tensile σ_f of 2,000 MPa was measured under a uniaxial tension–tension load. The fatigue endurance limit defined by the ratio of applied tensile amplitude stress (σ_a) to tensile fracture strength, σ_B, after the cycles of 10^7 is 0.24 for the Cu-based alloy, as shown in Fig. 1.14.[34] The fatigue limit is much higher than 0.14 for Zr–Al–Ni–Cu BMG[35] and 0.02 for Zr–Ti–Be–Ni–Cu BMG.[36] The Cu–Ti–Zr BMG also has high tensile stress amplitude of 480 MPa after 10^7 cycles,[34] and is comparable to that for a chrome-molybdenum steel (SCM435) and much higher than those for Ti-based crystalline alloys and Zr-based BMGs.[37] The fatigue crack initiated at a defect site located on the outer surface of the rod and propagated into the interior, accompanied by distinct striation patterns. The final fatigue fracture region consisted of a well-developed vein pattern. Although the fatigue strength of the Cu–Zr–Ti BMG is relatively high, it is expected that the elimination of surface defects caused by the decreases of inclusions and casting-introduced pores results in further improvement of fatigue strength. The fracture toughness of Cu–Zr–Ti BMG sheets was also evaluated by using a precracked fatigue test specimen which satisfies the ASTM E399 criterion[38] for the size and dimensions. The fracture toughness was measured to be ~68 MPam$^{1/2}$ which was slightly higher than that (40–60 MPam$^{1/2}$) for Zr-based BMGs.[39] Since Cu–Zr–Ti BMG exhibits high tensile σ_f, high ductility, high fatigue strength, and high fracture toughness, it is concluded that all the mechanical properties are superior to those for Zr-based BMGs.

By adding Ta, which is immiscible with Cu, to $Cu_{60}Hf_{25}Ti_{15}$ alloy a mixed phase alloy can be obtained that consists of a homogeneously dispersed bcc Ta-rich dendrite phase with a size of ~15 μm embedded in an amorphous matrix.[40] When the volume fraction of the bcc-Ta phase was ~11%, the dendrite-dispersed Cu-based alloy exhibited high yield strength of 2,100 MPa and large plastic elongation of 34%. This value was much larger than that (1.6%) for the monolithic alloy. A high density of shear bands was observed on the outer surface and fracture occurs along the maximum shear stress plane. The significant increase in compressive plasticity is presumably due

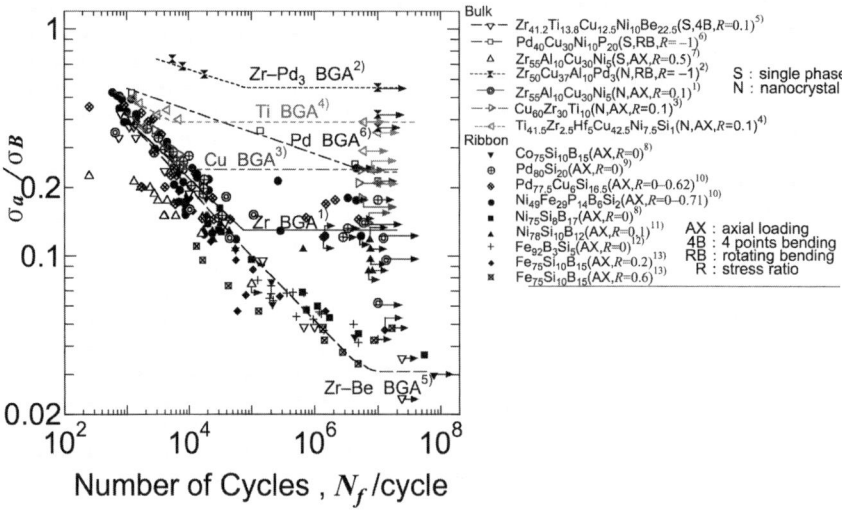

Fig. 1.14. S–N curves normalized by the tensile strength (σ_B) for bulk metallic glasses and amorphous alloy ribbons[49] (courtesy Japan Institute of Metals)

to easy generation of shear bands at the glass/dendrite interface through the increase in the stress concentration at the interface caused by the difference in the mechanical strength between the two phases.

1.5   Pd- AND Pt-BASED BMGS

For the noble metal-based BMGs, a $Pd_{40}Cu_{30}Ni_{10}P_{20}$ BMG was developed in 1996 with the lowest critical cooling rate for glass formation, R_c, of the order 0.1 K s^{-1}.[41] Very recently, another BMG with extremely high GFA was found in the Ni-free Pd–Pt–Cu–P alloy system.[42] In this alloy, the R_c was lower than 0.1 K s^{-1} and the maximum sample thickness was larger than 50 mm. This is the second alloy system in which BMGs have critical sizes larger than 50 mm and low R_c below 0.1 K s^{-1}.

1.6   CORRELATIONS AMONG FUNDAMENTAL PROPERTIES OF LTM-BASED BMGS

The relationship between tensile or compressive σ_f and Young's modulus, E, for the LTM-based BMGs is summarized in Fig. 1.15. Some data of conventional crystalline alloys are also included for comparison. One can see different good linear relation for BMGs and crystalline alloys. The σ_f and E increase in the order of Pt-, Pd-, Cu-, Ni-, Fe-, and Co-based alloys. The σ_f value at the same E is about three times higher than that for crystalline alloys.

The slope of the linear relation corresponding to elastic elongation limit is 2% which is about three times larger than that (0.65%) for crystalline alloys. Also, the deformation and fracture behaviors of the BMGs are independent of alloy component and strength level. The similar linear relation is also recognized between σ_f and T_g or T_l for the LTM-based BMGs, indicating that the strength is dominated by the random atomic configurations and bonding nature among the constituent elements.

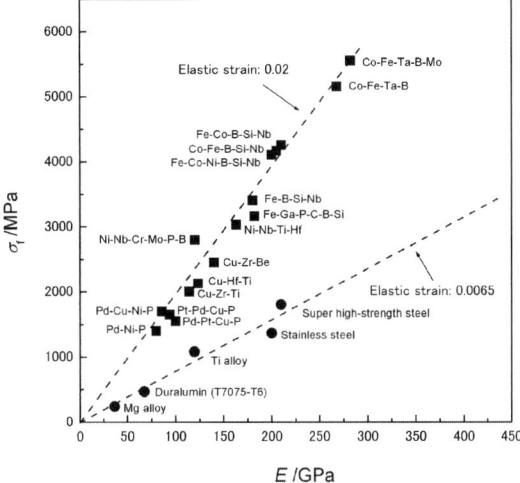

Fig. 1.15. Relationship between tensile or compressive fracture strength and Young's modulus for late transition metal-based BMGs. The data of conventional crystalline alloys are also shown for comparison[49] (courtesy Japan Institute of Metals)

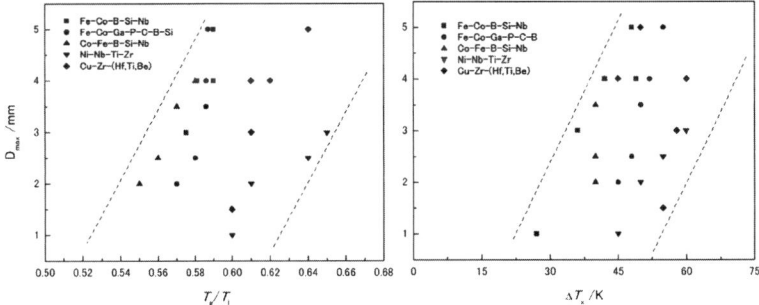

Fig. 1.16. Relationship between critical diameter and reduced glass transition temperature or supercooled liquid region for late transition metal-based BMGs[49] (courtesy Japan Institute of Metals)

As shown in Fig. 1.16, the relationship between D_{max} and T_g/T_l or ΔT_x for the LTM-based BMGs can be scarcely seen, though there are significant scatterings. This scarce relationship indicates that the high GFA is due to the combination of thermodynamic and kinetic factors, i.e., continuous increase in viscosity of supercooled liquid with decreasing temperature and high resistance of supercooled liquid against crystallization.

1.7   APPLICATIONS

The fundamental properties of the LTM-based BMGs are summarized in Table 1.5. The typical fundamental properties of Fe-, Co-, Ni-, and Cu-based BMGs were described in Sects. 1.3 and 1.4. In addition to these BMGs, Pt–Pd–Cu–P BMGs have a useful combination of a low T_g of ~500 K, a large ΔT_x of over 90 K, and high corrosion resistance. These properties are suitable for viscous flow working treatments on a nanometer scale in the supercooled liquid region. Because of this combination of unique features, LTM-based BMGs have already been used for various applications. Their applications will be briefly illustrated in this section.

To produce Fe-based BMG powders with sizes ranging from 0.1 to 2 mm, a mass-production type water atomization technique has been developed.

Table 1.5. The fundamental properties of the LTM-based BMG[49] (courtesy Japan Institute of Metals)

Fe-based	Soft magnetism (glass, nanocrystal)
	Hard magnetism (nanocrystal)
	High corrosion resistance
	High endurance against cycled impact deformation
Co-based	Soft magnetism (glass, nanocrystal)
	Hard magnetism (nanocrystal)
	High corrosion resistance
	High endurance against cycled impact deformation
Ni-based	High strength, high ductility
	High corrosion resistance
	High hydrogen permeation
Cu-based	High strength, high ductility (glass, nanocrystal)
	High fracture toughness, high fatigue strength
	High corrosion resistance
Pd-based	High strength
	High fatigue strength, high fracture toughness
	High corrosion resistance
Pt-based	Very low T_g
	Very low T_l
	High GFA
	High corrosion resistance
	Good nanoimprintability

The output of this technique has reached 20 tons per month. These Fe-based BMG powders have already been commercialized as shot-peening balls.[43] Shot peening generates a compressive residual stress field on the surface of material. This effect is far superior in BMGs compared to conventional crystalline shot-peening balls. In addition, Fe-based BMG balls have a significantly longer lifetime.

Stainless steel has traditionally been used for conventional pressure sensors. However, Ni–Nb–Ti–Zr–Cu–Co BMGs have much higher tensile σ_f, much lower E, and much better corrosion resistance. Therefore, Ni-based BMGs are expected to produce a new type of pressure sensor with higher sensitivity and better high-pressure properties.[44] Ni-based BMG diaphragms can be produced by the injection die-casting process. The strain gauge pattern on the surface of the diaphragms is made by a low temperature chemical vapor deposition technique. Pressure sensors using Ni- and Zr-based BMG diaphragms have been shown to exhibit 3.8 times higher sensitivity than that of conventional stainless steel diaphragms. This significantly higher sensitivity enables the miniaturization of the pressure sensor. The diameter of sensors can be reduced from 5.0 mm for conventional diaphragms to 2.5 mm for Ni-based BMG diaphragms. At present, one automobile manufacturer is using at least ten kinds of pressure sensors for injection control, oil pressure control, brake control, air conditioning, clogging monitor, etc. If fuel cell systems are adopted, the number of the pressure sensors is expected to double. Therefore, a significant increase in the pressure sensor market for automobiles is anticipated in the near future. To meet this demand, a mass-production line with a capacity of 50 million diaphragms per year is being prepared.

Microgeared motors with high rotating torques have been used in various engineering fields. The minimum size of the motors has decreased from 12 mm in 1980 to 7 mm in 2000 and is presently 2.4 mm. The durability of Ni-based BMG gears in a 2.4-mm-diameter motor increased by 313 times as compared with the tool steel gears. Even after 1,875 million revolutions, the Ni-based BMG gear kept its original shape, in contrast to the heavy wear found in steel gears after only 6 million revolutions. The world's smallest, 1.5 mm diameter, heavy load and high durability microgeared motors have been fabricated from high σ_f Ni-based BMG gear parts, as shown in Fig. 1.17.[45] The gears in a 1.5-mm-diameter motor cannot be made by conventional mechanical machining techniques. In fact, the tool steel gears in a 2.4-mm-diameter motor can be scarcely constructed by mechanical machining. The world's smallest microgeared motor with a diameter of 1.5 mm has been made from Ni-based BMG parts for the carrier shaft, planetary gears, and sun-gear carrier. This microgeared motor had high-rotating torques of 0.1 mN m for a two-stage stacked gear-ratio reduction

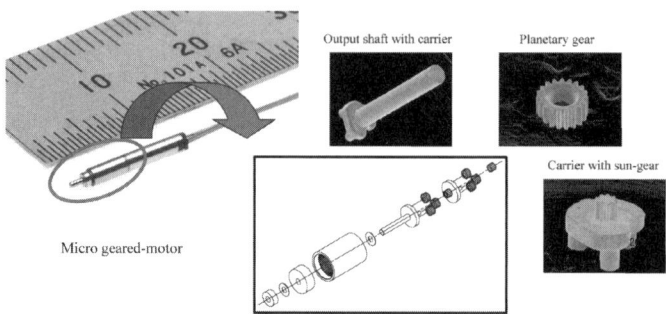

Fig. 1.17. Microgeared motor with the world's smallest size of 1.5 mm diameter constructed from Ni-based BMG alloy gears and an illustration of its construction[49] (courtesy Japan Institute of Metals)

system and 0.6 mN m for a three-stage system. These torques are 6–20 times higher than in a conventional 4.0-mm-diameter motor used in the vibration system of mobile telephones. These microgeared motors are expected to be used in advanced medical equipment such as endoscopes, micropumps, rotablators, and catheters for thrombus removal, precision optics, micro-industries, microfactories, etc.

Fe-based soft magnetic BMG alloys in Fe–Co–Ga–metalloid and Fe–Co–B–Si–Nb systems have been commercialized as consolidated magnetic cores for power supplies such as choke coils, common mode, and noise filter.[6,7] These commercialized magnetic cores exhibit very good soft magnetic properties, i.e., nearly constant relative permeability over a wide frequency range up to several MHz, good linear relationship between permeability and DC bias field, and much lower core losses as compared with Sendust and Permalloy. These Fe-based BMG Fe–Co–Ga–P–C–B–Si and Fe–Co–B–Si–Nb alloys exhibit high effective permeabilities above 100,000 at 1 kHz and high J_s of 1.2–1.3 T in a thick sheet form. These alloys have been tested as the yoke material in precision positioning linear actuators in which the wiring pitch for circuit process unit can be controlled on a scale of 10–30 nm. These linear actuators exhibit higher Lorentz force of 0.35 N in a frequency range of 20–40 Hz as compared with those for other soft magnetic alloys, indicating that they are appropriate for small, large force, fast drive, and energy-saving applications.

The recent rapid development of nano- and microscale working process techniques using focused ion beam (FIB), electron beam, and LIGA process etc. has significantly increased the importance of materials with homogeneous structures on a nanometer scale. BMG alloys have a homogeneous structure and can be deformed by working in the viscous flow regime at much lower temperatures than nanocrystalline alloys.

By applying FIB-based techniques, various complex patterns with smooth surfaces on a nanometer scale can be fabricated in BMGs.[46] These patterns cannot be obtained in conventional crystalline alloys. The controllable minimum size of complex patterns produced by the FIB technique can reach dimensions as small as 12 nm. Nanometer scale Pt-based BMG surfaces having highly functional characteristics that were fabricated by FIB-based methods are shown in Fig. 1.18.[47] The $Pt_{48.75}Pd_{9.75}Cu_{19.5}P_{22}$ BMG with low T_g of 502 K and large ΔT_x of 85 K can be regarded as an ideal viscous flow working material on a nanometer scale. The low T_g enables the use of polyimide dies supported on copper plates. Only one cycle viscous flow pressing against the polyimide die in the supercooled liquid region produced complex microinductors. The minimum pattern size caused by the viscous flow die-forging process reaches dimensions as small as 22 nm. This small size enables the fabrication of the imprinted nano data-pit patterns for the next generation of DVDs and data storage media, as shown in Fig. 1.19.

Fig. 1.18. Periodically nanostructured surfaces fabricated by superplastic nanoforging of Pt-based bulk metallic glass with focused ion beam machined dies of Zr-based metallic glass[49] (courtesy Japan Institute of Metals)

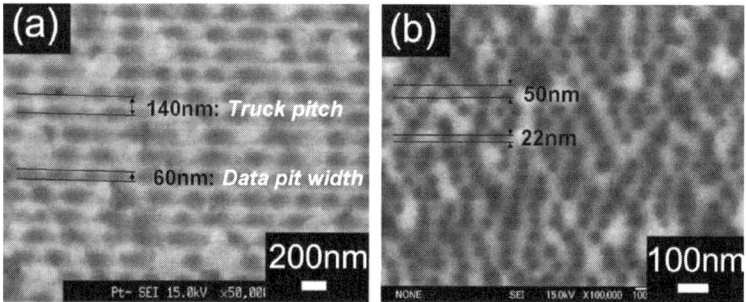

Fig. 1.19. Imprinted nano data-pit patterns for the next generation of DVDs[49] (courtesy Japan Institute of Metals)

Besides the applications described above, there have been several other applications that have been successfully commercialized,[48] including biomedical components, antiallergic jewelry, sporting equipment, knives, razor blades, cell phone and thumb drive cases, printer parts, as well as alloys for military and space applications.

1.8 CONCLUSIONS

The development of advanced metallic materials by use of the science and technology of supercooled metallic liquid started around 1990, and the new research field on the basis of this concept is believed to become more and more significant in the near future. The first LTM-based bulk glassy alloys were developed in Fe-based alloys in 1995. Since then, the late transition-based BMGs have been found in Fe–(Al,Ga)–(P,C,B,Si), Fe–(Cr,Mo)–(C,B), Fe–(Zr,Hf,Nb,Ta)–B, Fe–Ln–B (Ln = lanthanide metal), Fe–B–Si–Nb, and Fe–Nd–Al for Fe-based alloys, Co–(Ta,Mo)–B and Co–B–Si–Nb for Co-based alloys, Ni–Nb–(Ti,Zr)–(Co,Ni) for Ni-based alloys, and Cu–Ti–(Zr,Hf), Cu–Al–(Zr,Hf), Cu–Ti–(Zr,Hf)–(Ni,Co), and Cu–Al–(Zr,Hf)–(Ag,Pd) for Cu-based alloys.

These BMGs exhibit useful engineering properties of high mechanical strength, large elastic elongation, and high corrosion resistance. In addition, Fe- and Co-based bulk glassy alloys have good soft magnetic properties which cannot be obtained for conventional amorphous and crystalline type magnetic alloys. Fe- and Ni-based BMGs have already been used in some applications. These LTM-based BMGs are promising as new metallic engineering materials.

ACKNOWLEDGMENT

The authors thank the Japan Institute of Metals for permission to reproduce the figures and tables from Materials Transactions, 47 (2006) 1275[49] that are used in this chapter.

REFERENCES

1. A. Inoue, K. Ohtera, K. Kita, and T. Masumoto, New amorphous Mg–Ce–Ni alloys with high-strength and good ductility, *Jpn. J. Appl. Phys. Part 2* **27**(12), L2248–L2251 (1988).
2. A. Inoue, T. Zhang, and T. Masumoto, Al–La–Ni amorphous-alloys with a wide supercooled liquid region, *Mater. Trans. JIM* **30**(12), 965–972 (1989).

3. A. Inoue, T. Zhang, and T. Masumoto, Zr–Al–Ni amorphous-alloys with high glass-transition temperature and significant supercooled liquid region, *Mater. Trans. JIM* **31**(3), 177–183 (1990).

4. A. Peker and W. L. Johnson, A highly processable metallic glass-$Zr_{41.2}Ti_{13.8}Cu_{12.5}Ni_{10.0}Be_{22.5}$, *Appl. Phys. Lett.* **63**(17), 2342–2344 (1993).

5. A. Inoue, High-strength bulk amorphous-alloys with low critical cooling rates, *Mater. Trans. JIM* **36**(7), 866–875 (1995).

6. A. Inoue, Stabilization of metallic supercooled liquid and bulk amorphous alloys, *Acta Mater.* **48**(1), 279–306 (2000).

7. A. Inoue, Bulk amorphous and nanocrystalline alloys with high functional properties, *Mater. Sci. Eng. A* **304–306**, 1–10 (2001).

8. W. L. Johnson, Bulk glass-forming metallic alloys: Science and technology, *MRS Bull.* **24**(10), 42–56 (1999).

9. A. Inoue, Y. Shinohara, and J. S. Gook, Thermal and magnetic properties of bulk Fe based glassy alloys prepared by copper mold casting, *Mater. Trans. JIM* **36**(12), 1427–1433 (1995).

10. A. Inoue, T. Zhang, and A. Takeuchi, Bulk amorphous alloys with high mechanical strength and good soft magnetic properties in Fe–TM–B (TM=IV–VIII group transition metal) system, *Appl. Phys. Lett.* **71**(4), 464–466 (1997).

11. A. Inoue, A. Murakami, T. Zhang, and A. Takeuchi, Thermal stability and magnetic properties of bulk amorphous Fe–Al–Ga–P–C–B–Si alloys, *Mater. Trans. JIM* **38**(3), 189–196 (1997).

12. B. L. Shen, H. Koshiba, T. Mizushima, and A. Inoue, Bulk amorphous Fe–Ga–P–B–C alloys with a large supercooled liquid region, *Mater. Trans. JIM* **41**(7), 873–876 (2000).

13. A. Inoue and B. L. Shen, Soft magnetic bulk glassy Fe–B–Si–Nb alloys with high saturation magnetization above 1.5 T, *Mater. Trans.* **43**, 766–769 (2002).

14. A. Inoue, B. L. Shen, and C. T. Chang, Super-high strength of over 4000 MPa for Fe-based bulk glassy alloys in $[(Fe_{1-x}Co_x)_{0.75}B_{0.2}Si_{0.05}]_{96}Nb_4$ system, *Acta Mater.* **52**(14), 4093–4099 (2004).

15. A. Inoue and A. Katsuya, Multicomponent Co-based amorphous alloys with wide super-cooled liquid region, *Mater. Trans. JIM* **37**(6), 1332–1336 (1996).

16. A. Inoue, B. L. Shen, H. Koshiba, H. Kato, and A. R. Yavari, Cobalt-based bulk glassy alloy with ultrahigh strength and soft magnetic properties, *Nat. Mater.* **2**(10), 661–663 (2003).

17. B. L. Shen and A. Inoue, Enhancement of the fracture strength and glass-forming ability of CoFeTaB bulk glassy alloy, *J. Phys.: Condens. Matter.* **17**(37), 5647–5653 (2005).

18. X. M. Wang, I. Yoshii, A. Inoue, Y. H. Kim, and I. B. Kim, Bulk amorphous $Ni_{75-x}Nb_5M_xP_{20-y}B_y$ (M = Cr, Mo) alloys with large supercooling and high strength, *Mater. Trans. JIM* **40**(10), 1130–1136 (1999).

19. T. Zhang and A. Inoue, New bulk glassy Ni-based alloys with high strength of 3000 MPa, *Mater. Trans.* **43**, 708–711 (2002).

20. A. Inoue, W. Zhang, and T. Zhang, Thermal stability and mechanical strength of bulk glassy Ni–Nb–Ti–Zr alloys, *Mater. Trans.* **43**, 1952–1956 (2002).

21. A. Inoue, W. Zhang, T. Zhang, and K. Kurosaka, High-strength Cu-based bulk glassy alloys in Cu–Zr–Ti and Cu–Hf–Ti ternary systems, *Acta Mater.* **49**(14), 2645–2652 (2001).

22. A. Inoue, T. Zhang, K. Kurosaka, and W. Zhang, High-strength Cu-based bulk glassy alloys in Cu–Zr–Ti–Be system, *Mater. Trans.* **42**, 1800–1804 (2001).

23. A. Inoue, W. Zhang, T. Zhang, and K. Kurosaka, Cu-based bulk glassy alloys with good mechanical properties in Cu–Zr–Hf–Ti system, *Mater. Trans.* **42**, 1805–1812 (2001).

24. T. Zhang, K. Kurosaka, and A. Inoue, Thermal and mechanical properties of Cu-based Cu–Zr–Ti–Y bulk glassy alloys, *Mater. Trans.* **42**, 2042–2045 (2001).

25. W. Zhang and A. Inoue, Thermal stability and mechanical properties of Cu–Hf–Al base bulk glassy alloys with a large supercooled liquid region of over 100 K, *Mater. Trans.* **44**, 2346–2349 (2003).

26. A. Inoue and S. G. Kim, Japan Patent, P2000-345309A.

27. D. Turnbull, Under what conditions can a glass be formed? *Contemp. Phys.* **10**(5), 473–488 (1969).

28. K. I. Arai, N. Tsuya, M. Yamada, and T. Masumoto, Zero magnetostriction and extremely low residual magnetic loss in Fe–Co amorphous ribbons, *IEEE Trans. Magn.* **12**(6), 939–941 (1976).

29. T. Bitoh, A. Makino, and A. Inoue, Origin of low coercivity of Fe–(Al, Ga)–(P, C, B, Si, Ge) bulk glassy alloys, *Mater. Trans.* **44**, 2020–2024 (2003).

30. P. Sharma, H. Kimura, A. Inoue, E. Arenholz, and J. H. Guo, Temperature and thickness driven spin-reorientation transition in amorphous Co–Fe–Ta–B thin films, *Phys. Rev. B* **73**, 052401 (2006).

31. T. Shima, K. Takanashi, Y. K. Takahashi, and K. Hono, Preparation and magnetic properties of highly coercive FePt films, *Appl. Phys. Lett.* **81**(6), 1050–1052 (2002).

32. E. Matsubara, S. Sato, M. Imafuku, T. Nakamura, H. Koshiba, A. Inoue, and Y. Waseda, Structural study of amorphous $Fe_{70}M_{10}B_{20}$ (M = Zr, Nb and Cr) alloys by X-ray diffraction, *Mater. Sci. Eng. A* **312**, 136–144 (2001).

33. Y. Q. Zeng, N. Nishiyama, T. Wada, D. V. Louzguine-Luzgin, and A. Inoue, Ni-rich Ni–Pd–P glassy alloy with high strength and good ductility, *Mater. Trans.* **47**, 175–178 (2006).

34. K. Fujita, T. Hashimoto, W. Zhang, H. Kimura, and A. Inoue, Abstracts of 12th International Conference on Rapidly Quenched and Metastable Materials, 22–26 August 2005, Jeju, Korea, p. 147.

35. Y. Yokoyama, K. Fukaura, and A. Inoue, Effect of Ni addition on fatigue properties of bulk glassy $Zr_{50}Cu_{40}Al_{10}$ alloys, *Mater. Trans.* **45**, 1672–1678 (2004).

36. C. J. Gilbert, J. M. Lippmann, and R. O. Ritchie, Fatigue of a Zr–Ti–Cu–Ni–Be bulk amorphous metal: Stress/life and crack-growth behavior, *Scripta Mater.* **38**(4), 537–542 (1998).

37. *Metals Data Book* (Japan Institute of Metals, Maruzen, Tokyo, 2004), p. 139.

38. *Annual Book of ASTM* Standards (American Society for Testing and Materials, Philadelphia, 1994).

39. K. Fujita, T. Hashimoto, W. Zhang, H. Kimura, and A. Inoue, Abstracts of 12th International Conference on Rapidly Quenched and Metastable Materials, 22–26 August 2005, Jeju, Korea, p. 148.

40. C. L. Qin, W. Zhang, H. H. Kimura, and A. Inoue, Excellent mechanical properties of Cu–Hf–Ti–Ta bulk glassy alloys containing in-situ dendrite Ta-based BCC phase, *Mater. Trans.* **45**, 2936–2940 (2004).

41. A. Inoue, N. Nishiyama, and T. Matsuda, Preparation of bulk glassy $Pd_{40}Ni_{10}Cu_{30}P_{20}$ alloy of 40 mm in diameter by water quenching, *Mater. Trans. JIM* **37**, 181–184 (1996).

42. K. Takenaka, T. Wada, N. Nishiyama, H. Kimura, and A. Inoue, New Pd-based bulk glassy alloys with high glass-forming ability and large supercooled liquid region, *Mater. Trans.* **46**, 1720–1724 (2005).

43. A. Inoue, I. Yoshii, H. M. Kimura, K. Okumura, and J. Kurosaki, Enhanced shot peening effect for steels by using Fe-based glassy alloy shots, *Mater. Trans.* 44, 2391–2395 (2003).

44. N. Nishiyama, K. Amiya, and A. Inoue, *Mater. Sci. Eng. A* **449**, 79–83 (2007).

45. M. Ishida, H. Takeda, D. Watanabe, K. Amiya, N. Nishiyama, K. Kita, Y. Saotome, and A. Inoue, Fillability and imprintability of high-strength Ni-based bulk metallic glass prepared by the precision die-casting technique, *Mater. Trans.* **45**, 1239–1244 (2004).

46. P. Sharma, W. Zhang, K. Amiya, H. M. Kimura, and A. Inoue, Nanoscale patterning of Zr–Al–Cu–Ni metallic glass thin films deposited by magnetron sputtering, *J. Nanosci. Nanotechnol.* **5**, 416–420 (2005).

47. Y. Saotome, Abstracts of 12th International Conference on Rapidly Quenched and Metastable Materials, 22–26 August 2005, Jeju, Korea, p. 118.

48. http://www.liquidmetal.com/

49. A. Inoue, B. Shen, and A. Takeuchi, Developments and applications of bulk glassy alloys in late transition metal base system, *Mater. Trans.* **47**, 1275–1285 (2006).

Chapter 2

ATOMISTIC THEORY OF METALLIC LIQUIDS AND GLASSES

T. Egami

Department of Materials Science and Engineering and Department of Physics and Astronomy, University of Tennessee, Knoxville, TN 37996, and Oak Ridge National Laboratory, Oak Ridge, TN 37831, USA

2.1   INTRODUCTION

When we try to find out the structure–property relationships for metallic glasses, we feel lost and have trouble knowing where to begin, because describing the atomic structure of a glass is already a major challenge. The structure of glass is called *amorphous*, meaning shapeless, a term that refuses rigorous characterization. Actually it is a great accidental gift of nature that many substances are crystalline, so that we can discuss their structure in such a simple way, such as the lattice, symmetry, and the unit cell, even though there are so many, of the order of 10^{23} atoms cm^{-3} in a crystal. In this chapter, we start with the most basic question of how to describe the structure of liquids and glasses, and discuss how we can start constructing a theory that can describe the structure–property relationships of metallic glasses. We will take a local, rather than global, view of the structure, and consider how the local structure is related to the local properties.

The structure of liquids and glasses is usually described in terms of the atomic pair-density correlation function (PDF), $\rho_0 g(r)$, where ρ_0 is the atomic number density, or the radial distribution function (RDF), $4\pi r^2 \rho_0 g(r)$. The PDF describes the distribution of the distances between pairs of atoms, averaged over the volume and angle. It can be determined directly by X-ray, neutron, or electron diffraction experiments, by Fourier transforming the structure function, $S(Q)$, where Q is the momentum transfer of scattering

$(Q = 4\pi \sin\theta / \lambda$, θ, the diffraction angle; λ, the wavelength of the probe)[1,2] through

$$\rho_0 g(r) = \rho_0 + \frac{1}{2\pi^2 r} \int [S(Q) - 1] \sin Qr \cdot Q \, dQ, \qquad (2.1)$$

as shown in Figs. 2.1 and 2.2.[3] By analyzing the first peak of the PDF, we can determine the number of nearest neighbor atoms (coordination number, N_C) and the average and distribution of the nearest neighbor distances, a and Δa.

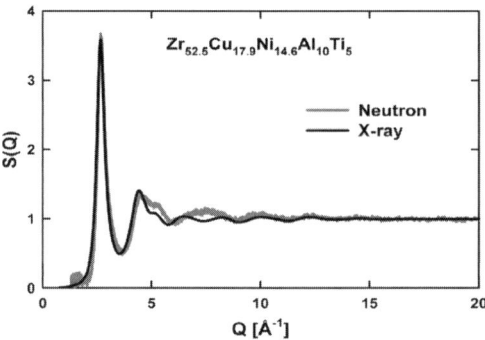

Fig. 2.1. Structure function $S(Q)$ of bulk metallic glass $Zr_{52.5}Cu_{17.9}Ni_{14.6}Al_{10}Ti_5$ determined by pulsed neutron and X-ray scattering[3]

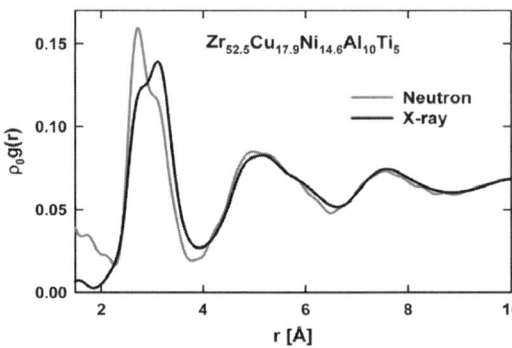

Fig. 2.2. Atomic pair-density function (PDF) of bulk metallic glass $Zr_{52.5}Cu_{17.9}Ni_{14.6}Al_{10}Ti_5$ obtained by Fourier transformation of $S(Q)$ in Fig. 2.1[3]

Since $S(Q)$ describes the scattering of particles by the structure, it can be related to, for instance, electrical resistivity by the Ziman formula,[4]

$$\rho = AS(2k_F), \qquad (2.2)$$

where k_F is the Fermi momentum of electrons and A is a constant. However, the PDF does not help in discussing other properties, such as atomic transport, since it is a one-dimensional function that describes only two-atom correlation. We need to know the real three-dimensional structure, and even when we know it, for instance by constructing a computer model of which the PDF agrees with the experimental one, it is not easy to characterize the three-dimensional structure in a useful manner. For this reason, we often resort to phenomenological concepts. The idea most frequently used in discussing atomic transport and deformation is *free-volume*. Free-volume is a space between atoms, and it is intuitively reasonable to assume that atoms need some space for moving around. The theory by Cohen and Turnbull[5–7] made this concept popular, and it was later applied to explain various properties. But, we have to be careful as its validity for metallic liquids was questioned from the beginning.[5]

In this chapter, we start with a review of the free-volume theory, and introduce an alternative, broader approach based upon the atomic bond topology. It is most important to recognize that the problem we are facing is a very difficult one, and many ideas may yet be discovered. We should not be satisfied with simply using old concepts such as the free-volume, but should actively seek and introduce new ideas that may facilitate deeper understanding of bulk metallic glasses. That is why this field is an exciting one for researchers who have courage to face the unknown and welcome new challenges.

2.2 FREE-VOLUME THEORY

The idea of free-volume goes back a long time. Nearly a century ago it was recognized that the viscosity of a liquid was strongly related to its volume.[8,9] But the success of Cohen and Turnbull[5–7] in quantifying the concept made this idea widely applicable. They started with recognizing the difference between liquid and gas. In the gaseous state atoms are free to move, but a liquid is condensed matter, and an atom is trapped in the "cage" of neighboring atoms. Since an atom is confined in the cage, a moving atom gets "backscattered" most of the time by the neighboring atoms and cannot move. Only when enough space, larger than v^*, happens to develop next to the atom, the atom can move into this space (Fig. 2.3). The diffusivity is then given by

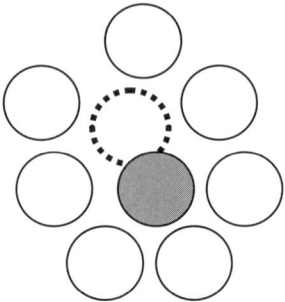

Fig. 2.3. Free-volume for an atom to move into

$$D = D(v^*) \int_{v^*}^{\infty} p(v)\mathrm{d}v = D(v^*)\exp\left(-\frac{\gamma v^*}{v_f}\right), \qquad (2.3)$$

where $p(v)$ is the probability distribution of space between atoms, given by

$$p(v) = \frac{\gamma}{v_f}\exp\left(-\frac{\gamma v^*}{v_f}\right), \qquad (2.4)$$

where γ is a constant of the order of unity and v_f is the total free-volume which depends linearly on temperature above the glass transition temperature T_g,[10]

$$v_f = \int v p(v)\mathrm{d}v = A + B(T - T_g). \qquad (2.5)$$

This expression yields the Vogel–Fulcher law for diffusivity.

For many molecular liquids, the magnitude of v^* is about 80% of the atomic volume, close to the atomic volume itself, but it is only about 10% for metallic liquids.[4] This already spells trouble for applying the original free-volume concept on metallic systems, but usually this important fact is conveniently ignored by the majority in the field of metallic glasses. Actually recent computer simulations of diffusion in atomic glasses by molecular dynamics (MD) clearly show that atomic diffusion does not happen as assumed in the free-volume theory (Fig. 2.4).[11–13] It is a more collective and diffusive process, where chain actions occur at many linearly connected atomic sites, almost like a process of billiards with linearly arranged balls. The reality of "free-volume" is closer to the "distributed free-volume" picture of Argon.[14]

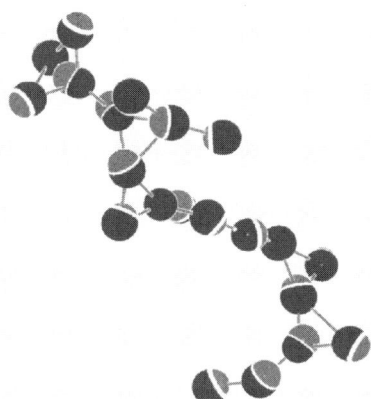

Fig. 2.4. Chain-like reaction for diffusion in a glass (reprinted from reference [12] with permission from Elsevier)

When a crystal is heated its volume expands because of vibrational anharmonicity, but the topology of the lattice structure remains unchanged unless there is a phase transformation. In contrast, the structure and the volume of a liquid are temperature dependent. The topology of the structure, for instance defined by the atomic connectivity, changes with temperature, and the configurational entropy due to topological disorder increases with temperature. The changes in the structure with temperature change occur quickly at high temperatures, but at low temperatures the kinetics of the change slows down. Below a certain temperature, the change becomes so sluggish that the structure appears to be frozen because of "kinetic arrest." This temperature is the glass transition temperature (Fig. 2.5). When a liquid is rapidly cooled, this kinetic arrest occurs at a higher temperature, and the system freezes above T_g (dashed line in Fig. 2.5), and volume, thus the value of v_f, remains high. The temperature at which the thermal arrest happens is called the *fictive temperature*, T_f, of that state.[15] If the frozen structure is warmed up (annealed) to a temperature below T_g, the structure slowly relaxes, and its fictive temperature and the volume come down. This phenomenon is called *structural relaxation*, and various other related properties change as a consequence.

Greer and Spaepen[16] and Van den Beukel and Radelaar[17] successfully extended the free-volume theory to account for the structural relaxation. But they noted that the process is second order, requiring the "sinks" for the free-volume, of which the nature is unclear. The apparent success of the free-volume theory in explaining the structural relaxation may have come from the fact that volume is the most convenient parameter to describe the fictive

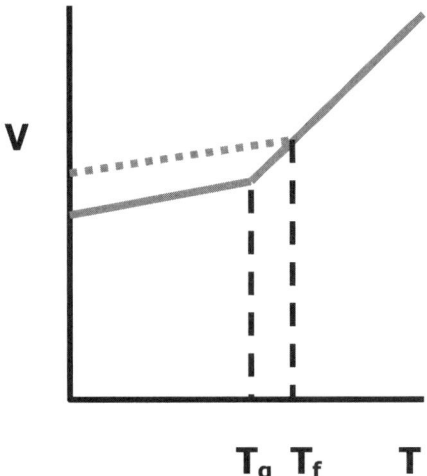

Fig. 2.5. Volume–temperature relation for metallic glasses cooled slowly (*solid line*) and rapidly (*dashed line*) from the liquid

temperature. The volume change is easy to measure accurately and provides a very good measure of the configurational entropy and the fictive temperature of the glass. We could have used the excess enthalpy or entropy instead of volume as the phenomenological parameter to describe the structure and relaxation, since they follow more or less the same kinetics. We should not confuse the success of the free-volume theory as a phenomenology with the atomistic reality of the free-volume concept.

Spaepen[18,19] also proposed the free-volume theory of plastic deformation. Free-volume, however, is similar to a lattice vacancy, and although its diffusion results in atomic transport, it does not produce shear deformation by itself, unless special geometry, such as dislocation, is assumed. The basis of the theory is that the shear process needs free space to operate, thus the free-volume controls the shear flow. However, this logic is valid only for hard-sphere models, and elastic bodies can undergo local shear without volume changes. As we will discuss below, the shear and diffusion processes are distinct, and require different mechanisms.

2.3 DESCRIBING THE STRUCTURE

The concept of free-volume is so simple and intuitive, and it is tempting to use it to explain various properties of metallic glasses. So what is wrong with this approach? Firstly, the concept of free-volume originates more or less from the model of dense random packing of hard spheres.[20,21] In this model, atoms cannot be compressed, so they need free space to move around. However, metallic glasses are quite remote from the hard-sphere models. For instance the packing density of the hard-sphere dense-random-packing (HS-DRP) model is lower than the fcc crystal by 15%.[20,21] But when the fcc metal melts, the volume increases only of the order of 1–2%. The interatomic potential in metals is more harmonic, and atoms can be squeezed, at some energy cost. We should take a more balanced view of the density fluctuation, which goes both positive and negative. For instance it is well known that the local density fluctuation in liquid, $\Delta\rho$, follows the thermodynamical law[22]

$$(\Delta\rho)^2 = \frac{T\rho^2}{B\Omega},\qquad(2.6)$$

where B is the bulk modulus, Ω is the volume of the system, and ρ is the number density of atoms. Note that thermal fluctuation itself does not change the total volume, since $\langle\Delta\rho\rangle = 0$! Only the anharmonicity relates the density fluctuation to volume expansion as discussed below.

The second problem is that the free space in the structure, "free-volume," cannot be described by the volume alone, and we have to consider the shape of the free-volume as well. The volume is the $\ell = 0$ quantity (ℓ, angular momentum), and we need higher-order terms, for instance in the spherical harmonics expansion. The extension of the free-volume theory to consider "shoving" is one such an attempt.[23,24] In this chapter, we introduce the general topological approach to describe the local atomic structure, and how knowing the local topology of atomic structure could facilitate understanding the properties of metallic glasses.

The first step is to consider the network of atomic connectivity or the topology of the atomic structure. This is natural for covalently bonded materials where atomic bonds are well defined. But they are reasonably well defined even in metallic liquids, since the interatomic potential has a negative curvature a little beyond the potential minimum, which tends to separate the second neighbor from the first neighbor.[5] So we define the nearest neighbors as neighbors within the first minimum in the atomic pair-density function (PDF) or the radial distribution function (RDF),[1,2] and assign metallic bonds between the central atom and the nearest neighbors. There are about a dozen nearest neighbors in the DRP structure as we discuss below. But the number of neighbors quickly increases as we go to the second neighbors and beyond; thus the detailed topology becomes less important, and we consider the effect of second neighbors and beyond in a continuum approximation. The topology of atomic connectivity is not necessarily static. In the liquid state at high temperatures, atomic bonds are cut and formed at a rate similar to the vibration of an atom. The local topology fluctuates with time and changes with temperature.

The most obvious way to characterize the local atomic connectivity is the local coordination number, N_C, or the number of nearest neighbors. Let us consider how the value of N_C is determined in a metallic glass. For simplicity we assume that the total energy of the system is given by a spherical pairwise interatomic potential

$$E_{\text{total}} = \sum_{i,j} \phi(r_{ij}),\qquad(2.7)$$

where r_{ij} is the distance between the ith and jth atoms. For simplicity let us assume that the potential is so short range that only the nearest neighbors interact with each other. Because any interatomic potential is harmonic near the potential minimum at $r = a$, we may expand (2.7) as

$$E_{\text{total}} = \sum_{i,j} \left[\phi(a) + \frac{1}{2} \frac{\partial^2 \phi}{\partial r^2}\bigg|_{r=a} (r_{ij} - a)^2 + \cdots \right]$$

$$= E_{\text{bond}} + E_{\text{elastic}} + \cdots,\qquad(2.8)$$

$$E_{\text{bond}} = N\phi(a)\langle N_C \rangle,$$

$$E_{\text{elastic}} = \frac{b}{2} \sum_{i,j} (r_{ij} - a)^2, \quad b = \frac{\partial^2 \phi}{\partial r^2}\bigg|_{r=a},\qquad(2.9)$$

where N is the total number of atoms and $\langle \cdots \rangle$ is the volume average. When N_C is small, the structure is open, and the nearest neighbors of an atom are not touching each other. Then E_{elastic} is zero, and the total energy is simply proportional to the average coordination number. But as $\langle N_C \rangle$ increases the neighbors become crowded, and the elastic energy increases as they push each other (Fig. 2.6). The minimum of E_{total} as a function of $\langle N_C \rangle$ is thus determined by the competition between the bonding energy and the elastic energy. Because atoms can be accommodated snugly without pushing each other in the fcc structure, the minimum of E_{total} must be for $\langle N_C \rangle$ greater than 12. In metallic glasses, the first peak of the PDF has some width, which is related to the local elastic energy at the atomic level.

We now relate the elastic energy to the local fluctuation in the value of N_C. The best way to imagine this is to consider a spherical atom with the radius r_A being embedded in the glass or liquid of atoms with the radius r_B. If the atomic size ratio, $x = r_A/r_B$, is small, N_C will be small. We may argue that N_C should be roughly proportional to $(1 + x)^2$, since near neighbor B atoms at the distance of $r_A + r_B = (1+x)r_B$ from the A atom would fill the

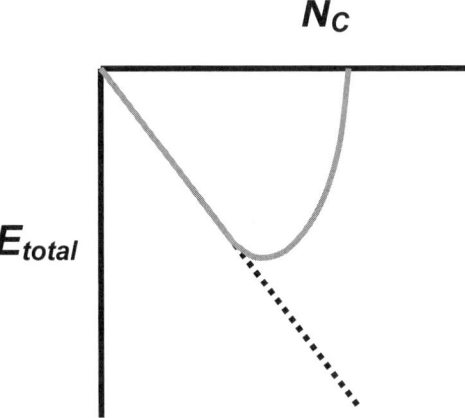

Fig. 2.6. Total energy of amorphous metal interacting via a pairwise potential (2.8) as a function of the coordination number, N_C

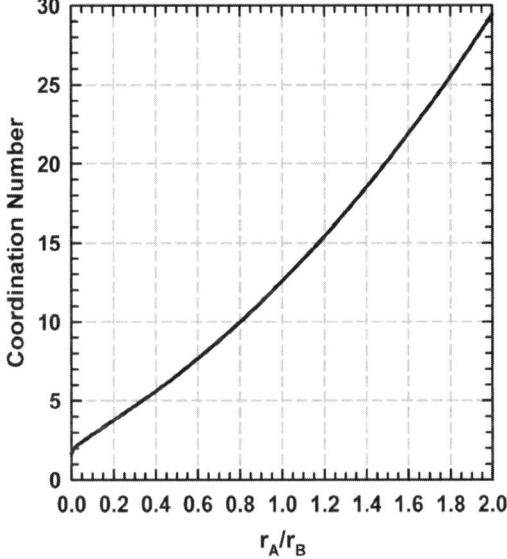

Fig. 2.7. The coordination number, N_C, of an element A in the liquid of element B, according to equation (2.10) as a function of $x = r_A/r_B$ (reprinted from reference [25] with permission from Elsevier)

surface of the A atom with a certain constant packing fraction. Indeed for a system with a short-range pairwise potential, it was possible to derive a quantitative expression for the average local coordination number N_C (Fig. 2.7),[25]

$$N_C(x) = 4\pi \left(1 - \frac{\sqrt{3}}{2}\right)(1+x)\left(1 + x + \sqrt{x(x+2)}\right). \tag{2.10}$$

Note that for $x = 1$, $N_C = 4\pi = 12.56$, which should give the minimum in Fig. 2.6. This equation was heuristically derived and confirmed by computer simulation,[25] but a more rigorous proof is yet to be formulated.

Now let us assume $r_A < r_B$, so that $x < 1$ and $N_C(x) < 4\pi$. Then, if we insert a B atom at the site of the A atom, because the radius of a B atom is larger, it would result in local pressure on the B atom. In other words, the fluctuation in the local coordination produces the fluctuation in the local atomic level pressure. If we define the local atomic level pressure of an ith atom, $p(i)$, as the local increase in the energy due to volume strain, then it is given by

$$p(i) = \frac{1}{V_i} \sum_j \mathbf{f}_{ij} \cdot \mathbf{r}_{ij}, \tag{2.11}$$

where V_i is the local atomic volume of the ith atom, $\mathbf{f}_{ij}$ is the two-body force, and $\mathbf{r}_{ij}$ is the separation, between the atoms i and j.[26] Note that for the system interacting with a pairwise interatomic potential $\phi(r)$

$$f_{ij} = -\frac{d\phi}{dr} = -b(r_{ij} - a) + \cdots, \tag{2.12}$$

where b is given in (2.9). Thus, a smaller value of N_C that requires smaller $\langle r_{ij} \rangle$ results in compression at that site. The local pressure thus defined is indeed correlated with the local coordination number, N_C, in the model structure.[27]

The local topology of the atomic bonds can be described not only by the number of the bonds, N_C, but also by the anisotropy of the connection. For instance a hoop of atoms in the x–y plane may be different from that in the x–z plane. The central atom may be bound tightly in the x–y plane, but loosely in the x–z plane. This gives rise to the local shear stress. In general, the local stress tensor can be defined by[26]

$$\sigma^{\alpha\beta}(i) = \frac{1}{V_i} \sum_j f_{ij}^\alpha r_{ij}^\beta, \tag{2.13}$$

where α and β are Cartesian coordinates. The atomic level stresses can be calculated also quantum mechanically from the first principles[28] using the Hellmann–Feynman force.[29] Similarly, the local elastic moduli, $C^{\alpha\beta\gamma\delta}$, can be defined,[27] and thus the local strain,

$$\varepsilon^{\alpha\beta}(i) = \frac{\sigma^{\gamma\delta}(i)}{C^{\alpha\beta\gamma\delta}(i)}. \tag{2.14}$$

Because the glass is macroscopically isotropic unless mechanically deformed, it is convenient to use the spherical coordinates and express them in terms of spherical harmonics. The stress will then be

$$\sigma_\ell^m(i) = \frac{1}{V_i}\sum_j f_{ij} r_{ij} Y_\ell^m\left(\frac{\mathrm{r}_{ij}}{r_{ij}}\right), \quad (\ell = 0, 2), \tag{2.15}$$

where $Y_\ell^m(\mathrm{n})$ are the spherical harmonics.[27] The term $\ell = 0$ is the hydrostatic pressure, $p(i)$, while the $\ell = 2$ terms ($m = -2,\dots,2$) are the five shear stress components. The total shear stress, $\tau(i)$, is given by

$$\tau^2(i) = \sum_m \left[\sigma_2^m(i)\right]^2. \tag{2.16}$$

Since the system is isotropic as a bulk, even though many local elastic moduli are needed to describe the local dynamics, only two elastic moduli survive volume averaging: the bulk modulus B and the shear modulus G.[27]

2.4 LOCAL DENSITY FLUCTUATION AND FREE-VOLUME

The distribution of the atomic level stresses was studied for a model generated by computer simulation.[30] Note that the volume average of the stresses, $\langle\sigma^{\alpha\beta}\rangle$, is equal to the external macroscopic stress, and is practically zero. So that the relevant quantities are the second- and higher-order moments. It was found that the distributions are mostly Gaussian, and according to the MD simulation, the second moment depends upon temperature. It has been suggested by the mean-field approximation[27] and proven by MD simulation[31] that at high temperatures the average of the local elastic energy due to local stress fluctuation is directly related to thermal energy

$$\frac{V\langle p^2\rangle}{2B} = \frac{1}{5}\frac{V\langle\tau^2\rangle}{2G} = \frac{kT}{4}, \tag{2.17}$$

where $\langle\cdots\rangle$ is thermal and temporal or ensemble average and V is the average atomic volume (Fig. 2.8). This means that the total potential energy, $3kT/2$, is equally divided among the elastic energies for the six stress components. Equivalently, the local strains that represent local topological fluctuations are given by

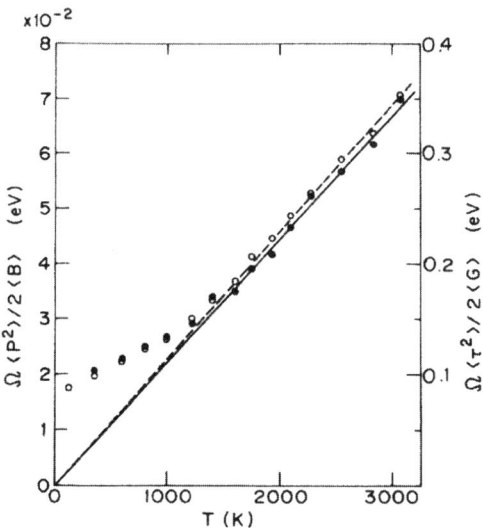

Fig. 2.8. Temperature dependence of the local elastic energy, $\Omega\langle p^2\rangle/2B$ and $\Omega\langle\tau^2\rangle/2G$, by molecular dynamics (MD) simulation, where Ω is the atomic volume (reprinted from reference [31] with permission from the American Physical Society)

$$\frac{VB}{2}\langle\varepsilon_v^2\rangle = \frac{VG}{10}\langle\varepsilon_s^2\rangle = \frac{kT}{4}. \qquad (2.18)$$

Equations (2.17) and (2.18) extrapolate to zero at $T = 0$, which means all the atomic bonds have to be the ideal length. However, it is impossible to achieve this in the dense-random-packed (DRP) structure because of topological frustration.[27,32] This means that the system will not be able to achieve thermal equilibrium and becomes nonergodic. In other words, it freezes into a glass structure. This is the glass transition. The glass transition is not a real thermodynamic phase transition, but is a kinetic phenomenon that depends upon the cooling rate. The higher the cooling rate is, the higher the value of $\langle p^2\rangle$ or $\langle\varepsilon_v^2\rangle$ the system freezes in. Now for small pressure the atomic level pressure is linearly related to the local volume by $\Delta V(i) = V(i)(p(i)/B(i))$, but the amplitude of the local pressure is large enough for the nonlinear terms as in (2.12) to be important as shown in Fig. 2.9:

$$\langle V\rangle = V_0\left(1 + \left\langle\frac{p^2}{B^2}\right\rangle\right) + \cdots. \qquad (2.19)$$

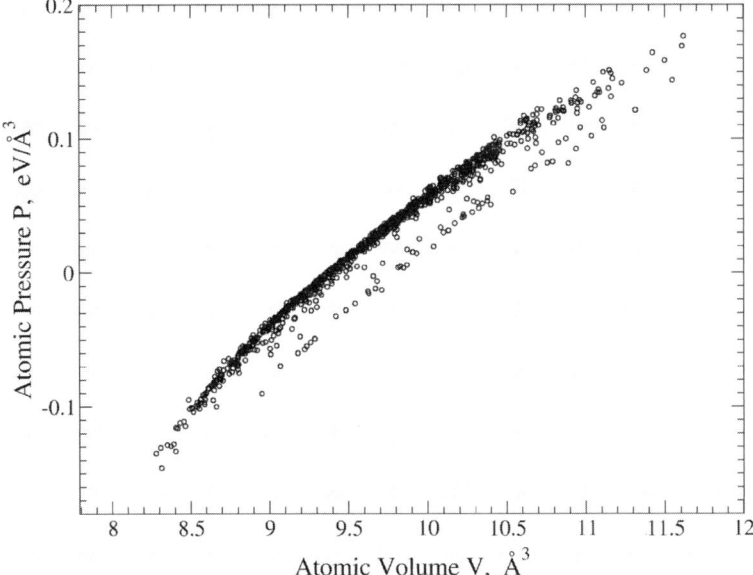

Fig. 2.9. The relationship between the local atomic volume and the local atomic level pressure. Due to anharmonicity the increase in the width of pressure distribution results in the increase in the total volume (reprinted from reference [33] with permission from EDP Sciences)

Note that $\langle p \rangle = 0$. Thus, the volume change and the change in $\langle p^2 \rangle$ are directly related through anharmonicity.[33] Thus the macroscopic free-volume is proportional to $\langle p^2 \rangle$. As we discuss below, we can define the atomic sites with large enough $p(I)$ as the free-volume sites. But the total volume is not proportional to the density of such sites. The change in the PDF due to structural relaxation observed by X-ray diffraction[34] confirms this view (Fig. 2.10). If the structural relaxation reduces free-volume, either the first peak in the PDF should shift to a shorter distance, or the outer portion of the first peak should shift the weight to the inner portion. Neither of these is observed, and instead the first peak became sharper, by eliminating both short and long extreme atomic distances. The result is consistent with reducing the distribution in $p(i)$. The change in the PDF assuming the narrowing in the pressure distribution explains the data very well as shown in Figs. 2.10 and 2.11.[35]

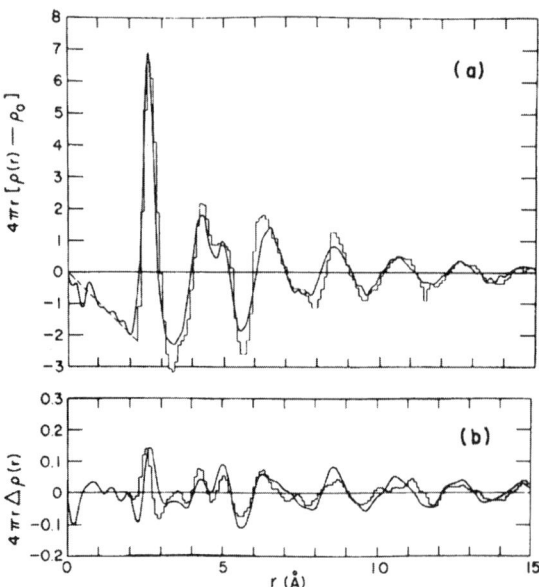

Fig. 2.10. (**a**) The X-ray PDF of glassy $Fe_{40}Ni_{40}P_{14}B_6$ (*smooth line*)[34] and the model PDF (*histogram*).[35] (**b**) Change in the PDF due to structural relaxation (*smooth line*)[34] and the change in the model PDF calculated assuming the narrowing of the pressure distribution (*histogram*) (reprinted from reference [35] with permission from the American Physical Society)

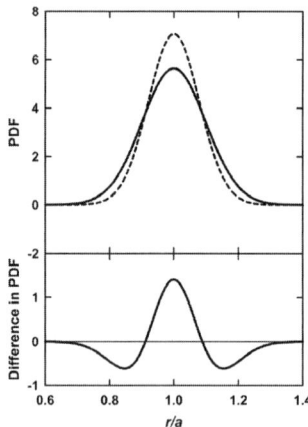

Fig. 2.11. Narrowing of a PDF peak due to structural relaxation (*above*) and the difference (*below*)

2.5 LOCAL TOPOLOGICAL INSTABILITY AND THE GLASS TRANSITION

Let us now consider when and how the glass transition happens. In (2.10), the coordination number, N_C, is a continuous function of the size ratio, x, while the actual coordination number at each atom has to be an integer. A nonintegral coordination number can be achieved as an ensemble average, or at high temperatures where the local coordination fluctuates fast with time. But as the liquid is cooled and the atomic motion is suppressed, the discrete nature of the coordination becomes a problem. When the particular coordination of an atom does not satisfy (2.10), that atom will be under the pressure due to the mismatch, and produces local elastic energy. Thus, the local energy landscape is periodic as a function of x, being minimum at values of x that give an integral value of N_C in (2.10) (Fig. 2.12).

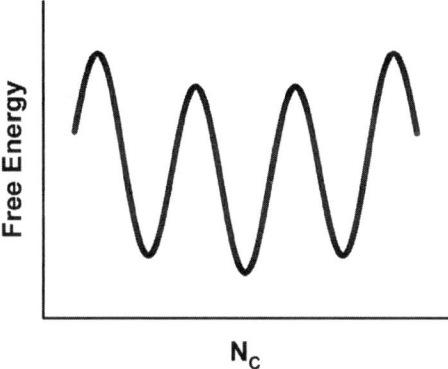

Fig. 2.12. The energy landscape of an atom as a function of $x = r_A/r_B$. The minima correspond to integral values of N_C

This means that the local topology is unstable at the point where x gives a half-integer. The amount of variation in x that corresponds to $1/2$ in N_C is given by

$$\Delta x = \frac{1/2}{\partial N_C(x)/\partial x}, \qquad (2.20)$$

where, from (2.8),

$$\frac{\partial N_C(x)}{\partial x} = 8\pi \left(1 - \frac{\sqrt{3}}{2}\right)\left[1 + x + \sqrt{x(x+2)} + \frac{1}{2\sqrt{x(x+2)}}\right]. \qquad (2.21)$$

The change in x can be caused either by changing the element or by uniformly expanding the system, for instance by thermal expansion. The homogeneous volume strain, $\langle \varepsilon_v \rangle$, is given by

$$\langle \varepsilon_v \rangle = (3/2)\Delta x / x, \tag{2.22}$$

since the same volume strain is shared by the central atom as well as the atoms in the neighboring shell. At $x = 1$, this gives the critical homogenous volume strain

$$\varepsilon_v^{crit}(H) = \frac{3}{2}\Delta x \Big|_{x=1} = \frac{6\sqrt{3} - 9}{8\pi} = 0.0554. \tag{2.23}$$

Thus when the total volume expands more than 6%, the structure should become unstable. Indeed this condition appears to be satisfied by many elements upon melting.[36] This condition was the basis for predicting the composition limit for glass formation for a binary system[37]

$$c_B^{min} = 0.1 \frac{V}{|\Delta V|}, \tag{2.24}$$

where c_B^{min} is the minimum concentration of the B element to form a glass when alloyed into the A matrix, and $\Delta V = V_A - V_B$, as shown in Fig. 2.13.

On the other hand if only the central atom has volume strain while others remain unchanged, the local volume strain, ε_v, is related to Δx by

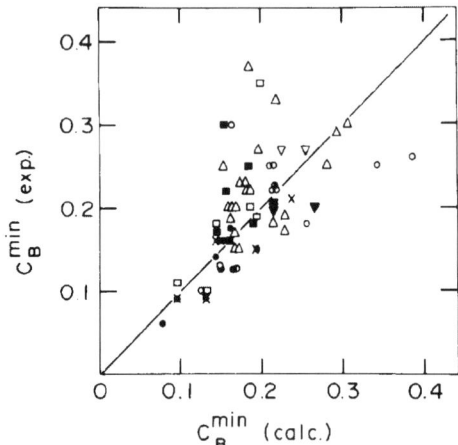

Fig. 2.13. The minimum composition of the alloying element B to obtain a glass in binary A–B alloys, c_B^{min}, calculated vs. experimental (reprinted from reference [37] with permission from Elsevier)

$$\varepsilon_v = 3\Delta x / x, \tag{2.25}$$

therefore at $x = 1$ the critical local volume strain is given by

$$\varepsilon_v^{crit}(L) = 3\Delta x \big|_{x=1} = \frac{6\sqrt{3} - 9}{4\pi} = 0.111. \tag{2.26}$$

Thus if the local volume strain is larger than 11%, the site is topologically unstable, and the local coordination number may change. This leads to the definition of the free-volume in terms of the critical local volume strain. We may define the site with the negative (dilatational) volume strain larger than 11% as the *free-volume site* or the *n*-type defect, and the sites with the positive (compressive) volume strain larger than 11% as the *antifree-volume site* or the *p*-type defect (Fig. 2.14).[30,33] Structural relaxation is explained in terms of the recombination of the *n*-type and *p*-type defects.[33,35] The defect sites are *liquid-like*, in a sense that they are topologically unstable, while the sites with the volume strain less than 11% are *solid-like*, since they are topologically stable. As shown below, the glass transition takes place by the percolation of the liquid-like sites, as envisaged by Cohen and Grest.[38]

It is interesting that the value of the critical local volume strain given in (2.24) is of the same order of magnitude as the "critical free-volume," v^*, considered by Cohen and Turnbull[5] for metallic liquids. This implies that the fluctuation in the topology of atomic environment does not require real

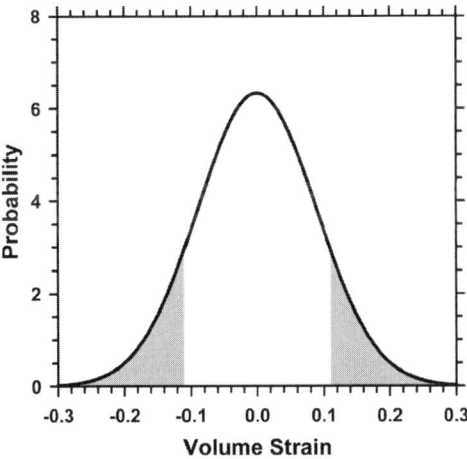

Fig. 2.14. Distribution of the atomic level volume strain, $\varepsilon_v = p/B$, and the free-volume (*n*-type defects) and the antifree-volume (*p*-type defects) defined by the topological instability condition (2.24)

free-volume as large as the atomic volume, but only the local dilatation of about 11% that allows the change in the coordination by unity.

Now, the atomic level volume strain ε_v has a Gaussian distribution.[30] Because the critical local volume strain, $\varepsilon_v^{\text{crit}}(L)$, is 11% as in (2.24), the density of the liquid-like sites, free-volume (n-type defects) and antifree-volume (p-type defects) sites, or the sites with $\varepsilon_v < -\varepsilon_v^{\text{crit}}(L)$ or $\varepsilon_v^{\text{crit}}(L) < \varepsilon_v$, is given by the complementary error function, CE(y),

$$p(\text{liq}) = \text{CE}(y_C) = \frac{2}{\sqrt{\pi}} \int_{y_C}^{\infty} e^{-y^2} dy, \qquad (2.27)$$

where

$$y_C = \frac{\varepsilon_v^{\text{crit}}(L)}{\sqrt{2}\langle \varepsilon_v^2 \rangle^{1/2}}. \qquad (2.28)$$

Because $\langle \varepsilon_v^2 \rangle$ is proportional to T as in (2.17), we plot the density of the liquid-like sites, $p(\text{liq})$, as a function of $\langle \varepsilon_v^2 \rangle$ in Fig. 2.15.

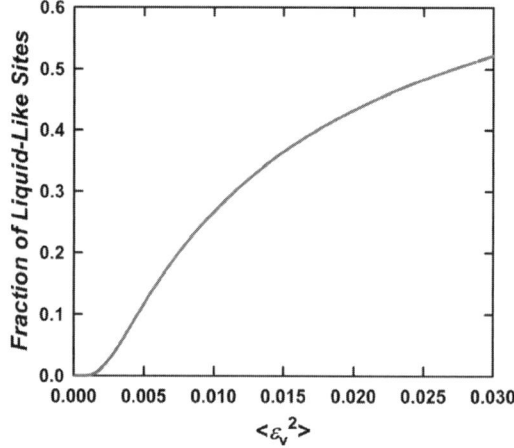

Fig. 2.15. Density of the liquid-like sites, $p(\text{liq})$, as a function of the second moment of the local volume strain, $\langle \varepsilon_v^2 \rangle$

Let us now go back to the freezing of the local density fluctuations with cooling (as shown in Fig. 2.7) because of the glass transition. The logic is that as the topological fluctuation becomes small with cooling the system starts to see the integral nature of the coordination, and becomes trapped in the local minima of the energy landscape in Fig. 2.12. Then, what we need to know is the barrier height of the local energy landscape to change the local coordination number. So we have to know the elastic energy due to local deformation. Now, the basis of (2.15) is that the atomic level stresses are totally localized, and the stresses at neighboring sites are uncorrelated. This assumption is valid at high temperatures as shown in Fig. 2.8, but when the system freezes this assumption is no longer valid. Since the atoms are all connected through the network of metallic bonds, one cannot just deform the environment of one atom, without affecting the other.

This coupling can be described in terms of a long-range stress field around the deformed site, calculated in the continuum approximation by Nabarro[39] and generalized by Eshelby.[40] The idea is that, to place an elastic sphere into a spherical hole in the elastic medium with a different radius, one has to first deform the sphere to match the hole. This volume strain is called the *transformation strain*, ε^T. Then one places the sphere in the hole, and lets the system relax. The strain in the sphere is given by

$$\varepsilon_1 = \frac{\varepsilon^T}{K_\alpha}, \tag{2.29}$$

with

$$K_\alpha = \frac{3(1-v)}{2(1-2v)}, \tag{2.30}$$

where v is the Poisson's ratio, and the total elastic energy is given by

$$E_{el} = \frac{BV}{2K_\alpha}(\varepsilon^T)^2. \tag{2.31}$$

Thus, the total elastic energy due to the local atomic level pressure is given by

$$E_v = \frac{\langle p^2 \rangle}{2BV} K_\alpha = \frac{BV}{2K_\alpha}\langle (\varepsilon_v^T)^2 \rangle. \tag{2.32}$$

In the extension of (2.17) this energy should be equal to $kT/4$ in the super-cooled liquid state just above the glass transition. At the glass transition temperature, $\langle (\varepsilon_v^T)^2 \rangle$ freezes into a constant value, so that the glass transition temperature should be given by

$$kT_g = 4E_v^{crit} = \frac{2BV}{K_\alpha}\left(\varepsilon_v^T(T_g)\right)^2. \tag{2.33}$$

As shown in Fig. 2.16, this equation agrees with the experimental data with impressive accuracy, with $\varepsilon_v^T(T_g) = 0.095$.[41]

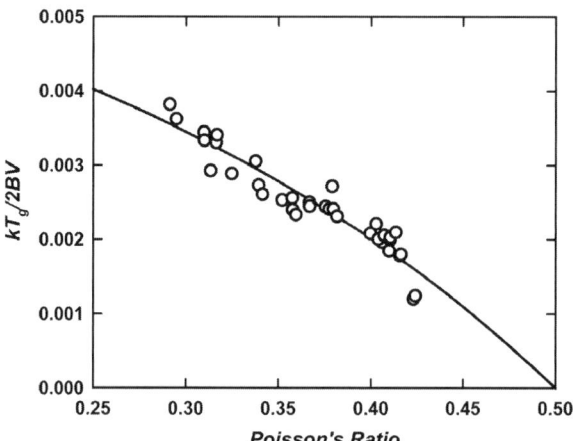

Fig. 2.16. Glass transition temperature divided by $2BV$, as (2.4), plotted as a function of Poisson's ratio v for various metallic glasses. The *solid line* indicates $(\varepsilon_v^T)^2 / K_\alpha$ with $\varepsilon_v^T = 0.095 \pm 0.004$, as in (2.33)[41]

Now, the critical local volume strain, $\varepsilon_v^{\text{crit}}(L)$, is 11% as in (2.24) and the local volume strain has a Gaussian distribution

$$f(\varepsilon_v) = \frac{1}{\sqrt{\pi}\sigma} \exp\left(-\frac{\varepsilon_v^2}{2\sigma^2}\right). \tag{2.34}$$

So for the standard deviation σ equal to $\varepsilon_v^T(T_g) = 0.095$, the critical value of $y = \varepsilon_v / \sqrt{2}\sigma$ is $y^{\text{crit}} = \varepsilon_v^{\text{crit}}(L) / \sqrt{2}\varepsilon_v^T(T_g) = 0.825$. This means that the total density of the *liquid-like* sites, p(liq), is given by

$$p(\text{liq}) = CE(y^{\text{crit}}) = 0.243, \tag{2.35}$$

as in (2.25) and Fig. 2.15. This value is close to the percolation limit for the DRP structure, which is estimated to be about $p_c = 0.2$ for $N_C = 12$.[42] This result implies that the glass transition occurs through percolation transition of the liquid-like states, as predicted by Cohen and Grest.[38]

A part of this small disagreement is due to the uncertainty of the percolation concentration in the DRP structure. Another is the kinetic nature of the transition that it slightly depends upon the cooling rate. Yet another reason is that, in estimating the critical value of ΔN_C ($= 0.5$), we assumed that the local energy landscape is purely a sinusoidal function. This is not realistic, since $N_C = 12.56$ is the overall minimum and we need a value of ΔN_C slightly larger to overcome the barrier. If we assume $p(\text{liq}) = CE(y^{\text{crit}}) = 0.2$, then

$y^{\text{crit}} = 0.905$, and for $\sigma = \varepsilon_v^T(T_g) = 0.095$ this gives $\varepsilon_v^{\text{crit}}(L) = 0.122$, a value only slightly above (2.24). If we assume $\Delta N_C = 0.55$, instead, we obtain $\varepsilon_v^{\text{crit}} = 0.122$, in perfect agreement with the experimental value. Thus the glass transition temperature is expressed by (2.31) virtually without input from the experiment, from only the atomic volume and elastic moduli, which can be calculated by the first-principle methods.

2.6 DEFORMATION OF METALLIC GLASSES

Free-volume theory is widely used also to explain the deformation of metallic glasses.[18,19] The argument is that the local shear transformation requires free-volume, since otherwise atoms cannot move. This implies a strong relationship between shear deformation and diffusion. If that is the case, the Stokes–Einstein relationship,

$$D = \frac{kT}{2\pi a \eta},\qquad(2.36)$$

where D is the diffusivity, η is the viscosity, and a is the diameter of the diffusing object, should be observed. In metallic glasses, however, it holds only with unrealistically small values of a, suggesting that many atomic processes contribute only to diffusion but not to viscous flow.[43] As we discussed above, the concept of free-volume is rooted on the hard-sphere model, whereas the hard-sphere model is not valid for metals. Density fluctuations can be both positive and negative, and shear action may occur even without free-volume. Actually, since the Poisson's ratio is less than 0.5 (0.25–0.4 for most metals), shear deformation reduces the volume and *creates* free-volume, so it may not require preexisting free-volume.

A more realistic approach is to consider deformation from the point of view of atomic bond rearrangement. If the structure is defined by the topology of atomic connectivity, deformation should involve changes in the bond arrangement. Because it is most likely that the total number of bonds is conserved during the rearrangement, deformation proceeds mostly by bond exchange, such as the one in Fig. 2.17.[44] When a static stress is applied, such bond exchange will result in the bond orientational anisotropy (BOA), which was actually observed by X-ray diffraction experiment.[44]

Bond rearrangements occur even during elastic deformation. The shear elastic constant of a glass is lower by 25–30% than the corresponding crystal of the same composition.[45] This contrasts to the bulk modulus, which is almost the same for both. A part of the reason for this softening is that the atomic response to the macroscopic shear stress is not uniform and non-collinear.[46] But the major reason is the bond rearrangements. No matter how small the shear strain is, the bond-exchange mechanism in Fig. 2.17 is activated, and local plastic deformation takes place, as shown in Fig. 2.18.[47] This is related to the fact that the local shear modulus has a distribution down to zero.[27] Therefore in glasses, shear deformation is inherently *anelastic*, at any stress level. However, because the density of the bond-exchange

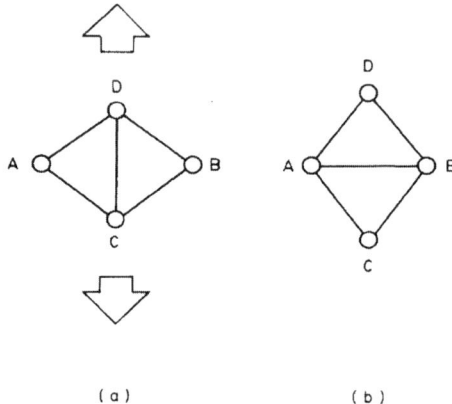

(a)   (b)

Fig. 2.17. The bond-exchange mechanism of shear deformation.[44] When a vertical tensile stress is applied the bond C–D is cut, and the new bond A–B is formed. The total number of bonds remains unchanged, but the distribution of orientation becomes anisotropic. Bond orientational anisotropy (BOA) is formed as a result of such a bond-exchange process (reprinted from reference [44] with permission from the American Physical Society)

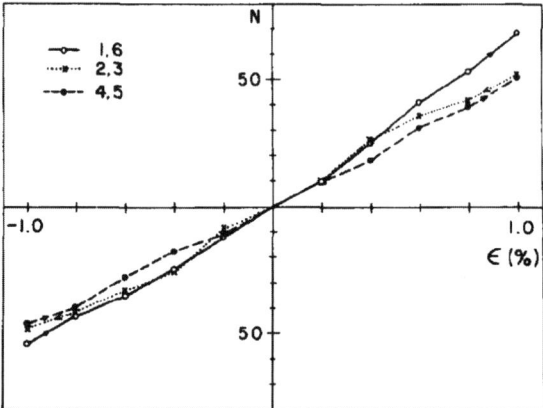

Fig. 2.18. The number of bonds cut or formed during the cyclic deformation (cycle shown by the number). No matter how small the strain is, always some bonds are exchanged, resulting in anelastic deformation (reprinted from reference [47] with permission from Elsevier)

incidents is constant for a unit strain increment, the response *appears* elastic. It is possible that these bond-exchange units are the origin of the two-level systems that gives rise to the linear specific heat at low temperatures due to tunneling.[48,49]

At high temperatures the viscosity of a liquid is strongly dependent on temperature, and the temperature dependence varies significantly among the glass-forming liquids. To describe this variation, Angell[50] introduced the concept of *liquid fragility* by plotting log $\eta(T)$, where $\eta(T)$ is the viscosity of a liquid at temperature T, against T_g/T (Angell plot). Liquids that show strong variation above T_g in this plot are called "fragile," while those that do not are called "strong." The slope of the Angell plot,

$$m = \frac{\partial \log \eta(T)}{\partial (T_g / T)}\bigg|_{T=T_g},$$

(2.37)

is called the *fragility coefficient*. Recently it was discovered that m is closely related to the Poisson's ratio and rapidly increases with v.[51] This claim is still controversial,[52,53] but it appears that the correlation is strong within a class of glasses, for instance for metallic glasses, and the relationship varies from class to class. It is known that the Poisson's ratio is closely related to the covalency of the bond. Now let us examine how we can explain this relationship between m and v in terms of the freezing of the local structure.

As shown above, the glass transition occurs because of the freezing of local density fluctuations. But how about the shear fluctuations? An indication of such a behavior was suggested by MD simulation.[31] The local shear stresses of the neighboring sites are correlated at low temperature due to the long-range Eshelby stress field, but this correlation persists above the glass transition temperatures, and disappears only at a temperature almost twice the glass transition temperature. This could be related to the freezing of the shear stresses. The temperature at which the local shear fluctuation freezes can be given by

$$kT_s = E_s^{crit} = \frac{GV}{2}\langle(\varepsilon_s^T)^2\rangle.$$

(2.38)

If we examine the Angell plot,[50] a plot of log η vs. T_g/T, the slope changes around ~1 poise, forming a knee.[54] This is a strong candidate for T_s, although we need confirmation by experiments or simulation. Now G and B are related by

$$\frac{G}{B} = \frac{2(1-2v)}{3(1+v)}.$$

(2.39)

From the condition that $T_s = T_g$ at the limit of v approaching 0.5, we obtain

$$\frac{(\varepsilon_s^T)^2}{(\varepsilon_v^T)^2} = \frac{B}{GK_\alpha} = \frac{1+v}{1-v}\Big|_{v=0.5} = 3, \qquad (2.40)$$

$$\varepsilon_s^{crit} = 0.159. \qquad (2.41)$$

Thus we get

$$\frac{T_s}{T_g} = \frac{3(1-v)}{1+v} \qquad (2.42)$$

and the liquid fragility coefficient is given by

$$m = \frac{\partial \log \eta(T)}{\partial(T_g/T)}\Big|_{T=T_g} = \frac{13}{1-(T_g/T_s)} = 13K_\alpha = \frac{39(1-v)}{2(1-2v)}. \qquad (2.43)$$

Equation (2.43) explains very well the close correlation between the Poisson's ratio and the liquid fragility.[51] In terms of the interatomic potential, the Poisson's ratio depends on the balance between the two-body term and the many-body volume-dependent term.[55] The two-body term represents covalency and decreases the Poisson's ratio. Thus more covalent glasses, such as silicates, have low values of v and show a strong liquid behavior. If the two-body term is absent and the elastic energy is only volume dependent, $G = 0$ and $v = 0.5$. This is the case for a fluid, such as water.

The theory described here can also form the microscopic basis for the mode-coupling theory which has been successful in depicting the dynamic behavior of a liquid.[56,57] The mode-coupling theory is a hydrodynamic theory extended by including the nonlinear coupling to slow viscous mode. Even though it has been successful, it is essentially a phenomenological theory that requires input from experiment for each composition, and thus does not have a predictive power until some experiment is performed.

Now the discussions above address homogeneous deformation, such as mechanical creep. But homogeneous deformation is observed only at high temperatures and/or low strain rates, and at room temperature deformation often occurs inhomogeneously through formation of localized shear band. If some stress concentration occurs, usually accidentally because of defects or surface irregularity, local deformation becomes accelerated if the local modulus is stress dependent and nonlinear, decreasing with stress. This leads to crack or shear band formation as discussed elsewhere in this book. While the details of the shear band formation are a question of mesoscopic mechanics, the nonlinearity of the stress–strain curve, which is the fundamental cause of the shear band formation, is an atomistic phenomenon. For instance recently it was observed that the ductility is primarily determined by the Poisson's ratio.[58] A possibility is that it is related to the behavior of liquid, because the local heating within the shear band is enough to make the temperature in the band exceed the glass transition temperature.[59] However, that is not likely to be the case, since a more recent observation shows that the heating is merely the consequence of local deformation, not the cause.[60] The atomistic mechanism of brittleness as it relates to the Poisson's ratio remains the subject of further research.

2.7 GLASS FORMABILITY

One of the central questions in the field of bulk metallic glasses is that at what compositions bulk metallic glasses form. Because the experimental process of discovering a new glass forming is a long and tedious one, the need for guiding principles is keenly felt by many. It has been known for a long time that the heat of formation and the atomic size difference are the key parameters for glass formation,[61] an inverse of the Hume–Rothery rule for stability of solid solution. The research group of A. Inoue at Tohoku University in Sendai, Japan utilized the wisdom beautifully to discover a large number of bulk metallic glass compositions.[62] Three principles they used are:

1. Use three or more elements.
2. The difference in the atomic size should be greater than 12%.
3. The heat of formation has to be negative.

The principles 1 and 2 are related to the composition criterion (2.24),[37] and represent the need of destabilizing the solid solution to form a glass. When we consider the question of glass formability, many tend to think of the stability of a glass itself. But a *glass* can be equally half-empty of half-full. More often than not glass stability is a consequence of the instability of a crystal. The criterion (2.24) represents the composition limit beyond which the crystalline solid solution of the same composition becomes topologically unstable. Then the glass is formed as a default, if the liquid is cooled fast enough not to allow diffusion.

But binary alloys are prone to form intermetallic compounds. Having three or more elements reduces this chance, and in return stabilizes a glass. This idea is often referred to as the "confusion principle." As far as the present author can recall this was first coined by H. H. Liebermann around 1976, when he was a graduate student at the University of Pennsylvania. It was then rediscovered by many authors. It nicely catches the idea of confusing the crystal so that it will not easily form, and increasing the entropy to stabilize the complex glass.

Unlike a crystal, a liquid does not have fixed topology of atomic connectivity. Atoms with different sizes can be easily accommodated by changing the nearest neighbor configurations, and thus the strain field due to size difference becomes quickly relaxed. While fluctuations in the local topology result in the local distortions and elastic energy, such as (2.32), in general atoms have a larger number of the nearest neighbors in the liquid state than in the crystalline state. Consequently the total energy (2.8) can be comparable to that of a solid solution, and the higher entropy of the liquid state gives the chance for the liquid state to be favored over the supersaturated solid solution. The free-energy balance, however, is very subtle, and requires more sophisticated theories to describe it accurately. This is another one of the fertile areas of research.

2.8   CONCLUSIONS

Because describing the structure of liquids and glasses meaningfully is already a major challenge, it is extremely difficult to create a theory that explains various properties of liquids and glasses. Many experimentalists still use the free-volume theory or at least the free-volume concept to explain the experimental observations. But the validity of the free-volume theory for metallic liquids is questionable, and its atomistic basis is challenged by recent MD simulations. It is more likely that the apparent success of the free-volume theory originates from the fact that the volume is the easiest property to measure that represents the fictive temperature of the system. In other words, the free-volume theory is successful as a phenomenology but not as a microscopic theory. A new theory has to be developed to describe the atomistic movements in glasses and liquids. In this chapter we introduce the theory of local topological fluctuations as an alternative theory. In this theory, topological fluctuations are represented by the atomic level stresses, and evolution of their distribution with temperature determines various thermal properties. Glass transition, structural relaxation, glass formation, and mechanical deformation have been well described by this theory. In particular, the importance of the Poisson's ratio, recognized by recent studies, emerges naturally from this theory. Although the details need to be worked out by further studies, this theory promises to be the one that could replace the free-volume theory in elucidating the complex behaviors of metallic glasses.

ACKNOWLEDGMENTS

The author gratefully acknowledges useful and insightful discussions with W. L. Johnson, S. J. Poon, A. L. Greer, C. T. Liu, J. M. Morris, P. K. Liaw,

E. George, and G. M. Stocks. This work was supported by the Division of Materials Science and Engineering, Office of Basic Energy Sciences (LLH), US Department of Energy under contract DE-AC05-00OR-22725 with UT-Battelle, LLC.

REFERENCES

1. B. E. Warren, *X-Ray Diffraction* (Dover, New York, 1969, 1990).
2. T. Egami and S. J. L. Billinge, *Underneath the Bragg Peaks: Structural Analysis of Complex Materials* (Pergamon, Elsevier Science, Oxford, 2003).
3. W. Dmowski, 2006, Unpublished.
4. J. M. Ziman, A theory of the electrical properties of liquid metals – The monovalent metals, *Philos. Mag. A* **6**, 1013–1034 (1961).
5. M. H. Cohen and D. Turnbull, Molecular transport in liquids and glasses, *J. Chem. Phys.* **31**, 1164–1169 (1959).
6. D. Turnbull and M. H. Cohen, Free-volume model of the amorphous phase: Glass transition, *J. Chem. Phys.* **34**, 120–125 (1961).
7. D. Turnbull and M. H. Cohen, On the free-volume model of the liquid–glass transition, *J. Chem. Phys.* **52**(6), 3038–3041 (1970).
8. A. J. Batschinski, Examination of the inner friction of liquids. I, *Z. Phys. Chem.* **84**, 643–706 (1913).
9. A. K. Doolittle, Studies in Newtonian flow. 2. The dependence of the viscosity of liquids on free-space, *J. Appl. Phys.* **22**, 1471–1475 (1951).
10. M. L. Williams, R. F. Landel, and J. D. Ferry, The temperature dependence of relaxation mechanisms in amorphous polymers and other glass forming liquids, *J. Am. Chem. Soc.* **77**, 3701–3707 (1955).
11. F. Faupel, W. Frank, M.-P. Macht, H. Mehrer, V. Naundorf, K. Rätzke, H. R. Schober, S. K. Sharma, and H. Teichler, Diffusion in metallic glasses and supercooled melts, *Rev. Mod. Phys.* **75**(1), 237–280 (2003).
12. H. R. Schober, Soft phonons in glasses, *Physica A* **201**(1–3), 14–24 (1993).
13. C. Donati, J. F. Douglas, W. Kob, S. J. Plimpton, P. H. Poole, and S. C. Glotzer, Stringlike cooperative motion in a supercooled liquid, *Phys. Rev. Lett.* **80**(11), 2338–2341 (1998).
14. A. S. Argon, Plastic deformation in metallic glasses, *Acta Metall.* **27**(1), 47–58 (1979).
15. G. W. Scherer, *Relaxation in Glass and Composites* (Wiley, New York, 1986).
16. A. L. Greer and F. Spaepen, Creep, diffusion, and structural relaxation in metallic glasses, *Ann. NY Acad. Sci.* **371**, 218–237 (1981).
17. A. Van den Beukel and S. Radelaar, On the kinetics of structural relaxation in metallic glasses, *Acta Metall.* **31**(3), 419–427 (1983).
18. F. Spaepen, A microscopic mechanism for steady state inhomogeneous flow in metallic glasses, *Acta Metall.* **25**(4), 407–415 (1977).
19. P. S. Steif, F. Spaepen, and J. W. Hutchinson, Strain localization in amorphous metals, *Acta Metall.* **30**(2), 447–455 (1982).
20. J. D. Bernal and J. Mason, Packing of spheres: Co-ordination of randomly packed spheres, *Nature* **188**, 910–911 (1960).
21. G. D. Scott, Packing of spheres: Packing of equal spheres, *Nature* **188**, 908–910 (1960).
22. L. D. Landau and E. M. Lifshitz, *Statistical Physics*, 2nd ed. (Addison-Wesley, Reading, 1969).

23. R. W. Hall and P. G. Wolynes, The aperiodic crystal picture and free energy barriers in glasses, *J. Chem. Phys.* **86**(5), 2943–2948 (1987).
24. J. C. Dyre, N. B. Olsen, and T. Christensen, Local elastic expansion model for viscous-flow activation energies of glass-forming molecular liquids, *Phys. Rev. B* **53**(5), 2171–2174 (1996).
25. T. Egami and S. Aur, Local atomic structure of amorphous and crystalline alloys: Computer simulation. *J. Non-Cryst. Solids* **89**(1–2), 60–74 (1987).
26. T. Egami, K. Maeda, and V. Vitek, Structural defects in amorphous solids: A computer simulation study, *Philos. Mag. A* **41**(6), 883–901 (1980).
27. T. Egami and D. Srolovitz, Local structural fluctuations in amorphous and liquid metals: A simple theory of glass transition, *J. Phys. F: Metal Phys.* **12**(10), 2414–2463 (1982).
28. O. H. Nielsen and R. M. Martin, Quantum-mechanical theory of stress and force, *Phys. Rev. B* **32**, 3780–3791 (1985).
29. R. P. Feynman, Forces in molecules, *Phys. Rev.* **56**, 340–343 (1939).
30. D. Srolovitz, K. Maeda, V. Vitek, and T. Egami, Structural defects in amorphous solids: Statistical analysis of a computer model. *Philos. Mag. A* **44**(4), 847–866 (1981).
31. S.-P. Chen, T. Egami, and V. Vitek, Local fluctuations and ordering in liquid and amorphous metals, *Phys. Rev. B* **37**, 2440–2449 (1988). http://link.aps.org/abstract/PRB/v37/p2440
32. D. R. Nelson, Order, frustration and defects in liquids and glasses, *Phys. Rev. B* **28**, 5515–5535 (1983).
33. T. Egami, K. Maeda, D. Srolovitz, and V. Vitek, Local atomic structure of amorphous metals, *J. Phys.* **41**(C8), 272–274 (1980).
34. T. Egami, Structural relaxation in amorphous $Fe_{40}Ni_{40}P_{14}B_6$ studied by energy-dispersive X-ray diffraction, *J. Mater. Sci.* **13**(12), 2587–2599 (1978).
35. D. Srolovitz, T. Egami, and V. Vitek, Radial distribution function and structural relaxation in amorphous solids, *Phys. Rev. B* **24**(12), 6936–6944 (1981). http://link.aps.org/abstract/PRB/v35/p2162
36. T. Egami, Universal criterion for metallic glass formation, *Mater. Sci. Eng. A* **226–228**, 261–267 (1997).
37. T. Egami and Y. Waseda, Atomic size effect on the formability of metallic glasses, *J. Non-Cryst. Solids* **64**(1–2), 113–134 (1984).
38. M. H. Cohen and G. Grest, Liquid–glass transition, a free-volume approach, *Phys. Rev. B* **20**(3), 1077–1098 (1979).
39. F. R. N. Nabarro, The strains produced by precipitation in alloys, *Proc. R. Soc. Lond. A* **175**, 519–538 (1940).
40. J. D. Eshelby, The determination of the elastic field of an ellipsoidal inclusion, and related problems, *Proc. R. Soc. Lond. A* **241**, 376–396 (1957).
41. T. Egami, S. J. Poon, Z. Zhang, and V. Keppens, Unpublished.
42. M. J. Powell, Site percolation in random networks, *Phys. Rev. B* **21**(8), 3725–3728 (1980).
43. A. L. Greer, Atomic transport and structural relaxation in metallic glasses, *J. Non-Cryst. Solids* **61–62**, 737–748 (1984).
44. Y. Suzuki, J. Haimovic, and T. Egami, Bond-orientational anisotropy in metallic glasses observed by X-ray diffraction. *Phys. Rev. B* **35**(5), 2162–2168 (1987). http://link.aps.org/abstract/PRB/v24/p6936
45. L. A. Davis, Mechanics of metallic glasses, in *Mechanics of Metallic Glasses. Rapidly Quenched Metals*, edited by N. J. Grant and B. C. Giessen (MIT, Cambridge, 1976), pp. 369–391.

46. D. Weaire, M. F. Ashby, J. Logan, and M. J. Weins, On the use of pair potentials to calculate the properties of amorphous metals, *Acta Metall.* **19**(8), 779–788 (1971).

47. Y. Suzuki and T. Egami, Shear deformation of glassy metals: Breakdown of Cauchy relationship and anelasticity, *J. Non-Cryst. Solids* **75**(1–3), 361–366 (1985).

48. R. C. Zeller and R. O. Pohl, Thermal conductivity and specific heat of noncrystalline solids, *Phys. Rev. B* **4**(6), 2029–2041 (1971).

49. P. W. Anderson, B. I. Halperin, and C. M. Varma, Anomalous low-temperature thermal properties of glasses and spin glasses, *Philos. Mag. A* **25**(1), 1–9 (1972).

50. C. A. Angell, Formation of glasses from liquids and biopolymers, *Science* **267**(5206), 1924–1935 (1995).

51. V. N. Novicov and A. P. Sokolov, Poisson's ratio and the fragility of glass-forming liquids, *Nature* **431**, 961–963 (2004).

52. L. Battezzati, Is there a link between melt fragility and elastic properties of metallic glasses? *Mater. Trans.* **46**(12), 2915–2919 (2005).

53. S. N. Yannopoulos and G. P. Johari, Poisson's ratio of glass and a liquid's fragility, *Nature* **442**(7102), E7–E8 (2006).

54. D. Kivelson, S. A. Kivelson, X. Zhao, Z. Nussinov, and G. Tarjus, A thermodynamic theory of supercooled liquids, *Physica A* **219**(1–2), 27–38 (1995).

55. M. I. Baskes, Many-body effects in fcc metals: Lennard–Jones embedded atom potential, *Phys. Rev. Lett.* **83**(13), 2592–2595 (1999).

56. W. Götze and L. Sjögren, Relaxation processes in supercooled liquids, *Rep. Prog. Phys.* **55**, 241–376 (1992).

57. P. Das, Mode-coupling theory and the glass transition in supercooled liquids, *Rev. Mod. Phys.* **76**, 785–851 (2004).

58. J. J. Lewandowski, W. H. Wang, and A. L. Greer, Intrinsic plasticity or brittleness of metallic glasses, *Philos. Mag. Lett.* **85**(2), 77–87 (2005).

59. J. J. Lewandowski and A. L. Greer, Temperature rise at shear bands in metallic glasses, *Nat. Mater.* **5**(1), 15–18 (2006).

60. A. L. Greer, Personal communication.

61. B. C. Giessen, Glass formation diagrams: A two-parameter presentation of readily glass forming binary alloy systems, in *Proceedings of 4th International Conference on Rapidly Quenched Metals*, edited by T. Masumoto and K. Suzuki (The Japan Institute of Metals, Sendai, 1982), pp. 213–216.

62. A. Inoue and A. Takeuchi, Recent progress in bulk metallic glasses, *Mater. Trans.* **43**(8), 1892–1906 (2002).

Chapter 3

MODELING: THE ROLE OF ATOMISTIC SIMULATIONS

Rachel S. Aga[1] and James R. Morris[1,2]

[1]*Materials Science and Technology Division, Oak Ridge National Laboratory, P.O. Box 2008, Oak Ridge, TN 37831-6115, USA*
[2]*Department of Materials Science and Engineering, University of Tennessee, Knoxville, TN 37996-2200, USA*

3.1 INTRODUCTION

Computational modeling is a very significant driving force in our current understanding of materials properties; many research efforts now incorporate theory and modeling to complement experimental investigations. Simulations of liquid and glasses date back to the earliest stages of simulation studies.[1–3] Simulations play several key roles in research, particularly in disordered systems. They provide key tests of theory, giving unbiased insights that lead to new theoretical ideas, and providing a key link between theory and experiment.

Despite this long history, many outstanding fundamental scientific issues in bulk metallic glasses (BMGs) remain, from the computational point of view. These challenges may be separated into two categories: those that concern generic properties not specific to a particular system; and challenges that address material specific properties that change from one material system to the next. Largely, current simulations focus on the genric properties, due to the difficulty of accurately modeling real materials. However, even for the generic properties, there are a number of challenges that must be met.

A major advantage of atomistic simulations is that a detailed picture of the model under investigation is available, and so they have been very instrumental in explaining the connection of macroscopic properties to the atomic scale. Simulations play a significant role in the development and testing of theories. For example, simulations have been extensively used to test the mode-coupling theory (MCT).[4-6] The theory predicts that at some critical temperature T_c, known as the *mode-coupling temperature*, the super-cooled liquid undergoes a structural arrest, prohibiting the system from accessing all possible states, thus, essentially undergoing an ergodic to nonergodic transition. It gives definite predictions on various correlation functions that can be calculated directly in simulations.[7,8] Simulations and MCT have played a tremendous role in elucidating a majority of what we now understand about the dynamics of glass-forming systems.

Simulations can also be used to compare with experimental results to validate the model, so that one can use simulation results to measure pro-perties not accessible to experiments. In many cases, as will be illustrated in the next sections, results of simulations motivate experimental investigations. Part of the goal of this chapter is to examine the contributions of atomic simulations to the current state of understanding of metallic glasses.

However, while simulations have provided key insights, they also face a number of limitations. Just as experimental techniques such as microscopy, calorimetry, and diffraction are critical for understanding materials by pro-viding important information, the techniques also have inherent limitations, including limited time- and length scales. Therefore, another goal of this chapter is to describe the limitations of simulations, as a cautionary note both to people performing simulations and for others reading simulation papers.

One key point is that a nonempirical electronic-structure-based approach for calculating energetics is currently limited, both in size and timescales, significantly more so than empirical potentials. Such calculations presently can accurately handle hundreds or possibly a couple of thousand atoms at most. The timescales are similarly short, due to the significant amounts of computational effort for calculating the energies and forces at each time step. Much more common is the use of classical, empirical potentials, ranging from extremely simple potentials such as hard spheres, and extending to significantly more complicated many-body potentials. The more complex potentials can correct many deficiencies common to pair potentials. How-ever, the empiricism required often makes the applicability to real materials somewhat questionable. When considering such approaches, it is important to recognize that nominally similar potentials may produce significantly different results, with occasionally unexpected behavior that is a result of the potential and not of the real material. This sensitivity is often not considered in any detail.

Even if the potentials are reasonably accurate (or at least are assumed to be acceptable for the problem at hand), there are important questions. As known experimentally, the primary influence of glass formation is the phase diagram. Thus, in addition to limitations concerning material specificity, timescale, and system size, the accuracy of the phase diagram of a multi-component model system is another challenge. Often, this is completely ignored; even a simple evaluation of the local region of the phase diagram is not performed. Section 3.2 discusses these limitations, as well as the various models used in studying properties related to BMGs.

In Sect. 3.3, we discuss the competition between nucleation and glass formation. This is a critical issue: Glass formation may simply be thought of as an absence of nucleation; this is the heart of the idea of a "critical cooling rate." We include some brief discussion on the very open questions of crystal nucleation from simulations, for both monatomic systems and alloys. We note that even for monatomic systems, the issues of nucleation and growth are an area of particular interest.[9,10] The questions of growth, phase separation, and other applications will be mentioned here.

Section 3.4 deals with the thermodynamic and dynamical properties between the melting temperature and the glass transition temperature. A number of changes in macroscopic behavior have been documented, with various suggestions as to the microscopic changes that lead to them. A number of issues have been raised directly using simulation, yet many questions remain.

In Sect. 3.5, we discuss work on the mechanical properties of BMGs. Less attention has been paid to these until recently, despite the fact that the promise of applications for these materials is limited by the usually poor ductility of these materials. Again, there are generic issues to be dealt with, such as the mechanisms of deformation, the nature of the formation, and propagation of shear bands, as well as material-specific issues. The study of mechanical properties makes clear the limitations of atomistic simulations, and particular attention will be paid to recent work[11] that demonstrates the difficulties of making definitive statements using simulations.

Section 3.6 briefly summarizes and will present a brief outlook for research directions and outstanding theoretical questions that may be particularly fruitful in the near future.

3.2 ATOMISTIC SIMULATIONS

Atomistic simulations require the use of a potential energy function that defines the interactions between atoms in the system.[12,13] In a Monte Carlo simulation, the interaction potential is used to calculate the change in energy with a change in configuration, so that the search for configurations that minimizes the free energy can be carried out. In a molecular dynamics (MD) simulation, the potential energy is used to calculate the force on an atom due to the presence of all other atoms. Solving a system of coupled equations of motions (Newton's law) then gives the positions and trajectories of atoms, a full description of the time evolution of the system. From these quantities, various thermodynamic parameters and correlation functions can be derived using statistical mechanics and thermodynamics.[12-14]

Despite the availability of various models for glass-forming systems, several questions remain to be considered. To what extent can the available interaction potentials model the real system? What about when we go beyond binary systems, what are the available interaction potentials? What information is available on melting properties and phase diagrams for a given potential?

3.2.1 Limitations

The results of atomistic simulations are dependent on the interaction potential used, and the degree to which simulations and experiments can be quantitatively compared depends on how well the interaction models the experimental system. Typically, BMGs consist of three or more components, and developing an accurate potential for such a multicomponent system is not a trivial task. Given the qualitative questions that are still being studied, many studies on glass formation avoid the issue of fitting potentials, and are instead carried out on pairwise interaction potentials such as hard spheres,[1,3,15-21] purely repulsive soft spheres,[22] and the Lennard-Jones (LJ) model.[23-34] More recently, many-body potentials are being implemented to simulate more realistic systems (see below). With any model used, there is always the question of how closely the results may be compared to measure properties. Material specificity is a major drawback of the current status of atomistic simulations. However, many properties that are generic to glass formation have been obtained, and the contributions of simulations to understanding BMGs cannot be neglected.

The phase diagram is often not known for model systems. In the absence of thermodynamic information, the degree of undercooling or the driving force for crystallization cannot be calculated. Indeed, until recently, the calculation of the melting temperature for model monatomic systems was not straightforward. The advent of coexistence simulations, combined with faster computers, has changed this,[35-37] and now even first-principles calculations are being used to examine the melting points.[38-40] For binary or higher-order systems, simulations often simply ignore the phase diagram, even though techniques exist to calculate the binary phase diagram.[41-44] Thus, another challenge is the need for accurate phase diagrams of multicomponent model systems.

Simulations are also affected by timescale limitations. A typical time step in a classical MD simulation is on the order of a femtosecond. This means that to let the system evolve on the order of a nanosecond, $\sim 10^6$ MD steps are necessary. Each step requires the calculation of interatomic interactions, which also puts a limit on the system size that can be practically simulated. Present computational power allows for easy access to simulations of 1–10 ns and system sizes of several thousands to a few million atoms. A decade ago, simulations performed at the low-temperature supercooled region consisted of only 10^5 time steps, which at that time, were considered very long runs, and the system typically consisted of a few thousand atoms.[45]

The size of typical time steps poses a corresponding limitation on the cooling rate. Cooling down a system by 1,000 K over 10^6 steps at 1 fs per step gives a cooling rate of 10^{12} K s^{-1}. This is orders of magnitude higher than experimental cooling rates used to form BMGs, which range from 10^{-3} to 500 K s^{-1}. At such high simulation cooling rates, even monatomic systems form glasses.[45–47] Indeed, it becomes a challenge for examining nucleation at moderate undercoolings.[9,48] The resulting structure and resulting mechanical properties may be different at these extreme rates.[11] This is particularly true for multicomponent systems, as the local chemical equilibrium will depend on the diffusion rates. At very fast quenches, the degree of chemical ordering will be dramatically different than at slower rates.

3.2.2 Simulation models

For many basic questions regarding the nature of the glass transition, simple simulation models or interatomic potentials are used. These models include pairwise potentials including hard spheres, purely repulsive soft spheres, and LJ potentials. While hard spheres were mostly used to describe packing in the amorphous state,[49] purely repulsive soft spheres were used in some of the first studies to look at dynamics-related behavior such as hopping[50] and localization of instantaneous normal modes.[51,52] With the formulation of MCT, the interest in testing MCT predictions by simulation emerged, using lattice-gas[53] and LJ[54] models.

Simulations that require modeling of more specific properties require more complicated interactions. For metallic systems, for example, many-body interactions should be accounted for. First-principles calculations may also be implemented, and these are particularly useful in examining stable and metastable phases, and constructing phase diagrams.

Lattice-gas model: Kob–Andersen

One approach to studying the fundamental glass transition is to simplify further, to a lattice model. This limits the number of available states, allowing for a more detailed study. This approach has been used to examine the nature of the glass transition, particularly whether there are quantities that change discontinuously or diverge at the transition, hallmarks of a true phase transition. Kob and Andersen introduced a kinetically constrained lattice-gas model, now known as the Kob–Andersen (KA) model,[53] to test a simple version of MCT, where activated hopping processes are excluded. The KA model is simple and has been a system of interest in theoretical studies that involve understanding behavior associated with glass formation.[55,56] In the model, the possible positions of the particles are those of the lattice sites of a simple cubic lattice. The particles essentially interact via a hard-core repulsion in that each site can be occupied by a maximum of only one particle. At each time step, a particle move is attempted by randomly choosing a particle and one of its six nearest neighbor lattice sites. The particle moves to the site if three conditions are all satisfied (1) the site is unoccupied, (2) the particle has m or less nearest neighbors, and (3) the site has $m + 1$ or fewer nearest neighbor sites that are occupied. If the system is *self-averaging*, i.e., a sufficiently long simulation produces thermally average quantities, it is said to be *ergodic*. A sufficient region of the phase space is sampled during the simulation for average quantities to be well defined. The breakdown of ergodicity is characteristic of a glass. For the KA model to have an ergodic to nonergodic transition at high densities, $m = 3$ is chosen. The conditions above introduce kinetic constraints preventing the particle from moving when it has the maximum number of neighbors, therefore incorporating the formation of "cages" (long-lived groups of neighboring atoms) in high-density liquids.

The system qualitatively agrees with the relaxation behavior expected in glass-forming systems from MCT. It exhibits a power-law density dependence of the diffusion coefficient, $D \sim (\rho_c - \rho)^\gamma$, where the critical density ρ_c is 0.881 particles per lattice site, and the exponent γ is 3.1. We note that similar types of models have been used to explore more complex behavior as well, including the nature of fragility in liquids.[57]

Pair potentials: Hard spheres and Lennard-Jones

The hard-sphere system is the simplest meaningful pairwise interaction potential for describing liquids and glasses.[58] The system consists of particles with hard cores, where the potential energy is infinite when cores overlap, and is zero otherwise. Early work has used the model to describe atomic packing in amorphous metals, suggesting that the structure is that of a dense random packing of hard-spheres liquid.[49] The model is useful, as the packing density where the glass transition occurs is known,[59] and is reasonably well approximated by monodisperse colloidal systems, which may be studied more directly than the related atomic systems. Such studies have been carried out to understand the dynamics of both glass transformations[60–62] and crystal nucleation.[63] Binary systems of hard spheres have also been used to look at atomic size effects.[64,65] However, due to the lack of a finite energy scale, there is no meaningful definition of temperature for the hard-sphere systems. Most studies have therefore used more realistic potentials.

One of the widely used models for glass-forming systems is the binary Lennard-Jones used by Kob and Andersen.[7,8,66,67] Results from this model will be discussed in later sections of this chapter, so we introduce it here. The model is described by the Lennard-Jones potential

$$V_{\alpha\beta}(r) = 4\varepsilon_{\alpha\beta}\left[\left(\frac{\sigma_{\alpha\beta}}{r}\right)^{12} - \left(\frac{\sigma_{\alpha\beta}}{r}\right)^{6}\right], \tag{3.1}$$

where $\varepsilon_{AA} = 1.0$, $\sigma_{AA} = 1.0$, $\varepsilon_{AB} = 1.5$, $\sigma_{AB} = 0.8$, $\varepsilon_{BB} = 0.5$, and $\sigma_{BB} = 0.88$, with 80% A particles. These parameters were chosen to improve glass formability, and are similar to simple models of $Ni_{80}P_{20}$,[68,69] with ε_{AA} and σ_{AA} setting the energy and length scales, respectively. This incorporates the "rule of thumb" for glasses that there should be a positive heat of mixing: A–B atom pairs can have significantly lower energies than either A–A or B–B pairs. The mode-coupling temperature for the system (where self-correlation times diverge) is $T_c = 0.435\varepsilon_{AA}$.[26]

Another widely used binary model for glass-forming systems is a 50/50 Lennard-Jones with the following parameters: $\sigma_{AA}/\sigma_{BB} = 1.2$ and $\varepsilon_{AA} = \varepsilon_{BB}$, with the masses $m_A/m_B = 2$. The usual Lorentz–Berthelot mixing rules are used for the interaction of atoms with different types, namely $\sigma_{AB} = \frac{1}{2}(\sigma_{AA} + \sigma_{BB})$ and $\varepsilon_{AB} = (\varepsilon_{AA}\varepsilon_{BB})^{1/2}$. The model was first used by Wahnström,[70] where he studied the dynamics and compared with MCT predictions. The mode-coupling temperature obtained is $T_c = 0.574\varepsilon_{AA}$.[71,72] Results based on this potential will be discussed later in this chapter.

Many-body potentials

In metallic systems, the details of the electronic structure on the interaction between atoms become important. In this case, the use of a many-body potential interaction is desirable. Pair potentials produce a number of un-physical effects, particularly on the elastic constants and vacancy energies. Terms beyond the pair potential form clearly affect the properties of metallic glasses, whose elasticity and open regions ("free volume") are of particular interest.

There are a number of many-body interactions in use, many of which use the same mathematical form. These describe the energy of an atom as a function of a fictitious "electron density" from the surrounding atoms, plus a pair term. These methods use the following form for the total energy

$$E = \sum_i F(\rho_i) + \frac{1}{2}\sum_{j\neq i}\phi(r_{ij}), \tag{3.2}$$

where F_i is the embedding energy experienced by atom i due to the sum of electron densities ρ^a from all other atoms. In the case of a spherically symmetric electron density, the density is written as:

$$\rho_i = \sum_{j \neq i} \rho_j^a(r_{ij}).$$ (3.3)

The methods that use this form include the embedded atom model (EAM),[73] effective medium theory,[74] the Rosato–Guillope–Legrand potentials,[75] the Glue potentials,[76] and the Sutton–Chen potentials.[77] We will refer to these as EAM potentials, for convenience. Baskes[78] has extended this to nonspherically symmetric charge densities.

An extension of the EAM method has also been proposed, where a many-body term is added to the traditional Lennard-Jones pair potential to investigate the effects of many-body interactions in fcc metals.[79] A variant of this potential is being implemented[80] to look at the effects in including many-body interactions to the pairwise modified Johnson potential for Fe, $\phi_{MJ}(r_{ij})$.[81] Following Baskes,[79] the total energy is written as

$$E = \sum_i F(\overline{\rho}_i) + \frac{1}{2} \sum_{j \neq i} \phi(r_{ij}).$$ (3.4)

The embedding energy has the form

$$F(\overline{\rho}) = \frac{A}{2} F_0 \overline{\rho}[\ln(\overline{\rho}) - 1].$$ (3.5)

The parameter A determines the strength of the many-body potential; in the limit that $A = 0$, the energy reduces to a pair potential. The embedding energy is a functional of the superposition of individual contributions $\rho(r)$ from neighboring atoms

$$\overline{\rho}_i = \frac{1}{\rho_0} \sum_{j \neq i} \rho(r_{ij}),$$ (3.6)

where $\rho(r_{ij}) = \exp[\beta(r/r_0 - 1)]$, β describes the decay of the electron density, and F_0, ρ_0, and r_0 are chosen such that the equilibrium energy and lattice spacing do not change as the many-body interaction is introduced. The pair term is given by

$$\phi(r_{ij}) = \phi_{MJ}(r_{ij}) - \frac{2}{F_0} F(\rho(r)).$$ (3.7)

In the model described, the bulk modulus is essentially preserved, while the shear modulus varies as $A\beta^2$ at $T = 0$. Morris and coworkers are using the model to study glass properties in systems with different Poisson's ratios.

Ab initio calculations

Ab initio calculations are the most accurate techniques for describing atomistic energies and forces, as they explicitly include electronic contributions to the energies. These calculations are termed "ab initio" because they utilize no empirical parameters (though there are several approximations, depending upon the method being used). While accurate, they are also the most limited in terms of numbers of atoms and timescales for simulations.

One application of these is to supplement phase diagram information, by energetics of both stable and metastable phases.[82–90] This is particularly useful for ternary systems where experimental information is limited; however, such systems require a large number of calculations. They have also been used to study local structures and dynamics in liquid metals and alloys,[40,91–95] as well as to predict the type of atomic additions needed to suppress nucleation.[83] We also note that such calculations are often important for developing interatomic potentials, particularly for alloy systems where there is limited experimental data on metastable phases. Without proper care, empirical potentials may produce very different energetics of competing crystal structures or defect energies (including interstitials, vacancies, and surface energies), and in the worst case, they can produce an incorrect ground state. By calculating these quantities accurately and fitting the empirical potential to these quantities, the potential may be optimized to avoid unphysical behavior.

3.3 NUCLEATION VS. GLASS TRANSITION

Localized density fluctuations in an undercooled liquid result in the formation of crystalline clusters. Depending on the Gibbs free energy difference (or driving force) and the interfacial free energy (or work of formation of the interface) between the liquid and the solid phases, a cluster may shrink or grow spontaneously if it has reached the critical nucleus size. A simple and widely used description of the process is classical nucleation theory (CNT).[96] However, the applicability of CNT to crystal nucleation from the liquid remains an open question. Studies have shown that even for monatomic systems, the classical description of nucleation is insufficient. Transient effects, diffuseness of the interface, and temperature dependence of the interfacial free energy are some of the factors that need to be considered.[9,48,97–99]

Using MD simulations of an EAM of aluminum,[9] Aga and coworkers[9] have demonstrated that an accurate prediction of crystal-nucleation times obtained from MD simulations can be made in the large undercooling limit, where transient nucleation dominates. In their study, all parameters – including the thermodynamic driving force, the temperature-dependent diffusion rate, and the interfacial free energy – were determined from separate simulations, so there are no fitting parameters in the comparison between theory and simulation. They show that it is necessary to include transient nucleation into the classical description of nucleation to explain the simulation results. In addition, an effective temperature-dependent interfacial free energy is introduced. The study included an examination of the system size dependence of nucleation. They find that the occurrence of nucleation in MD simulations is limited by the system size; nucleation is typically observed only if the average expected number of critical nuclei in the simulated system is at least unity.

In the absence of nucleation, glass formation may occur. If the system is cooled fast enough for the system to avoid the formation of a structure with long-range order, the system will continue to be in the supercooled liquid state. Glass transition occurs when, upon cooling, the system reaches an effectively "frozen" disordered state. More discussion on the definition of the glass transition is given in Sect. 3.4.

Of particular interest in BMG studies is glass-forming ability. Simulations are useful in understanding factors that affect glass formation, since they can provide information of the atomic level structure allowing one to directly observe the phase changes and the phase separation that occur. Simulation studies are also ideal for isolating effects of certain factors that are otherwise difficult to study experimentally. For example, by analyzing the structure of a model for a metallic glass former, it has been demonstrated that the atomic size ratio has a direct effect on whether crystallization, glass formation, or phase separation will occur.[100] For slightly different sizes, crystallization was observed. A decrease in the size ratio favored glass formation, and further decrease resulted to phase separation.

Of equal importance to studying glass-forming ability is understanding nucleation. In addition to atomic size effects, the multicomponent nature of BMGs points to the need to consider compositional effects, which give rise to phase separation and chemical ordering. Experiments have shown the onset of phase separation prior to crystallization.[101] Even in the absence of nucleation, phase separation has also been reported to occur in metallic glasses.[102–104] Atomistic simulations and other theoretical approaches have not yet played a significant role in understanding compositional effects in nucleation and glass formation.

3.4 PROPERTIES OF THE SUPERCOOLED LIQUID

The definition of a glass transition temperature is not unique, in the sense that it exhibits dependence on timescale and thermal history. From the thermodynamic point of view, it can be defined as the temperature at which the temperature dependence of the volume or enthalpy changes, typically measured on heating. Experimentally determining T_g from calorimetry, for example, involves identifying the temperature at which an abrupt change in the heat capacity occurs.

Alternatively, the kinetic definition of T_g may be used. A glass is a low-temperature liquid that has a very long relaxation time exceeding experimental timescales, such that it appears to be "frozen," and in that sense, is solid-like, but has a disordered configuration. Because of the marked slowdown in the atomic motion as T_g is approached, a definition based on transport coefficients is very reasonable. A widely accepted definition for T_g is that by Angell,[105] who proposed that T_g is the temperature at which the shear viscosity is 10^{13} poise.

To frame the microscopic issues that differentiate the liquid from the glass, we enumerate a number of changes in behavior that occur or have been proposed to occur, as the system is cooled from a high-temperature liquid (well above the melting point) to the melting temperature T_m, and the lower glass transition temperature T_g to a low-temperature glass. These changes include (but are not limited to) the following:

- A crossover from single-particle diffusion to "cage diffusion" where neighboring bonds become long-lived relative to times for motion of atoms on the length scale of interatomic separations (sometimes referred to as the *mode-coupling transition*)[26,54]
- The development of correlated diffusion, resulting in reduced dimensionality of the diffusion paths, with chains of particles moving in a coherent fashion[66,106]
- A "shear-freezing" temperature where fluctuations in atomic level shear stresses become static, possibly different from the temperature where atomic level compressive/tensile stresses become static[107,108]
- A "nucleation temperature" where the nucleation rate peaks. Alternately, this may be thought of as the "nose" of a time–temperature–transformation (TTT) diagram, the temperature where the nucleation rate reaches a maximum
- A breakdown in the Stokes–Einstein relation, relating diffusion and viscosity[21,57,109–111]
- The "Kauzmann" temperature, where the entropy of the super-cooled liquid drops below that of the crystal phase[112,113]

Having identified these, one is immediately faced with the key issues: generically, the relationship between these changes and the glass transition temperature is not at all understood; even the order of these is not clear.

Material-specific questions may be asked as well. One very broad and important question relates to the idea of a "fragile" liquid. The simplest model of diffusion, coupled with the Stokes–Einstein relation, leads to a temperature-dependent shear viscosity of the form

$$\eta(T) = \eta_0 \exp(-E_\eta / k_B T). \tag{3.8}$$

In general, however, this only holds at high temperatures, and many glass-forming liquids exhibit significant deviations of this near the glass transition temperature.[105,114–117] Liquids that have only small deviations from this behavior are known as "strong" glasses, while those that deviate significantly are known as "fragile" glasses. The fragility is defined as

$$m = \frac{\partial \ln \eta(T_g)}{\partial(T_g / T)}. \tag{3.9}$$

How does the fragility depend on interatomic potential or composition? Does the change in fragility correlate with or result from changes in the above temperatures? This remains an active experimental and theoretical question, not simply for metallic glasses,[118–120] as we will discuss below.

Fragility has its own thermodynamic and kinetic definitions as well. According to the definition of thermodynamic fragility, fragile liquids are those that, on cooling, have a more rapid approach to $\Delta S \to 0$ as a function of (T/T_m). From the kinetic definition, fragile liquids are those that have a faster approach to $\eta \to \eta(T_g)$ as a function of (T_g/T), compared with less fragile liquids.

The remainder of this section is focused on the dynamics above T_g. The dynamics as the glass transition is approached is a rich subject due to anomalous behaviors that are different from that of the normal high-temperature liquid behavior, as observed in both simulations and experiments. The changes in the temperature dependence of dynamic quantities occur at viscosities orders of magnitude lower than the 10^{13} poise value at the glass transition temperature, giving rise to various "crossover" temperatures between T_m and T_g. These include the temperature related to the change in scaling behavior of the Stokes–Einstein relation T_{SE}, the crossover to activated crossings in the potential energy landscape (PEL) T_x, the temperature T_s where the shear viscosity changes in behavior for fragile liquids, and the mode-coupling temperature T_c.

As will be seen in the discussions to follow, all these crossover temperatures seem to occur close to each other. It is therefore reasonable to ask, what are the microscopic mechanisms that relate these dynamical properties characteristic of the different crossover temperatures? This challenge has raised considerable interest in current studies of supercooled liquids and glasses. Most of these studies may have been influenced by early hypotheses by Adam and Gibbs[121] and Goldstein.[122] The idea of cooperative motion has been suggested by Adam and Gibbs, who argue that the flow in glass-forming liquids involves motion of cooperatively rearranging regions, whose size determines the temperature dependence of relaxation times.[121] Goldstein[122] describes the role of potential energy barriers to viscous flow. He states that, at shear relaxation times greater than a nanosecond, activated hopping between local potential energy barriers starts to dominate the flow process, qualitatively changing liquid viscous behavior.

The shear viscosity and diffusion coefficient give a straightforward measure of the change in the dynamics of a system as a function of temperature. The relationship between these two transport properties is given by the Stokes–Einstein relation[123]

$$D\eta = \frac{k_B T}{Ca}.$$ (3.10)

Here, a is the effective hydrodynamic diameter of a solute with translational diffusion coefficient D in a solvent of shear viscosity η, C is a constant (2π or 3π) that depends on the boundary condition imposed at the surface of the solute, k_B is Boltzmann's constant, and T is the temperature. As originally formulated, the diameter is a constant; however, in practice, the value of $D\eta/T$ is not independent of temperature, resulting in an effective temperature-dependent diameter $a(T)$.

We have calculated[111] the effective hydrodynamic diameter from MD simulations using a modified Johnson potential for Fe.[108,124–126] The scaling behavior suggested by the Stokes–Einstein relation does not hold: from $T = 1,300$ K to $T = 900$ K, the hydrodynamic diameter decreases by approximately an order of magnitude, indicating a change in behavior in the liquid. Above 1,300–3,000 K, the diameter only weakly depends on temperature.

This change in the scaling behavior was observed in earlier experimental investigations of the dynamics in supercooled organic and ionic liquids.[127] More recently, other studies – including experimental investigations on $Zr_{46.7}Ti_{8.3}Cu_{7.5}Ni_{10}Be_{27.5}$[109] and simulations using the LJ Kob–Andersen model,[110] a two-dimensional Yukawa liquid,[128,129] and the hard-sphere system[21] – find a "breakdown" in the Stokes–Einstein relation as well. The microscopic mechanism for the deviation is not known, and attempts to explain the temperature dependence are being made.[57,130,131] The change in the diffusion–viscosity scaling is just one of the many interesting issues that stimulate current investigations of supercooled liquid dynamics near T_g.

MCT[6] treats the glass transition problem from a dynamical point of view. In its idealized version, the shear viscosity diverges at the mode-coupling temperature T_c as given by

$$\eta = \eta_0 \left[\frac{T - T_c}{T_c} \right]^\gamma . \tag{3.11}$$

The theory predicts a purely dynamical phase transition at T_c, where the system undergoes structural arrest and becomes nonergodic. However, the structure may not be "frozen" yet at T_c due to the presence of activated hopping processes that are neglected in the ideal form. These contribute to the flow in the liquid and they could possibly restore ergodicity in the system. Thus, the mode-coupling temperature T_c marks a crossover between two types of dynamic behavior: one that is based on single-particle diffusion and one that is hopping dominated. It will become evident in later discussions that the hopping processes are strongly cooperative, involving correlated motion of a number of neighboring atoms.

An important early study by Rössler[127] showed that the breakdown in the Stokes–Einstein relation for various molecular liquids occurs close to T_c, indicating a change in the diffusion mechanism. Here, it was demonstrated that a graph of τ/η vs. $1/T$, where τ is the rotational diffusion correlation time, gives a positive slope at high temperatures. At low temperatures the slope is negative, and the crossover between these regimes is T_c.

In the plot presented by Angell,[105] where $\log_{10}(\eta)$ is plotted against T_g/T, fragile liquids show a slowly increasing function on the high-temperature side. Near a certain temperature T_s, the shear viscosity shifts to a rapidly increasing function of temperature. The shoulder in the plot where the shift occurs has been said to occur at the same viscosity values where T_c is normally observed.[7,8,70] Once again, this illustrates that the mode-coupling temperature is a crossover between two dynamic regimes. But the question remains, what is the underlying mechanism that gives rise to these two dynamic behaviors? Are the activated hopping processes related to the observed strong temperature dependence of viscosity? If so, how exactly do such processes affect macroscopic transport properties? How is the change in scaling factor of the Stokes–Einstein relation affected by this type of motion?

Because of the role of dynamics in glass transition, and the significance of the dynamical phase transition predicted by MCT in supercooled liquids, studies devoted to acquiring a detailed understanding of atomic motions below T_g are of particular interest. One of the first studies to directly show the existence of strongly correlated jump motion or hopping was performed on a simulation of a binary soft-sphere mixture.[50] The simulations were done just above the glass transition temperature. Although no direct measurement of the activation energy was made, it was found that a group of four to ten dynamically linked nearest neighbor atoms participate in the hopping motions.

The presence of particles participating in coordinated jump motions gives rise to what is now known as dynamical heterogeneities, where the super-cooled liquid consists of regions of slow-moving particles and clusters of "fast" or "mobile" particles. In a simulation of a binary LJ system, Kob[54,132] showed that as the temperature is decreased, the fastest particles become correlated, forming clusters that increase in size as the temperature decreases. It was also shown that the fast particles had significantly shorter relaxation times than the rest of the particles in the system. This supports the presence of heterogeneous relaxation, which could be giving rise to the nonexponential loss of correlation in supercooled liquids.[133]

A study by Weeks and coworkers[62] gave a direct observation of the structural relaxation in supercooled liquids through confocal microscopy imaging of the three-dimensional dynamics of a colloidal suspension. In their experiments, they were able to identify clusters of fast-moving parti-cles providing an experimental verification of the existence of dynamical heterogeneities.

Attempts are being made to explain the relationship between the dynamics of particles in the supercooled liquid and its PEL. The work by Stillinger and Weber[134] has been a motivation in understanding supercooled liquid dynamics in terms of the transitions between "inherent structures" and local minima in the PEL. Wahnström[70] investigated in detail the dynamics of a binary LJ mixture. The study correlates cooperative motion with activated hopping in the PEL, providing a clue of the significance of the role of PEL in supercooled liquid dynamics. It was demonstrated that cooperative motion begins when the timescale for structural relaxations is about a nanosecond. Recalling the proposed mechanism by Goldstein,[122] Wahnström states that activated jumps dominate shear flow when the shear relaxation time is greater than a nanosecond. The study therefore suggests that cooperative motion and activated processes in the PEL may be strongly correlated.

Using the same binary LJ model as Wahnström, Schrøder and coworkers[71] observed that the cooperative motion of particles is in fact associated with transitions between local minima. In agreement with the proposal by Goldstein,[122] the study shows that there exists a crossover temperature T_x below which transitions between inherent structures occur, and above which vibrations around inherent structures occur. The temperature T_x is found close to the MCT temperature T_c. Relating PEL work to the breakdown in Stokes–Einstein relation, the change in dynamics is proposed to be due to the transition from a hydrodynamically governed to a landscape-dominated dynamics.[71,135–138]

3.5   MECHANICAL PROPERTIES AND STRUCTURE

Ultimately, interest in the application of BMG materials resides in their unusual mechanical properties and their temperature dependence.[139] Their high strength is due to a lack of plastic deformation modes at normal temperatures; this lack of modes also typically leads to poor ductility. Commonly, the failure mechanisms are through shear banding, a localized deformation that leads to cracking. The shear-banding process is still poorly understood, in particular the role of heat generation during deformation, and the change in structure in the shear band. Stability analysis for continuum materials can provide important insights into the onset and dynamics of shear banding,[140,141] but cannot provide insights into the changes that may occur in both structure and material response in the region of a shear band. Instead, atomistic information should provide inputs to longer length scale modeling, connecting local structure with constitutive behaviors such as strain softening.

Simulations play a role in understanding generic mechanical properties of glass-forming systems, allowing for an atomistic description of the process. However, this area remains poorly understood. In this section, we will not review the "standard" models of "shear transition zones" proposed by Argon[142] or the "free volume theory" by Spaepen,[143] but limit ourselves to recent work on simulations. This section is kept brief, due to the rapidly changing information in this area.

In principle, studying deformation using molecular dynamics is straight-forward: from a liquid, quench the system to form a glass, and then apply the deformation and "see what happens." Many simulations follow this approach. However, this is an ideal case to demonstrate the subtlety of such simulations. As a first step, one must choose the type of deformation to perform. For example, a simple shear simulation can examine the change from elastic deformation to Newtonian flow.[33] On the other hand, a "tension" experiment more closely matches experimental conditions.[144,145] If resolved shear stresses were the critical issues, then these would give identical results. However, simulations of systems in tension do not produce shear bands; instead, a "necking" instability occurs.[144,145] This is consistent with continuum predictions that show that necking should always be initiated prior to shear banding.[140] Strikingly, the simulated elongation can be quite large, over 40%. Experimentally, the elongational strain of metallic glasses at failure is typically only a couple of percent at most, and is reduced by large strain rates. The strain rates in simulation are orders of magnitude faster than simulation, typically $10^8 \, s^{-1}$ or higher. Not surprisingly, the simulated strain at which failure occurs is altered by "notching" the sample to produce a local stress concentration, but still the resultant strain is surprisingly large.[144,145] This notch also produces localized deformation, but

it is unclear whether this has the same character as homogeneous deformations that produce localized strains.

In simple strain, at low temperatures, a very small strain results in an
elastic deformation with a proportional stress. However, under constant strain-
rate conditions, a "steady state" Newtonian flow can be achieved with an
essentially constant stress state. Again, such simulations appear to be straight-
forward, and work using this approach has shown that this can result in shear
localization.[146] However, the results can be sensitive to "sample preparation": The history of the glass being deformed affects the resultant deformation. One important demonstration of this[33] using the Kob–Anderson
potential showed that different initial temperatures of the liquid, subsequently
quenched to the same temperature and deformed under identical conditions,
can produce *either* localized flow or homogeneous flow. The structure of the
glass, as measured by the pair distribution function, was nearly the same
in all cases; however, a three-body correlation function which measures
the degree of bond alignment showed differences, and also changed under
deformation. Presumably, both the density of the systems and the chemical
order are also affected by the initial temperature of the liquid, and these
quantities certainly may affect these results. The observed change in behavior
suggests that the initial temperature of the liquid can affect the mechanical
properties of the glass; however, given the extremely rapid quench rates in
simulation, it is not clear that the conclusion carries over to experimental
quench rates.

Even for identical initial structures, the history of the system plays a role.
Again using the Kob–Andersen potential,[26] Rottler and Robbins[32] examined
the role of "aging." Instead of immediately deforming the quenched structure,
they performed simulations of the system with no deformation prior to
starting the deformation, and showed that the shear yield stress increased
logarithmically with aging time. In analogy with friction models, they demonstrate that the effects of both waiting time and strain rate may be combined
into a single universal function.

While at low temperatures, one expects a finite stress in response to a
small strain, in the liquid phase small static shears should only produce a
transient stress that decays to zero as a function of time. Recent work[34] has
examined the temperature dependence of the shear yield stress for the Kob–
Andersen LJ model, and shown that the results appear to follow a MCT
behavior, with a yield stress for T below T_c behaving as

$$\sigma_y(T) - \sigma_y(T_c) = \sigma_0 \left(1 - \frac{T}{T_c} \right)^{0.5}. \tag{3.12}$$

The fitted value of T_c is found to be slightly lower than the previously calculated MCT.[26] This may be due to difficulties of measuring the low strain-rate limit of the yield stress near the critical temperature.

3.6   CONCLUSIONS

Rather than reviewing the previous sections, we conclude by examining a few of the many interesting fundamental issues in BMGs, particularly those where we see key opportunities. Although the current focus of many researchers is the discovery of new alloys and the improvement of existing ones (in terms of both critical cooling rate and size of cast system), there are a number of interesting scientific questions that are raised or may be studied with more care using these alloys.

In Sect. 3.3, we discussed results on nucleation of crystals from liquid or amorphous phases. This is important, as one view of glass formation is that it is the *avoidance* of crystal nucleation. Most theoretical studies of nucleation[97,99,147,148] have emphasized systems that do not change composition. However, many BMGs are based on alloy systems with equilibrium intermetallic phases that have limited compositional ranges; as a result, the nucleation requires a change of composition. This "coupled" problem of compositional change and crystal formation is a very interesting area that remains to be fully addressed.[149–151] The same issues occur during devitrification on heating. Simultaneous time-resolved small-angle scattering and diffraction experiments on BAM-11[101] show the coupling of these, and suggest that chemical phase separation occurs just prior to the formation of the crystalline phase. Chemical phase separation on length scales of tens of nanometers can occur in metallic glasses even in the absence of nucleation.[102–104] In that case, nucleation is occurring in a chemically *inhomogeneous* phase, and this will obviously affect the nucleation dynamics. The role of chemical changes can play a role, even when the glass composition is identical to the stable crystalline phase. Amorphous Zr_2Pd devitrifies first into a metastable quasicrystalline phase[152] with a composition closer to Zr_3Pd,[153,154] and subsequently into the stable C16 ($MoSi_2$) phase. This is suggested to occur due to local icosahedral order in the amorphous phase,[155–158] lowering the crystal/amorphous interfacial free energy; however, this still neglects the compositional changes necessary for the transition. Similar issues arise for other systems.[159] It is also interesting that this does not occur in the chemically similar Zr_2Cu system,[157,158,160] but even small amounts (~1 at.%) of Pd can produce the quasicrystalline phase.[152–157,160–162] The role of small chemical additions is an important, yet poorly understood, issue affecting many metallic glasses.[163–165] We also note that chemical phase separation in the liquid may assist in the formation of nanocrystals, formed directly from the quench.[166–169]

To date, the debate on whether the glass transition is a thermodynamic phase transition or a purely kinetic transition is still unresolved. In Sect. 3.4, we discussed the thermodynamic and kinetic properties associated with the glass transition. Many interesting changes in the dynamics occur prior to glass transition, as we discussed in Sect. 3.4. Rapid changes in shear viscosity, from 10^{13} poise to ~1 poise, occur over a relatively small temperature range, from the glass transition temperature T_g to the melting temperature T_m. Over this temperature range, a crossover from high temperature to supercooled liquid behavior exists. Several crossover temperatures, all appearing to be close to each other, have been associated with the change in dynamical behavior. Although this topic has received a considerable amount of attention in the literature, the relationship between the identified crossover temperatures is not yet clear.

In Sect. 3.5, we discussed simulations of deformation of metallic glasses. Clearly, there is much to do in this area. Experimentally, the recent work demonstrating that compressive deformation results in strain-induced softening, consistent with softer shear-band boundaries,[170] indicates that shear banding changes the structure in the glass, and that such softening is not due solely to localized heating during the shear-band formation. This is supported by the observation that the change in response can be reversed by annealing the sample. However, the details of the structural changes at shear band, and why they result in softening, are not clear. The simulations of Albano and Falk[33] show the difficulty in making definitive statements: Identical compositions and very similar structures can result in either localized deformation or homogeneous flow under identical simulation conditions.

Less is known about the role of nanocrystals in the deformation process. Clearly, the size and morphology of nanocrystals play a role: Dendritic-type morphologies appear to improve the mechanical response more than compact nanocrystals.[171,172] Also important may be the mechanical behaviors of the nanocrystals. Most nanocrystals occurring in BMGs are hard, brittle intermetallic phases, adding strength and resistance to shear banding. However, a ZrCu metallic glass can form ZrCu B2 nanocrystals; this phase is a shape-memory alloy similar to NiTi,[173–178] with a high-temperature B2 phase. The B2-based shape-memory alloys can exhibit significant ductilities; this may contribute to the unusual mechanical properties reported in this system.[179–181]

In summary, we see many fundamental issues concerning the behavior of BMGs, and opportunities for broader studies enabled by the study of BMGs. Nucleation and growth of the crystal phase are still poorly understood, either during the quench or during devitrification on heating. The role of composition is partially understood, but the role of minor alloying additions is still a major question, particularly when such additions make qualitative changes in behavior (such as different devitrification pathways or dramatically improved glass formability). The relationship between fragility, elasticity, and plasticity has some tantalizing experimental connections, with some theoretical suggestions but many remaining questions. Plastic deformation is perhaps the most difficult of the subjects to address, particularly at the atomistic level, due to the separation of time- and length scales between experiment and theory, as well as the absence of careful experimental work that can differentiate between different theoretical descriptions.

ACKNOWLEDGMENTS

This work stems from many useful discussions with J. Eckert, T. Egami, Y. Gao, E. P. George, L. Granasy, J. J. Hoyt, K. Kelton, Z. P. Lu, M. Kramer, M. K. Miller, R. Napolitano, D. Sordelet, X. L. Wang, and Y. Y. Ye. We particularly thank T. Egami and G. S. Painter for critically reviewing the manuscript. We acknowledge support from the Division of Materials Science and Engineering, Office of Basic Energy Sciences, US Department of Energy under contract DE-AC05-00OR-22725 with UT-Battelle.

REFERENCES

1. B. J. Alder and T. E. Wainwright, Phase transition for a hard sphere system, *J. Chem. Phys.* **27**(5), 1208–1209 (1957).
2. B. J. Alder and T. E. Wainwright, Studies in molecular dynamics. II. Behavior of a small number of elastic spheres, *J. Chem. Phys.* **33**(5), 1439–1451 (1960).
3. W. W. Wood and J. D. Jacobson, Preliminary results from a recalculation of the Monte Carlo equation of state of hard spheres, *J. Chem. Phys.* **27**(5), 1207–1208 (1957).
4. U. Bengtzelius, W. Gotze, and A. Sjolander, Dynamics of supercooled liquids and the glass-transition, *J. Phys. C: Solid State* **17**(33), 5915–5934 (1984).
5. E. Leutheusser, Dynamical model of the liquid–glass transition, *Phys. Rev. A* **29**(5), 2765–2773 (1984).
6. W. Gotze and L. Sjogren, Relaxation processes in supercooled liquids, *Rep. Prog. Phys.* **55**(3), 241–376 (1992).
7. W. Kob and H. C. Andersen, Testing mode-coupling theory for a supercooled binary Lennard-Jones mixture. I. The Van Hove correlation-function, *Phys. Rev. E* **51**(5), 4626–4641 (1995).
8. W. Kob and H. C. Andersen, Testing mode-coupling theory for a supercooled binary Lennard-Jones mixture. II. Intermediate scattering function and dynamic susceptibility, *Phys. Rev. E* **52**(4), 4134–4153 (1995).
9. R. S. Aga, J. R. Morris, J. J. Hoyt, and M. Mendelev, Quantitative parameter-free prediction of simulated crystal-nucleation times, *Phys. Rev. Lett.* **96**(24), 245701 (2006).

10. F. H. Streitz, J. N. Glosli, and M. V. Patel, Beyond finite-size scaling in solidification simulations, *Phys. Rev. Lett.* **96**(22), 225701 (2006).

11. X. Y. Fu, M. L. Falk, and D. A. Rigney, Sliding behavior of metallic glass. Part II. Computer simulations, *Wear* **250**, 420–430 (2001).

12. M. P. Allen and D. J. Tildesley, *Computer Simulation of Liquids* (Oxford University Press, New York, 1987).

13. D. Frenkel and B. Smit, *Understanding Molecular Simulation*, 2nd ed. (Academic, London, 2001).

14. K. Binder, J. Horbach, W. Kob, W. Paul, and F. Varnik, Molecular dynamics simulations, *J. Phys.: Condens. Matter* **16**(5), S429–S453 (2004).

15. W. G. Hoover and F. H. Ree, Melting transition and communal entropy for hard spheres, *J. Chem. Phys.* **49**(8), 3609–3717 (1968).

16. X. C. Zeng and D. W. Oxtoby, Density functional theory for freezing of a binary hard sphere liquid, *J. Chem. Phys.* **93**(6), 4357–4363 (1990).

17. R. Agrawal and D. A. Kofke, Thermodynamic and structural properties of model systems at solid–fluid coexistence. I. Fcc and bcc soft spheres, *Mol. Phys.* **85**(1), 23–42 (1995).

18. T. S. Grigera and G. Parisi, Fast Monte Carlo algorithm for supercooled soft spheres, *Phys. Rev. E* **63**(4), 045102 (2001).

19. L. Gránásy, T. Pusztai, G. Tóth, and Z. Jurek, Phase field theory of crystal nucleation in hard sphere liquid, *J. Chem. Phys.* **119**(19), 10376–10382 (2003).

20. G. Parisi and F. Zamponi, The ideal glass transition of hard spheres, *J. Chem. Phys.* **123**(14), 144501 (2005).

21. S. K. Kumar, G. Szamel, and J. F. Douglas, Nature of the breakdown in the Stokes–Einstein relationship in a hard sphere fluid, *J. Chem. Phys.* **124**(21), 214501 (2006).

22. B. B. Laird and A. D. J. Haymet, The crystal liquid interface of a body-centered-cubic-forming substance – Computer-simulations of the R-6 potential, *J. Chem. Phys.* **91**(6), 3638–3646 (1989).

23. J. P. Hansen and L. Verlet, Phase transitions of the Lennard-Jones system, *Phys. Rev.* **184**(1), 151–161 (1969).

24. J. Q. Broughton, G. H. Gilmer, and K. A. Jackson, Crystallization rates of a Lennard-Jones liquid, *Phys. Rev. Lett.* **49**(20), 1496–1500 (1982).

25. E. Burke, J. Q. Broughton, and G. H. Gilmer, Crystallization of fcc (111) and (100) crystal–melt interfaces: A comparison by molecular dynamics for the Lennard-Jones system, *J. Chem. Phys.* **89**(2), 1030–1041 (1988).

26. W. Kob and H. C. Andersen, Scaling behavior in the beta-relaxation regime of a supercooled Lennard-Jones mixture, *Phys. Rev. Lett.* **73**(10), 1376–1379 (1994).

27. R. Agrawal and D. A. Kofke, Thermodynamic and structural properties of model systems at solid–fluid coexistence. II. Melting and sublimation of the Lennard-Jones system, *Mol. Phys.* **85**, 43–59 (1995).

28. Y. C. Shen and D. W. Oxtoby, bcc symmetry in the crystal–melt interface of Lennard-Jones fluids examined through density functional theory, *Phys. Rev. Lett.* **77**(17), 3585–3588 (1996).

29. H. E. A. Huitema, B. van Hengstum, and J. P. van der Eerden, Simulation of crystal growth from Lennard-Jones solutions, *J. Chem. Phys.* **111**(22), 10248–10260 (1999).

30. M. A. Barroso and A. L. Ferreira, Solid–fluid coexistence of the Lennard-Jones system from absolute free energy calculations, *J. Chem. Phys.* **116**(16), 7145–7150 (2002).

31. J. R. Morris and X. Song, The anisotropic free energy of the Lennard-Jones crystal–melt interface, *J. Chem. Phys.* **119**(7), 3920–3925 (2003).

32. J. Rottler and M. O. Robbins, Unified description of aging and rate effects in yield of glassy solids, *Phys. Rev. Lett.* **95**(22), 225504 (2005).

33. F. Albano and M. L. Falk, Shear softening and structure in a simulated three-dimensional binary glass, *J. Chem. Phys.* **122**(15), 154508 (2005).

34. F. Varnik and O. Henrich, Yield stress discontinuity in a simple glass, *Phys. Rev. B* **73**(17), 174209 (2006).

35. J. R. Morris, C. Z. Wang, K. M. Ho, and C. T. Chan, Melting line of aluminum from simulations of coexisting phases, *Phys. Rev. B* **49**(5), 3109–3115 (1994).

36. J. R. Morris and X. Song, The melting lines of model systems calculated from coexistence simulations, *J. Chem. Phys.* **116**(21), 9352–9358 (2002).

37. J. B. Sturgeon and B. B. Laird, Adjusting the melting point of a model system via Gibbs–Duhem integration: Application to a model of aluminum, *Phys. Rev. B* **62**(22), 14720–14727 (2000).

38. D. Alfe, Melting curve of MgO from first-principles simulations, *Phys. Rev. Lett.* **94**(23), 235701 (2005).

39. L. Vocadlo, D. Alfe, G. D. Price, and M. J. Gillian, Ab initio melting curve of copper by the phase coexistence approach, *J. Chem. Phys.* **120**(6), 2872–2878 (2004).

40. D. Alfe, First-principles simulations of direct coexistence of solid and liquid aluminum, *Phys. Rev. B* **68**(6), 064423 (2003).

41. H. Ramalingam, M. Asta, A. van de Walle, and J. J. Hoyt, Atomic-scale simulation study of equilibrium solute adsorption at alloy solid–liquid interfaces, *Interface Sci.* **10**(2–3), 149–158 (2002).

42. M. R. Hitchcock and C. K. Hall, Solid–liquid phase equilibrium for binary Lennard-Jones mixtures, *J. Chem. Phys.* **110**(23), 11433–11444 (1999).

43. R. Sibug-Aga and B. B. Laird, Structure and dynamics of the interface between a binary hard-sphere crystal of NaCl type and its coexisting binary fluid, *Phys. Rev. B* **66**(14), 144106 (2002).

44. R. Sibug-Aga and B. B. Laird, Simulations of binary hard-sphere crystal–melt interfaces: Interface between a one-component fcc crystal and a binary fluid mixture, *J. Chem. Phys.* **116**(8), 3410–3419 (2002).

45. J. J. Ullo and S. Yip, Dynamical transition in a dense fluid approaching structural arrest, *Phys. Rev. Lett.* **54**(14), 1509–1512 (1985).

46. U. Bengtzelius, Dynamics of a Lennard-Jones system close to the glass-transition, *Phys. Rev. A* **34**(6), 5059–5069 (1986).

47. M. I. Mendelev, J. Schmalian, C. Z. Wang, J. R. Morris, and K. M. Ho, Interface mobility and the liquid–glass transition in a one-component system described by an embedded atom method potential, *Phys. Rev. B* **74**(10), 104206 (2006).

48. H. E. A. Huitema, J. P. van der Eerden, J. J. M. Janssen, and H. Human, Thermodynamics and kinetics of homogeneous crystal nucleation studied by computer simulation, *Phys. Rev. B* **62**(22), 14690–14702 (2000).

49. G. S. Cargill, Dense random packing of hard spheres as a structural model for noncrystalline metallic solids, *J. Appl. Phys.* **41**(5), 2248–2250 (1970).

50. H. Miyagawa, Y. Hiwatari, B. Bernu, and J. P. Hansen, Molecular-dynamics study of binary soft-sphere mixtures – Jump motions of atoms in the glassy state, *J. Chem. Phys.* **88**(6), 3879–3886 (1988).

51. S. D. Bembenek and B. B. Laird, Instantaneous normal modes and the glass transition, *Phys. Rev. Lett.* **74**(6), 936–939 (1995).

52. B. B. Laird and H. R. Schober, Localized low-frequency vibrational-modes in a simple-model glass, *Phys. Rev. Lett.* **66**(5), 636–639 (1991).

53. W. Kob and H. C. Andersen, Kinetic lattice-gas model of cage effects in high-density liquids and a test of mode-coupling theory of the ideal-glass transition, *Phys. Rev. E* **48**(6), 4364–4377 (1993).

54. M. Nauroth and W. Kob, Quantitative test of the mode-coupling theory of the ideal glass transition for a binary Lennard-Jones system, *Phys. Rev. E* **55**(1), 657–667 (1997).

55. S. Franz, R. Mulet, and G. Parisi, Kob–Andersen model: A nonstandard mechanism for the glassy transition, *Phys. Rev. E* **65**(2), 021506 (2002).

56. V. Lecomte, C. Appert-Rolland, and F. van Wijland, Chaotic properties of systems with Markov dynamics, *Phys. Rev. Lett.* **95**(1), 010601 (2005).

57. Y. J. Jung, J. P. Garrahan, and D. Chandler, Excitation lines and the breakdown of Stokes–Einstein relations in supercooled liquids, *Phys. Rev. E* **69**(6), 061205 (2004).

58. J. D. Weeks, D. Chandler, and H. C. Andersen, Role of repulsive forces in determining the equilibrium structure of simple liquids, *J. Chem. Phys.* **54**(12), 5237–5247 (1971).

59. P. N. Pusey and W. Vanmegen, Phase-behavior of concentrated suspensions of nearly hard colloidal spheres, *Nature* **320**(6060), 340–342 (1986).

60. W. Vanmegen, S. M. Underwood, and P. N. Pusey, Nonergodicity parameters of colloidal glasses, *Phys. Rev. Lett.* **67**(12), 1586–1589 (1991).

61. W. Vanmegen and P. N. Pusey, Dynamic light-scattering study of the glass-transition in a colloidal suspension, *Phys. Rev. A* **43**(10), 5429–5441 (1991).

62. E. R. Weeks, J. C. Crocker, A. C. Levitt, A. Schofield, and D. A. Weitz, Three-dimensional direct imaging of structural relaxation near the colloidal glass transition, *Science* **287**(5453), 627–631 (2000).

63. U. Gasser, E. R. Weeks, A. Schofield, P. N. Pusey, and D. A. Weitz, Real-space imaging of nucleation and growth in colloidal crystallization, *Science* **292**, 258–262 (2001).

64. M. D. Eldridge, P. A. Madden, P. N. Pusey, and P. Bartlett, Binary hard-sphere mixtures – A comparison between computer-simulation and experiment, Mol. Phys. **84**(2), 395–420 (1995).

65. P. Jalali and M. Li, Atomic size effect on critical cooling rate and glass formation, *Phys. Rev. B* **71**(1), 014206 (2005).

66. C. Donati, S. C. Glotzer, P. H. Poole, W. Kob, and S. J. Plimpton, Spatial correlations of mobility and immobility in a glass-forming Lennard-Jones liquid, Phys. Rev. E **60**(3), 3107–3119 (1999).

67. T. B. Schrøder and J. C. Dyre, Hopping in a supercooled binary Lennard-Jones liquid, *J. Non-Cryst. Solids* **235–237**, 331–334 (1998).

68. T. A. Weber and F. H. Stillinger, Interactions, local order, and atomic-rearrangement kinetics in amorphous nickel-phosphorous alloys, Phys. Rev. B **32**(8), 5402–5411 (1985).

69. T. A. Weber and F. H. Stillinger, Local order and structural transitions in amorphous metal–metalloid alloys, *Phys. Rev. B* **31**(4), 1954–1963 (1985).

70. G. Wahnström, Molecular-dynamics study of a supercooled 2-component Lennard-Jones system, *Phys. Rev. A* **44**(6), 3752–3764 (1991).

71. T. B. Schrøder, S. Sastry, J. C. Dyre, and S. C. Glotzer, Crossover to potential energy landscape dominated dynamics in a model glass-forming liquid, J. Chem. Phys. **112**(22), 9834–9840 (2000).

72. S. C. Glotzer, V. N. Novikov, and T. B. Schrøder, Time-dependent, four-point density correlation function description of dynamical heterogeneity and decoupling in supercooled liquids, *J. Chem. Phys.* **112**(2), 509–512 (2000).

73. M. S. Daw and M. I. Baskes, Semiempirical, quantum mechanical calculation of hydrogen embrittlement in metals, *Phys. Rev. Lett.* **50**(17), 1285–1288 (1983).

74. K. W. Jacobsen, J. K. Norskov, and M. J. Puska, Interatomic interactions in the effective-medium theory, *Phys. Rev. B* **35**(14), 7423–7442 (1987).

75. V. Rosato, M. Guillope, and B. Legrand, Thermodynamical and structural-properties of fcc transition metals using a simple tight-binding model, *Philos. Mag. A* **59**(2), 321–336 (1989).

76. F. Ercolessi and J. B. Adams, Interatomic potentials from first-principles calculations: The force-matching method, *Europhys. Lett.* **26**(8), 583–588 (1994).

77. A. P. Sutton and J. Chen, Long-range Finnis Sinclair potentials, Philos. Mag. Lett. **61**(3), 139–146 (1990).

78. M. I. Baskes, Modified embedded-atom potentials for cubic materials and impurities, *Phys. Rev. B* **46**(5), 2727–2742 (1992).

79. M. I. Baskes, Many-body effects in fcc metals: A Lennard-Jones embedded-atom potential, *Phys. Rev. Lett.* **83**(13), 2592–2595 (1999).

80. J. R. Morris, R. S. Aga, T. Egami, and V. Levashov (2006) (in preparation).

81. D. Srolovitz, V. Vitek, and T. Egami, An atomistic study of deformation of amorphous metals, *Acta Metall.* **31**(2), 335–352 (1983).

82. M. C. Gao, N. Unlu, G. J. Shiflet, M. Mihalkovic, and M. Widom, Reassessment of Al–Ce and Al–Nd binary systems supported by critical experiments and first-principles energy calculations, *Metall. Mater. Trans. A* **36**(12), 3269–3279 (2005).

83. M. Mihalkovic and M. Widom, Ab initio calculations of cohesive energies of Fe-based glass-forming alloys, *Phys. Rev. B* **70**(14), 144107 (2004).

84. S. H. Zhou and R. E. Napolitano, Phase equilibria and thermodynamic limits for partitionless crystallization in the Al–La binary system, *Acta Mater.* **54**(3), 831–840 (2006).

85. R. G. Hennig, A. E. Carlsson, K. F. Kelton, and C. L. Henley, Ab initio Ti–Zr–Ni phase diagram predicts stability of icosahedral TiZrNi quasicrystals, Phys. Rev. B **71**(14), 144103 (2005).

86. R. G. Hennig, K. F. Kelton, A. E. Carlsson, and C. L. Henley, Structure of the icosahedral Ti–Zr–Ni quasicrystal, *Phys. Rev. B* **67**(13), 134202 (2003).

87. W. J. Golumbfskie, R. Arroyave, D. Shin, and Z.-K. Liu, Finite-temperature thermo-dynamic and vibrational properties of Al–Ni–Y compounds via first-principles calcula-tions, *Acta Mater.* **54**(8), 2291–2304 (2006).

88. R. Arroyave, A. van de Walle, and Z. K. Liu, First-principles calculations of the Zn–Zr system, *Acta Mater.* **54**(2), 473–482 (2006).

89. W. Golumbfskie and Z. K. Liu, CALPHAD/first-principles re-modeling of the Co–Y binary system, *J. Alloy Compd.* **407**(1–2), 193–200 (2006).

90. Y. Zhong, M. Yang, and Z. K. Liu, Contribution of first-principles energetics to Al–Mg thermodynamic modeling, *Calphad* **29**(4), 303–311 (2005).

91. M. Asta, V. Ozolins, J. J. Hoyt, and M. van Schilfgaarde, Ab initio molecular-dynamics study of highly nonideal structural and thermodynamic properties of liquid Ni–Al alloys, *Phys. Rev. B* **64**(2), 020201 (2001).

92. S. Y. Wang, C. Z. Wang, F. C. Chuang, J. R. Morris, and K. M. Ho, Ab initio molecular dynamics simulation of liquid $Al_{88}Si_{12}$ alloys, *J. Chem. Phys.* **122**(3), 034508 (2005).

93. S. Y. Wang, C. Z. Wang, F. C. Chuang et al., Ab initio molecular dynamics simulation of liquid Al_xGe_{1-x} alloys, *Phys. Rev. B* **70**(22), 224205 (2004).

94. G. Kresse and J. Hafner, Ab initio molecular dynamics for liquid metals, *Phys. Rev. B* **47**(1), 558–561 (1993).

95. J.-D. Chai, D. Stroud, J. Hafner, and G. Kresse, Dynamic structure factor of liquid and amorphous Ge from ab initio simulations, *Phys. Rev. B* **67**(10), 104205 (2003).

96. D. Kashiev, *Nucleation: Basic Theory with Applications* (Butterworth-Heinemann, Oxford, 2000).

97. L. Granasy, T. Pusztai, G. Toth, Z. Jurek, M. Conti, and B. Kvamme, Phase field theory of crystal nucleation in hard sphere liquid, *J. Chem. Phys.* **119**(19), 10376–10382 (2003).

98. L. Granasy and T. Pusztai, Diffuse interface analysis of crystal nucleation in hard-sphere liquid, *J. Chem. Phys.* **117**(22), 10121–10124 (2002).

99. L. Granasy, T. Borzsonyi, and T. Pusztai, Nucleation and bulk crystallization in binary phase field theory, *Phys. Rev. Lett.* **88**(20), 206105 (2002).

100. H. J. Lee, T. Cagin, W. L. Johnson, and W. A. Goddard III, Criteria for formation of metallic glasses: The role of atomic size ratio, *J. Chem. Phys.* **119**(18), 9858–9870 (2003).

101. X. L. Wang, J. Almer, C. T. Liu, Y. D. Wang, J. K. Zhao, A. D. Stoica, D. R. Haeffner, and W. H. Wang, In situ synchrotron study of phase transformation behaviors in bulk metallic glass by simultaneous diffraction and small angle scattering, *Phys. Rev. Lett.* **91**(26), 265501 (2003).

102. S. C. Glade, J. F. Loffler, S. Bossuyt, W. L. Johnson, and M. K. Miller, Crystallization of amorphous $Cu_{47}Ti_{34}Zr_{11}Ni_8$, *J. Appl. Phys.* **89**(3), 1573–1579 (2001).

103. M. K. Miller, D. J. Larson, R. B. Schwarz, and Y. He, Decomposition in $Pd_{40}Ni_{40}P_{20}$ metallic glass, *Mater. Sci. Eng. A* **250**(1), 141–145 (1998).

104. M. K. Miller, T. D. Shen, and R. B. Schwarz, Atom probe tomography study of the decomposition of a bulk metallic glass, *Intermetallics* **10**(11–12), 1047–1052 (2002).

105. C. A. Angell, Formation of glasses from liquids and biopolymers, *Science* **267**(5206), 1924–1935 (1995).

106. C. Donati, J. F. Douglas, W. Kob, S. J. Plimpton, P. H. Poole, and S. C. Glotzer, Stringlike cooperative motion in a supercooled liquid, *Phys. Rev. Lett.* **80**(11), 2338–2341 (1998).

107. T. Egami, Universal criterion for metallic glass formation, *Mater. Sci. Eng. A* **226–228**, 261–267 (1997).

108. T. Egami and D. Srolovitz, Local structural fluctuations in amorphous and liquid-metals – A simple theory of the glass-transition, *J. Phys. F* **12**(10), 2141–2163 (1982).

109. U. Geyer, W. L. Johnson, S. Schneider, Y. Qiu, T. A. Tombrello, and M. P. Macht, Small atom diffusion and breakdown of the Stokes–Einstein relation in the supercooled liquid state of the $Zr_{46.7}Ti_{8.3}Cu_{7.5}Ni_{10}Be_{27.5}$ alloy, *Appl. Phys. Lett.* **69**(17), 2492–2494 (1996).

110. P. Bordat, F. Affouard, M. Descamps, and F. Muller-Plathe, The breakdown of the Stokes–Einstein relation in supercooled binary liquids, *J. Phys.: Condens. Matter* **15**(32), 5397–5407 (2003).

111. R. S. Aga, J. R. Morris, V. Levashov and T. Egami, "Local structural fluctuations in a supercooled liquid," Bulk Metallic Glasses, edited by P. K. Liaw and R. A. Buchannan (TMS Society, Warrendale, Pennsylvania, 2006), p. 59–64.

112. W. Kauzmann, The nature of the glassy state and the behavior of liquids at low temperatures, *Chem. Rev.* **43**(2), 219–256 (1948).

113. K. Ito, C. T. Moynihan, and C. A. Angell, Thermodynamic determination of fragility in liquids and a fragile-to-strong liquid transition in water, *Nature* **398**(6727), 492–495 (1999).

114. R. Bohmer, K. L. Ngai, C. A. Angell, and D. J. Plazek, Nonexponential relaxations in strong and fragile glass formers, *J. Chem. Phys.* **99**(5), 4201–4209 (1993).

115. C. A. Angell, Relaxation in liquids, polymers and plastic crystals – Strong fragile patterns and problems, *J. Non-Cryst. Solids* **131**, 13–31 (1991).
116. C. A. Angell, K. L. Ngai, G. B. McKenna, P. F. McMillan, and S. W. Martin, Relaxation in glass-forming liquids and amorphous solids, *J. Appl. Phys.* **88**(6), 3113–3157 (2000).
117. L. M. Martinez and C. A. Angell, A thermodynamic connection to the fragility of glass-forming liquids, *Nature* **410**(6829), 663–667 (2001).
118. V. N. Novikov and A. P. Sokolov, Poisson's ratio and the fragility of glass-forming liquids, *Nature* **431**(7011), 961–963 (2004).
119. V. N. Novikov, Y. Ding, and A. P. Sokolov, Correlation of fragility of supercooled liquids with elastic properties of glasses, *Phys. Rev. E* **71**(6), 061501 (2005).
120. S. N. Yannopoulos and G. P. Johari, Glass behaviour – Poisson's ratio and liquid's fragility, *Nature* **442**(7102), E7–E8 (2006).
121. G. Adam and J. H. Gibbs, On temperature dependence of cooperative relaxation properties in glass-forming liquids, *J. Chem. Phys.* **43**(1), 139–146 (1965).
122. M. Goldstein, Viscous liquids and glass transition – A potential energy barrier picture, *J. Chem. Phys.* **51**(9), 3728–3739 (1969).
123. N. H. March and M. P. Tosi, *Introduction to Liquid State Physics* (World Scientific, Singapore, 2002).
124. R. A. Johnson, Interstitials and vacancies in alpha iron, *Phys. Rev. A* **134**(5), 1329–1336 (1964).
125. S. P. Chen, T. Egami, and V. Vitek, Orientational ordering of local shear stresses in liquids – A phase-transition, *J. Non-Cryst. Solids* **75**(1–3), 449–454 (1985).
126. S. P. Chen, T. Egami, and V. Vitek, Local fluctuations and ordering in liquid and amorphous metals, *Phys. Rev. B* **37**(5), 2440–2449 (1988).
127. E. Rössler, Indications for a change of diffusion mechanism in supercooled liquids, *Phys. Rev. Lett.* **65**(13), 1595–1598 (1990).
128. M. Broccio, D. Costa, Y. Liu, and S.-H. Chen, The structural properties of a two-Yukawa fluid: Simulation and analytical results, *J. Chem. Phys.* **124**(8), 084501 (2006).
129. B. Liu, J. Goree, and O. S. Vaulina, Test of the Stokes–Einstein relation in a two-dimensional Yukawa liquid, *Phys. Rev. Lett.* **96**(1), 015005 (2006).
130. Y. J. Jung, J. P. Garrahan, and D. Chandler, Excitation lines and the breakdown of Stokes–Einstein relations in supercooled liquids, *Phys. Rev. E* **69**(6), 061205 (2004).
131. F. H. Stillinger and P. G. Debenedetti, Alternative view of self-diffusion and shear viscosity, *J. Phys. Chem. B* **109**(14), 6604–6609 (2005).
132. W. Kob, C. Donati, S. J. Plimpton, P. H. Poole, and S. C. Glotzer, Dynamical hetero-geneities in a supercooled Lennard-Jones liquid, *Phys. Rev. Lett.* **79**(15), 2827–2830 (1997).
133. K. Schmidt-Rohr and H. W. Spiess, Nature of nonexponential loss of correlation above the glass-transition investigated by multidimensional NMR, *Phys. Rev. Lett.* **66**(23), 3020–3023 (1991).
134. F. H. Stillinger and T. A. Weber, Dynamics of structural transitions in liquids, *Phys. Rev. A* **28**(4), 2408–2416 (1983).
135. L. Angelani, G. Ruocco, M. Sampoli, and F. Sciortino, General features of the energy landscape in Lennard-Jones-like model liquids, *J. Chem. Phys.* **119**(4), 2120–2126 (2003).
136. L. Angelani, G. Parisi, G. Ruocco, and G. Viliani, Potential energy landscape and long-time dynamics in a simple model glass, *Phys. Rev. E* **61**(2), 1681–1691 (2000).

137. S. Sastry, P. G. Debenedetti, F. H. Stillinger, T. B. Schrøder, J. C. Dyre, and S. C. Glotzer, Potential energy landscape signatures of slow dynamics in glass forming liquids, *Physica A* **270**(1–2), 301–308 (1999).

138. M. Vogel, B. Doliwa, A. Heuer, and S. C. Glotzer, Particle rearrangements during transitions between local minima of the potential energy landscape of a binary Lennard-Jones liquid, *J. Chem. Phys.* **120**(9), 4404–4414 (2004).

139. M. F. Ashby and A. L. Greer, Metallic glasses as structural materials, *Scripta Mater.* **54**(3), 321–326 (2006).

140. R. Hill, A general theory of uniqueness and stability in elastic–plastic solids, *J. Mech. Phys. Solids* **6**, 236–249 (1958).

141. A. Needleman, Material rate dependence and mesh sensitivity in localization problems, *Comput. Methods Appl. Mech. Eng.* **67**(1), 69–85 (1988).

142. A. S. Argon, Plastic-deformation in metallic glasses, *Acta Metall.* **27**(1), 47–58 (1979).

143. F. Spaepen, Microscopic mechanism for steady-state inhomogeneous flow in metallic glasses, *Acta Metall.* **25**(4), 407–415 (1977).

144. N. P. Bailey, J. Schiotz, and K. W. Jacobsen, Atomistic simulation study of the shear-band deformation mechanism in Mg–Cu metallic glasses, *Phys. Rev. B* **73**(6), 064108–064112 (2006).

145. Q. K. Li and M. Li, Molecular dynamics simulation of intrinsic and extrinsic mechanical properties of amorphous metals, *Intermetallics* **14**(8–9), 1005–1010 (2006).

146. F. Varnik, L. Bocquet, J. L. Barrat, and L. Berthier, Shear localization in a model glass, *Phys. Rev. Lett.* **90**(9), 095702 (2003).

147. S. Auer and D. Frenkel, Prediction of absolute crystal nucleation rate in hard-sphere colloids, *Nature* **409**(6823), 1020–1023 (2001).

148. A. Cacciuto, S. Auer, and D. Frenkel, Solid–liquid interfacial free energy of small colloidal hard-sphere crystals, *J. Chem. Phys.* **119**(14), 7467–7470 (2003).

149. K. F. Kelton, Kinetic model for nucleation in partitioning systems, *J. Non-Cryst. Solids* **274**(1–3), 147–154 (2000).

150. K. F. Kelton, Time-dependent nucleation in partitioning transformations, *Acta Mater.* **48**(8), 1967–1980 (2000).

151. K. F. Kelton and K. L. Narayan, Time-dependent nucleation in partitioning systems, in *Materials Research Society Symposium Proceedings* **481** (Phase Transformations and Systems Driven Far from Equilibrium), 1998, pp. 107–112.

152. J. Kramer and D. J. Sordelet, Polymorphism in the short-range order of $Zr_{70}Pd_{30}$ metallic glasses, *J. Non-Cryst. Solids* **351**(19–20), 1586–1593 (2005).

153. B. S. Murty, D. H. Ping, M. Ohnuma, and K. Hono, Nanoquasicrystalline phase formation in binary Zr–Pd and Zr–Pt alloys, *Acta Mater.* **49**(17), 3453–3462 (2001).

154. B. S. Murty and K. Hono, On the criteria for the formation of nanoquasicrystalline phase, *Appl. Phys. Lett.* **84**(10), 1674–1676 (2004).

155. B. S. Murty, D. H. Ping, K. Hono, and A. Inoue, Icosahedral phase formation by the primary crystallization of a Zr–Cu–Pd metallic glass, *Scripta Mater.* **43**(2), 103–107 (2000).

156. B. S. Murty and K. Hono, Nanoquasicrystallization of Zr-based metallic glasses, *Mater. Sci. Eng. A* **312**(1–2), 253–261 (2001).

157. J. Saida, M. Matsushita, and A. Inoue, Direct observation of icosahedral cluster in $Zr_{70}Pd_{30}$ binary glassy alloy, *Appl. Phys. Lett.* **79**(3), 412–414 (2001).

158. J. Saida, M. Matsushita, C. Li, Formation of the icosahedral quasicrystalline phase in $Zr_{70}Pd_{30}$ binary glassy alloy, *Philos. Mag. Lett.* **81**(1), 39–44 (2001).

159. M. W. Chen, A. Inoue, T. Sakurai, E. S. K. Menon, R. Nagarajan, and I. Dutta, Redistribution of alloying elements in quasicrystallized $Zr_{65}Al_{7.5}Ni_{10}Cu_{7.5}Ag_{10}$ bulk metallic glass, *Phys. Rev. B* **71**(9), 092202 (2005).

160. J. Saida, M. Matsushita, and A. Inoue, Nano icosahedral quasicrystals in Zr-based glassy alloys, *Intermetallics* **10**(11–12), 1089–1098 (2002).

161. J. Saida, E. Matsubara, and A. Inoue, Nanoquasicrystallization in metallic glasses, *Mater. Trans.* **44**(10), 1971–1977 (2003).

162. J. Saida, M. Matsushita, and A. Inoue, Stability of supercooled liquid and transformation behavior in Zr-based glassy alloys, *Mater. Trans.* **43**(8), 1937–1946 (2002).

163. V. Ponnambalam, S. J. Poon, and G. J. Shiflet, Fe-based bulk metallic glasses with diameter thickness larger than one centimeter, *J. Mater. Res.* **19**(5), 1320–1323 (2004).

164. Z. P. Lu, C. T. Liu, J. R. Thompson, and W. D. Porter, Structural amorphous steels, *Phys. Rev. Lett.* **92**(24), 245503 (2004).

165. Z. P. Lu and C. T. Liu, Role of minor alloying additions in formation of bulk metallic glasses: A review, *J. Mater. Sci.* **39**(12), 3965–3974 (2004).

166. Z. G. Sun, W. Loser, J. Eckert, K. H. Muller, and L. Schultz, Phase separation in $Nd_{60-x}Y_xFe_{30}Al_{10}$ melt-spun ribbons, *Appl. Phys. Lett.* **80**(5), 772–774 (2002).

167. E. Pekarskaya, J. F. Loffler, and W. L. Johnson, Microstructural studies of crystallization of a Zr-based bulk metallic glass, Acta Mater. 51(14), 4045–4057 (2003).

168. Y. T. Wang, X. K. Xi, Y. K. Fang, D. Q. Zhao, M. X. Pan, B. S. Han, W. H. Wang, and W. L. Wang, Tb nanocrystalline array assembled directly from alloy melt, *Appl. Phys. Lett.* **85**(24), 5989–5991 (2004).

169. J. M. Li, Bulk nanostructure formation directly from the multicomponent alloy melt, *Appl. Phys. Lett.* **84**(3), 347–349 (2004).

170. H. Bei, S. Xie, and E. P. George, Softening caused by profuse shear banding in a bulk metallic glass, *Phys. Rev. Lett.* **96**(10), 105503 (2006).

171. G. He, J. Eckert, W. Loser, and L. Shultz, Novel Ti-base nanostructure-dendrite composite with enhanced plasticity, *Nat. Mater.* **2**(1), 33–37 (2003).

172. J. Eckert, J. Das, K. B. Kim, F. Baier, M. B. Tang, W. H. Wang, and Z. F. Zhang, High strength ductile Cu-base metallic glass, *Intermetallics* **14**(8–9), 876–881 (2006).

173. E. M. Carvalho and I. R. Harris, Constitutional and structural studies of the intermetallic phase, ZrCu, *J. Mater. Sci.* **15**(5), 1224–1230 (1980).

174. Y. N. Koval, G. S. Firstov, J. V. Humbeeck, L. Delaey, and W. Y. Jang, B2 intermetallic compounds of Zr. New class of the shape memory alloys, *J. Phys. IV* **5**(C8), 1103–1108 (1995).

175. Y. N. Koval, G. S. Firstov, and A. V. Kotko, Martensitic-transformation and shape memory effect in ZrCu intermetallic compound, *Scripta Metall. Mater.* **27**(11), 1611–1616 (1992).

176. D. Schryvers, SAED and HREM results suggest a NiTi B19′ based superstructure for CuZr martensite, *J. Phys. IV* **5**(C8), 1047–1051 (1995).

177. D. Schryvers, G. S. Firstov, J. W. Seo, J. Van Humbeeck, and Y. N. Koval, Unit cell determination in CuZr martensite by electron microscopy and X-ray diffraction, *Scripta Mater.* **36**(10), 1119–1125 (1997).

178. G. S. Firstov, J. van Humbeeck, Y. N. Noval, and R. G. Vitchev, Alloying of ZrCu-based high temperature shape memory alloys, *J. Phys. IV* **112**, 1075–1078 (2003).

179. Y. F. Sun, B. C. Wei, Y. R. Wang, W. H. Li, T. L. Cheung, and C. H. Shek, Plasticity-improved Zr–Cu–Al bulk metallic glass matrix composites containing martensite phase, *Appl. Phys. Lett.* **87**(5), 051905 (2005).

180. J. Das, M. B. Tang, K. B. Kim, R. Theissman, F. Baier, W. H. Wang, and J. Eckert, "Work-hardenable" ductile bulk metallic glass, *Phys. Rev. Lett.* **94**(20), 205501 (2005).

181. M. Chen, A. Inoue, W. Zhang, and T. Sakurai, Extraordinary plasticity of ductile bulk metallic glasses, *Phys. Rev. Lett.* **96**(24), 245502 (2006).

Chapter 4

EVALUATION OF GLASS-FORMING ABILITY

Z. P. Lu[1], Y. Liu[2], and C. T. Liu[2]

[1]*Materials Science and Technology Division, Oak Ridge National Laboratory, Oak Ridge, TN 37831-6115, USA*
[2]*The University of Tennessee, Department of Materials Science and Engineering, Knoxville, TN 37996-2200, USA*

4.1   INTRODUCTION

The emergence of synthetic bulk metallic glasses (BMGs) as a prominent class of functional and structural materials with a unique combination of properties has been an important part of the materials science scene over the past two decades. To date, a number of BMGs have been successfully developed and commercialized for engineering applications utilizing their exceptional properties. However, one of the biggest stumbling blocks for the use of these noncrystalline alloys is still the low glass-forming ability (GFA) of many systems, which is a long-standing problem that is far from being adequately solved. Understanding the nature of glass formation and GFA is the key to developing new BMGs with improved properties and economic manufacturability for industrial applications.

GFA, as related to the ease of devitrification, can be directly evaluated by the critical cooling rate (R_c) or the maximum attainable size (D_{max}) for glass formation. The smaller R_c or the larger D_{max} is, the higher GFA of the system should be. However, R_c is difficult to measure experimentally, and D_{max} strongly depends on the fabrication method used. An alternative approach is to establish reliable criteria for glass formation from its underlying physical insights and mechanisms, and then derive simple gauges to reflect relative GFA of various alloy systems. In the past, a great deal of effort has been devoted to this area; as a result, a variety of schemes, such as structural models,[1] nearly free electron theory,[2] chemical factors (e.g., electronegativity, electron transfer, bond strength, and ionization),[3] phase diagram features,[4] minimum volume criterion,[5] atomic size criterion,[6] solid solution model,[7] etc., have been proposed to assess GFA of metallic glasses. However, these approaches are only concerned with identifying the structural and thermodynamic factors which determine whether a glass will be formed when cooled from the liquid state, and the kinetics of glass formation was not taken into account. In addition, these criteria are very difficult to quantify in practice, and thus cannot be utilized as a guideline to search for better glass-forming compositions. Therefore, several simple GFA parameters have been deduced for various metallic systems from the consideration of the kinetic processes, viz., the crystal growth rate, the nucleation rate, or transformation kinetics (see references [8, 9] for details). Nevertheless, none of these parameters performed satisfactorily, and significant progress was not made in this area until the γ parameter was established recently.[10–12]

Appreciable progress has been made in understanding the physical insights of bulk glass formation, such as the crystallization mechanisms, alloying effects, liquid fragility, etc., and the macro- and microdeformation mechanisms of BMGs.[13–23] In this chapter, the focus is placed on how to effectively quantify and represent relative GFA of different glass-forming systems. Previous work on the known GFA indicators and a comprehensive review of recent developments in this area are summarized. One of the main emphases is the establishment of the γ parameter and the demonstration of its better reliability and applicability over all previous GFA indicators. In particular, underlying mechanisms and physical insights of the effective γ criterion will be analyzed in detail. Future directions in understanding and measuring GFA of metallic alloys will also be surmised. Specifically, this chapter contains the following sections:

1. Brief introduction of previous well-known GFA parameters
2. The γ indicator and its reliability
3. Summary of other recently developed GFA criteria/indicators
4. Limitations of all the newly developed GFA parameters
5. Prospective directions

4.2   BRIEF INTRODUCTION OF PREVIOUS WELL-KNOWN GFA PARAMETERS

Scientific efforts in searching for a proper GFA criterion/gauge for metallic glasses can be dated to the time when the first quenched Au–Si metallic glass was reported.[24] Subsequently, a variety of schemes have been proposed to attain an understanding of why some systems can be vitrified and others cannot, and what determines the composition ranges over which glasses can be made.[1–7] However, these approaches not only are extremely difficult to be quantified in practice but also lack the consideration of kinetic effects.

Alternatively, a few simple parameters have been suggested based on characteristic temperatures and other physical properties of the metallic glasses. Among them, the most famous is the reduced glass transition temperature T_{rg} (i.e., the ratio of the glass transition temperature T_g and the liquidus temperature T_l), proposed by Turnbull with an assumption that the nucleation frequency and crystal growth of a melt scales inversely with viscosity of the liquid.[25] It is important to point out that the T_{rg} indicator was originally developed based on a monoatomic system, which, to a great degree, limits its applicability to complex multicomponent BMGs. Later on, Lu et al.[26,27] confirmed that T_{rg} computed by T_l/T_g shows a better correlation with GFA than that given by T_g/T_m for different multicomponent alloy systems, where T_m is the melting point. Since the late 1980s, more and more BMGs have been discovered in various systems. A new representative GFA indicator, i.e., the supercooled liquid region ΔT_{xg} (the temperature difference between the onset crystallization temperature T_x and the glass transition temperature T_g), has been proposed for the easy glass formers by Inoue et al.[28] based on the considerations of supercooled liquid stability against crystallization.

Although both ΔT_{xg} and the T_g/T_l ratio were commonly used as the GFA indicators for BMGs in the 1990s, they did show contrasting trends vs. GFA in many alloy systems. For example, Waniuk et al.[29] confirmed that T_g/T_l correlated well with GFA in Zr–Ti–Cu–Ni–Be alloys whereas the supercooled liquid range ΔT_{xg} has no relationship with GFA. The glassy compositions with the largest ΔT_{xg} are actually the poorest glass formers in the system. Inoue et al.[30,31] also proved that the GFA is more closely associated with T_g/T_l values in Cu–Zr–Ti and Cu–Hf–Ti ternary systems rather than ΔT_{xg}. On the other hand, it was found that the ratio T_g/T_l is not reliable enough to infer relative GFA in $Pd_{40}Ni_{40-x}Fe_xP_{20}$ ($20 \geq x \geq 0$),[32] Fe–(Co,Cr,Mo,Ga,Sb)–P–B–C,[33] and $Mg_{65}Cu_{15}M_{10}Y_{10}$ (M = Ni, Al, Zn, and Mn)[34] alloy systems. On the contrary, ΔT_{xg} was claimed to be a reliable and useful gauge for the optimization of bulk glass formation in these systems. Thus, all of these observations suggest that a better criterion is urgently needed for effectively reflecting the GFA of newly emerged BMGs.

4.3 THE γ INDICATOR AND ITS RELIABILITY

Glass formation is always a competing process between the liquid phase and the resulting crystalline phases. GFA is specified as the ease by which a glass-forming liquid can be cooled to form an amorphous material without appreciable formation of crystalline phases. In this regard, GFA should include two components:[10,11,35]

1. How stable the liquid phase is.
2. How difficult it is to form competing crystalline phases (i.e., the resistance to crystallization).

The above two aspects are related but different properties. Liquid phase stability is related mainly to the short-range chemical and structural ordering of atoms in the molten state and the thermodynamic stability of the liquid which is normally expressed, for example, by minima of free energy at certain chemical compositions. The resistance to crystallization is primarily determined by two factors including (1) the relative stability of the solid amorphous phase as compared to crystalline phases (i.e., the driving force for crystallization) and (2) the kinetic stability determined by the nucleation and growth of competing crystalline phases.[35] If the liquid phase is stable upon cooling (i.e., high liquid phase stability) and/or the competing crystalline phases are difficult to precipitate (i.e., high resistance to crystallization), then glass formation from the melt would be facilitated. Both of them have to be taken into account as far as the GFA is concerned.

The time–temperature–transformation (TTT) diagram contains all the information needed to predict the formability and stability of any given glasses, as shown in Fig. 4.1. Starting with the work of Turnbull,[25] Uhlmann,[36] and Davies,[37] numerous attempts have been undertaken to describe glass formation kinetics by theoretically constructing TTT curves for different alloys. Weinberg et al.[38] and Clavaguera[39] have further modified this kinetic treatment for nonequilibrium crystallization and improved the "Nose method" of calculating critical cooling rate, R_c. Nevertheless, the application of the above kinetic description involves prior knowledge of a great number of the physical and thermal properties such as viscosity and heat capacity over a wide range of temperature, which is tedious and difficult to be measured experimentally. On the other hand, direct experimental determination of TTT curves was not assessable until the discoveries of multicomponent BMGs possessing high thermal stability with respect to crystallization (e.g., the $Pd_{40}Cu_{30}Ni_{10}P_{10}$ BMG[40]). As expected, such experimental work is extremely time consuming.

From Fig. 4.1, it can be easily understood that to form an amorphous solid material, the liquid must be cooled fast enough from above the liquidus temperature through the glass transition temperature without intersecting the TTT curve. The minimum cooling rate required to form a glass (i.e., critical cooling rate) is the cooling rate needed to bypass the nose of the TTT curve, as depicted by R_c in Fig. 4.1. Therefore, the GFA of a liquid is directly related to the location of the TTT curve in the time–temperature coordinates, i.e., the position of the TTT curve along the temperature axis and the time axis. In fact, the average position of the TTT curve along the temperature scale can be indicated by the line of $(1/2)(T_g + T_l)$, as shown in Fig. 4.1.

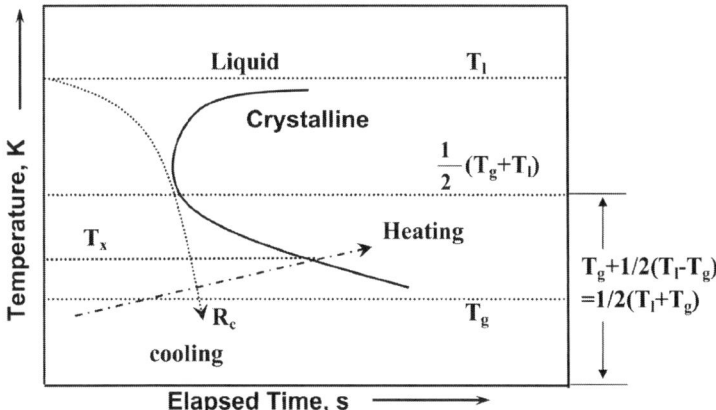

Fig. 4.1. Time–temperature–transformation (TTT) diagram. Crystallization occurs between T_l and T_g, and can be avoided by sufficiently cooling the liquid (R_c); when the amorphous solids are isochronally heated at a constant heating rate, the sample starts to crystallize at an onset temperature denoted as T_x (reprinted from reference [11] with permission from Elsevier)

From a physical point of view, liquid phase stability should be specified as the nature of the molten state (without referring to all kinetic factors). Liquid phase stability for glass-forming liquids should include two aspects (1) the stability of the liquid at the equilibrium state (i.e., stable state) and (2) the stability of the liquid during undercooling (i.e., metastable state). If two glass-forming liquids have the same T_g but different T_l, their relative liquid phase stability is then dominated by the stability of their stable states (i.e., the values of T_l). The lower the value of T_l is, the higher the liquid phase stability will be. In the case that two liquids have the same T_l but different T_g, their relative liquid phase stability is then dominated by the stability of their metastable states (i.e., the T_g values). The lower the value of T_g is, the higher will be the liquid phase stability. If two liquids have different T_l and T_g, then their liquid phase stability has to be measured by $(1/2)(T_g + T_l)$, which is the average of the stability of the liquids at equilibrium and metastable states. In general, a glass-forming liquid having a smaller value of $(1/2)(T_g + T_l)$ should have a relatively higher liquid phase stability. On the other hand, Thornburg[41] and Clavaguera et al.[42] pointed out that it is possible to determine experimentally a portion of the lower part of the TTT curve from rate-dependent thermograms upon reheating, and at temperatures below the nose, the onset times to crystallization measured on the reheated amorphous samples coincide with those measured on samples cooled from above T_l (i.e., the TTT curves at below nose temperatures are indeed the same in both cases).[43,44] When an amorphous solid is isochronally heated from the

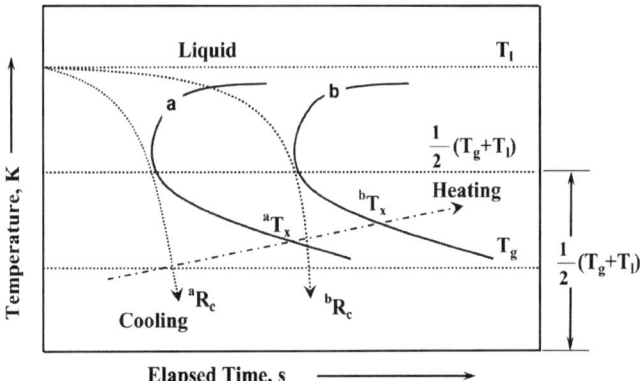

Fig. 4.2. TTT curves showing the effect of T_x measured upon continuous heating for different liquids with similar T_l and T_g; liquid b with higher onset crystallization temperature ^bT_x ($^aT_x < {}^bT_x$) shows a lower critical cooling rate ^bR_c ($^bR_c < {}^aR_c$) (reprinted from reference [12] with permission from the American Physical Society)

temperature below T_g at a low heating rate, the sample will start to crystallize at an onset temperature T_x, as illustrated in Fig. 4.1. If all liquids have the same liquid phase stability, then the GFA of a liquid can be reflected by quantity of T_x alone, as shown in Fig. 4.2, which schematically illustrates the T_x effect on GFA. The materials with a higher T_x likely have a longer onset time and a higher resistance to crystallization. Compared with liquid "a," liquid "b" has a higher onset crystallization temperature ^bT_x ($^aT_x < {}^bT_x$) and a longer onset time, thus consequently a lower critical cooling rate ^bR_c ($^bR_c < {}^aR_c$). Therefore, the onset crystallization temperature T_x measured upon continuous reheating alone can assess the GFA under the condition that the liquids have the same liquid phase stability. However, in real cases, glass-forming systems always have different liquid phase stability. To manifest the relative GFA among those liquids, T_x should be normalized to the average position of the TTT curve along the temperature axis (e.g., $(1/2)(T_g + T_l)$) such that all liquids have the same stability. Hence, the normalized T_x, denoted as γ, can be used as a gauge for GFA, which can be expressed as[10–12]

$$\gamma \propto T_x \left[\frac{1}{2(T_g + T_l)} \right] \propto \frac{T_x}{T_g + T_l}. \tag{4.1}$$

A summary of ΔT_{xg} ($T_x - T_g$), T_{rg} (T_g/T_l), and γ calculated based on recent results of BMGs, together with the critical cooling rate R_c and the critical section thickness D_{max} for glass formation in these alloy systems, is presented in Table 4.1.[10,12,26,27] The corresponding values for several conventional

Table 4.1. Summary of ΔT_{xg} $(T_x - T_g)$, T_{rg} (T_g/T_l), γ $(T_x/(T_g + T_l))$, critical cooling rate R_c, and maximum attainable size D_{max} for typical BMGs[10,12,26,27]

Alloy	$T_x - T_g$	T_g/T_l	$T_x/(T_g + T_l)$	R_c (K s^{-1})	D_{max} (mm)
$Mg_{80}Ni_{10}Nd_{10}$	16.3	0.517	0.353	1,251.4	0.6
$Mg_{75}Ni_{15}Nd_{10}$	20.4	0.570	0.379	46.1	2.8
$Mg_{70}Ni_{15}Nd_{15}$	22.3	0.553	0.373	178.2	1.5
$Mg_{65}Ni_{20}Nd_{15}$	42.1	0.571	0.397	30.0	3.5
$Mg_{65}Cu_{25}Y_{10}$	54.9	0.551	0.401	50.0	7.0
$Zr_{66}Al_8Ni_{26}$	35.6	0.537	0.368	66.6	
$Zr_{66}Al_8Cu_7Ni_{19}$	58.4	0.552	0.387	22.7	
$Zr_{66}Al_8Cu_{12}Ni_{14}$	77.4	0.559	0.401	9.8	
$Zr_{66}Al_9Cu_{16}Ni_9$	79.5	0.561	0.403	4.1	
$Zr_{65}Al_{7.5}Cu_{17.5}Ni_{10}$	79.1	0.562	0.403	1.5	16.0
$Zr_{57}Ti_5Al_{10}Cu_{20}Ni_8$	43.3	0.591	0.395	10.0	10.0
$Zr_{38.5}Ti_{16.5}Ni_{9.75}Cu_{15.25}Be_{20}$	48.0	0.628	0.415	1.4	
$Zr_{39.88}Ti_{15.12}Ni_{9.98}Cu_{13.77}Be_{21.25}$	57.0	0.625	0.420	1.4	
$Zr_{41.2}Ti_{13.8}Cu_{12.5}Ni_{10}Be_{22.5}$	49.0	0.626	0.415	1.4	50.0
$Zr_{42.63}Ti_{12.37}Cu_{11.25}Ni_{10}Be_{23.75}$	89.0	0.589	0.424	5.0	
$Zr_{44}Ti_{11}Cu_{10}Ni_{10}Be_{25}$	114.0	0.518	0.404	12.5	
$Zr_{45.38}Ti_{9.62}Cu_{8.75}Ni_{10}Be_{26.25}$	117.0	0.503	0.397	17.5	
$Zr_{46.25}Ti_{8.25}Cu_{7.5}Ni_{10}Be_{27.5}$	105.0	0.525	0.402	28.0	
$La_{55}Al_{25}Ni_{20}$	64.3	0.521	0.388	67.5	3.0
$La_{55}Al_{25}Ni_{15}Cu_5$	67.6	0.526	0.394	34.5	
$La_{55}Al_{25}Ni_{10}Cu_{10}$	79.8	0.560	0.420	22.5	5.0
$La_{55}Al_{25}Ni_5Cu_{15}$	60.9	0.523	0.389	35.9	
$La_{55}Al_{25}Cu_{20}$	38.9	0.509	0.366	72.3	3.0
$La_{55}Al_{25}Ni_5Cu_{10}Co_5$	76.6	0.566	0.421	18.8	9.0
$La_{66}Al_{14}Cu_{20}$	54.0	0.540	0.399	37.5	2.0
$Pd_{40}Cu_{30}Ni_{10}P_{20}$	74.0	0.685	0.458	0.1	72.0
$Pd_{42.5}Cu_{30}Ni_{7.5}P_{20}$	86.0	0.688	0.469	0.067	
$Pd_{42.5}Cu_{27.5}Ni_{10}P_{20}$	81.0	0.670	0.457	0.083	
$Pd_{40}Cu_{32.5}Ni_{7.5}P_{20}$	86.0	0.609	0.436	0.133	
$Pd_{40}Cu_{25}Ni_{15}P_{20}$	72.0	0.655	0.444	0.15	
$Pd_{45}Cu_{25}Ni_{10}P_{20}$	80.0	0.673	0.456	0.1	
$Pd_{45}Cu_{30}Ni_5P_{20}$	82.0	0.670	0.458	0.083	
$Pd_{37.5}Cu_{30}Ni_{12.5}P_{20}$	75.0	0.616	0.431	0.133	
$Pd_{81.5}Cu_2Si_{16.5}$	37.0	0.577	0.387		2.0
$Pd_{79.5}Cu_4Si_{16.5}$	40.0	0.585	0.392	500.0	0.75
$Pd_{77.5}Cu_6Si_{16.5}$	41.0	0.602	0.400	100.0	1.5
$Pd_{77}Cu_6Si_{17}$	44.0	0.569	0.388	125.0	2.0
$Pd_{73.5}Cu_{10}Si_{16.5}$	40.0	0.568	0.385		2.0
$Pd_{71.5}Cu_{12}Si_{16.5}$	28.0	0.565	0.377		2.0
$Pd_{40}Ni_{40}P_{20}$	63.0	0.585	0.409	1.57	25.0

(Continued)

Table 4.1. (*cont.*)

Alloy	$T_x - T_g$	T_g/T_1	$T_x/(T_g + T_1)$	R_c (K s^{-1})	D_{max} (mm)
$Nd_{60}Al_{15}Ni_{10}Cu_{10}Fe_5$	45.0	0.552	0.393		5.0
$Nd_{61}Al_{11}Ni_8Co_5Cu_{15}$	24.0	0.598	0.394		6.0
$Cu_{60}Zr_{30}Ti_{10}$	50.0	0.619	0.409		4.0
$Cu_{54}Zr_{27}Ti_9Be_{10}$	42.0	0.637	0.412		5.0
$Ti_{34}Zr_{11}Cu_{47}Ni_8$	28.8	0.597	0.389	100	4.5
$Ti_{50}Ni_{24}Cu_{20}B_1Si_2Sn_3$	74.0	0.554	0.393		1.0

metallic glasses are tabulated in Table 4.2.[10,26,27] For BMGs, the γ value is in the range from 0.350 to 0.500, whereas ΔT_{xg} ranges from 16.3 to 117 K and T_{rg} varies from 0.503 to 0.690. The relationships between the γ value and the critical cooling rate (Fig. 4.3a) and the critical section thickness (Fig. 4.3b) for glass formation in representative metallic glasses are shown in Fig. 4.3. By regression, the following fitting equations can be obtained:

$$R_c = 2.1 \times 10^{21} \exp(-114.8\gamma), \tag{4.2}$$

$$D_{max} = 2.80 \times 10^{-7} \exp(41.7\gamma). \tag{4.3}$$

To reveal how closely the estimated values for the regression line correspond to the actual experimental data, the statistical correlation parameter (R^2) computed using a common regression program ranges in value from 0 to 1. The higher the R^2 value is, the more reliable the regression line should be. As is clear in Fig. 4.3a, the R^2 value is as high as 0.90, suggesting that there is a solid correlation between the critical cooling rate R_c and the parameter γ. The predicted error band obtained at 95% confidence interval is also shown in Fig. 4.3a as two dashed lines. This prediction interval, which was also computed by the common regression program, describes the range where the data values will fall a percentage of the time for repeated measurements. A narrower band at a fixed confidence level (normally 95%) implies less scatter of the experimental data and a stronger correlation between independent variables. Compared with Fig. 4.3a, the data in Fig. 4.3b are more scattered. The widely spread data result in a lower R^2 value of 0.57 and a larger prediction band, which presumably attributes to the large variations in experimental processes during determining these maximum attainable sizes for different alloys.

As mentioned earlier, γ values for BMGs vary from 0.350 to 0.500. Substituting the maximum value of 0.500 into (4.3), we can predicate the maximum section size for BMGs to be larger than 300 mm, which corresponds to a critical cooling rate of 2.5×10^{-4} K s^{-1}. Additionally, the γ value for pure Ni is about 0.198 (see Table 4.2), its maximum size is

Table 4.2. Summary of ΔT_{xg} ($T_x - T_g$), T_{rg} (T_g/T_l), γ ($T_x/(T_g + T_l)$), and critical cooling rate R_c for some conventional metallic glasses[10,26,27]

Alloy	$T_x - T_g$	T_g/T_l	$T_x/(T_g + T_l)$	R_c (K s^{-1})
Ni	–	0.246	0.198	3.00×10^{10}
$Fe_{91}B_9$	–	0.369	0.269	2.60×10^7
$Pd_{95}Si_5$	–	0.383	0.277	5.00×10^7
$Pd_{75}Si_{25}$	–	0.488	0.328	1.00×10^6
$Zr_{65}Be_{35}$	–	0.503	0.335	1.00×10^7
$Ti_{63}Be_{37}$	–	0.497	0.332	6.30×10^6
$Pd_{82}Si_{18}$	–	0.605	0.377	1.80×10^3
$Mg_{77}Ni_{18}Nd_5$	7.8	0.484	0.332	4.90×10^4
$Mg_{90}Ni_5Nd_5$	22.8	0.464	0.334	5.30×10^4
$Au_{77.8}Si_{8.4}Ge_{13.8}$	–	0.466	0.318	3.00×10^6

determined to be 1 μm. This is in good agreement with the fact that no glass can be formed in pure Ni even by the melt-spinning technique.

In comparison, the relationship between T_{rg} and GFA for all metallic glasses listed in Tables 4.1 and 4.2 is shown in Fig. 4.4. The solid line in Fig. 4.4a is the best fit of the interrelationship between R_c and T_{rg}. The corresponding equation of the trend line and the resultant R^2 value are indicated in the graph. Although R_c is somewhat dependent on T_{rg}, however, compared with the correlation displayed in Fig. 4.3a, the current regression demonstrates a lower R^2 value of 0.73 and a larger prediction band, implying that the parameter γ correlates better with the critical cooling rate R_c than T_{rg}. Similarly, Fig. 4.4b depicts the relationship between T_{rg} and the critical section thickness D_{max}. Compared with Fig. 4.3b, it is clear that the parameter γ also has a better correlation with D_{max} than T_{rg}. This is in accordance with a lower R^2 value of 0.32 and a wider prediction band observed for the current $T_{rg} - D_{max}$ correlation. It is thus concluded that the indicator γ has a stronger correlation with GFA than T_{rg}.

The ratio T_g/T_l was introduced for purely kinetic reasons associated with the need to avoid crystallization.[25,37] In the first place, T_g is typically assumed to be less dependent on composition, while T_l often decreases more strongly with the solute concentration. The interval between T_l and T_g thus generally decreases and the value of T_{rg} increases with increasing alloying concentration, so that the probability of being cooled down without crystallization is enhanced, i.e., GFA is increased.[45] This is probably reliable for conventional binary alloy systems. However, T_l and T_g differ significantly for multicomponent systems by taking recent data into account (see Table 4.1). In this sense, T_{rg} values might not be able to judge the temperature interval $T_l - T_g$ for all systems. Secondly, T_{rg} theory arises from the requirement that viscosity must be large at temperatures between T_l and T_g.[46] Generally, the

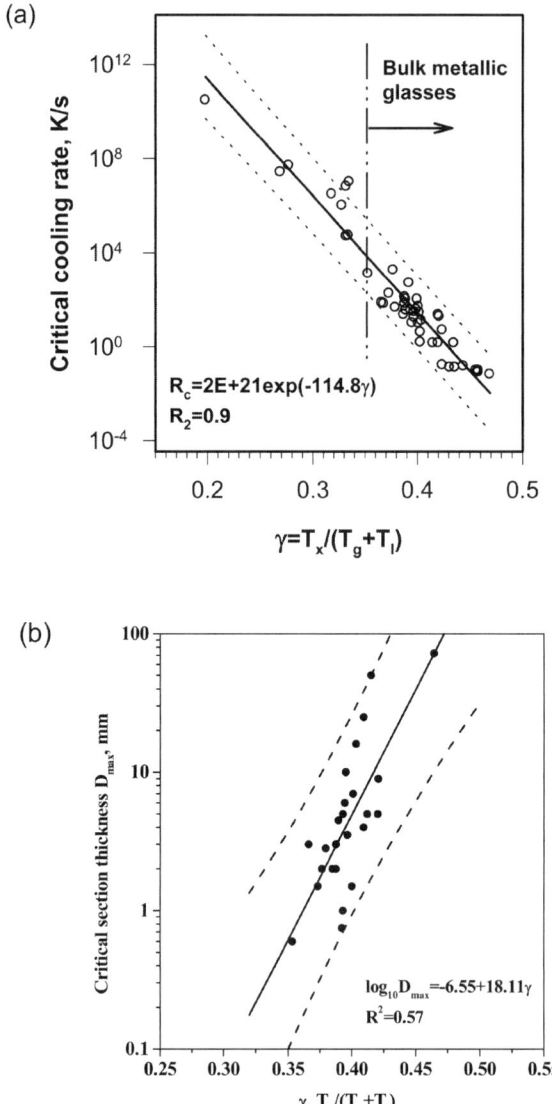

Fig. 4.3. The correlation between the γ parameter and (**a**) the critical cooling rate R_c and (**b**) the critical section thickness D_{max} for representative metallic glasses (reprinted from reference [11] with permission from Elsevier)

viscosity of glasses at T_g is 10^{12} Pa s; the higher the ratio T_{rg} is, the more viscous the melt becomes before it is ever undercooled and the more difficult the crystallization will be, thus enhancing GFA. The temperature variation of viscosity is different from system to system, depending on the classification (fragility concept) as defined by Angell.[47,48] T_g alone does not give any information about the temperature–viscosity relationship and hence the crystallization tendency. Therefore, T_g/T_l theory might not hold for some systems.

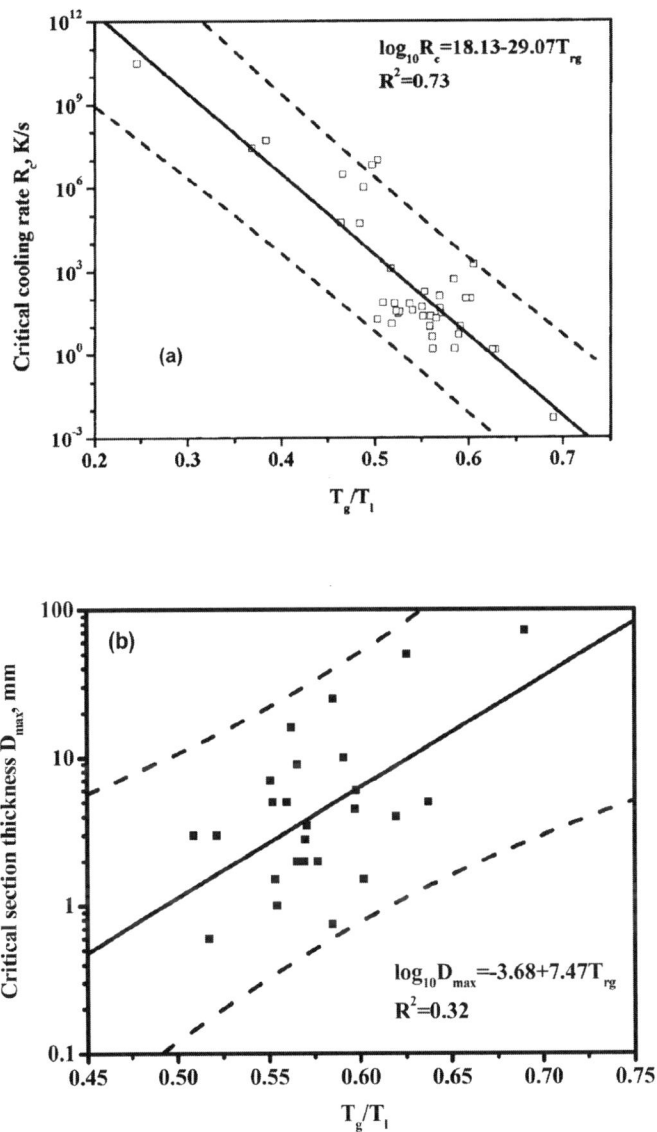

Fig. 4.4. (a) Critical cooling rate and (b) critical section thickness as a function of T_{rg} (T_g/T_l) for metallic glasses (reprinted from reference [10] with permission from Elsevier)

Figure 4.5 shows the relationship between ΔT_{xg} and GFA for BMGs. ΔT_{xg} values for some alloys even exceed 100 K, implying those glasses are rather stable upon reheating. As a whole, GFA nevertheless shows a very weak dependence on ΔT_{xg}, particularly the critical section thickness D_{max} as plotted in Fig. 4.5b. Apparently, ΔT_{xg} $(T_x - T_g)$ is a quantitative measure of glass stability, which is defined as the resistance of glasses toward devitrification upon reheating above T_g. However, GFA is specified as the ease by which melts can be cooled to form amorphous alloys without any crystal formation. It is well known that GFA and glass thermal stability are related but independent properties. Weinberg[49] demonstrated theoretically that an increasing GFA is not always accompanied by enhanced stability as measured by a difference $(T_x - T_g)$ of the same magnitude. Therefore GFA and thermal stability are akin concepts but they can be different for some systems. It is more likely that $T_x - T_g$ is just a reflection, or a corollary, rather than a cause of GFA. As such, it is inappropriate to utilize ΔT_{xg} alone as a gauge of GFA for BMGs.

As elaborated above, the γ parameter is statistically more reliable and effective than those well-known GFA indicators. This observation is further confirmed by experimental work reported recently in various alloy systems including Pd–Si, Ce–Ni–Cu–Al, Fe–Y–(Zr,Co)–(Mn,Mo,Ni)–Nb–B, Er–Co–Y–Al, Ca–Mg–Zn, Cu–Ti–Zr–Ni–Nb–Si, Cu–Zr–Al–Ag, Cu–Zr–Ti–Sn, Ni–Zr–Ti–Si, Nd–Fe–Co–Al, Ti–Cu–Ni–Sn–Be–Zr, and Mg–Cu–Gd alloys.[9,50–69] In particular, Nascimento et al.[9] have compared the reliability of all available GFA parameters in stoichiometric glass-forming oxides. They have demonstrated that, among all parameters studied, γ showed the best correlation with GFA in these oxide glasses. Moreover, experimental data revealed that γ is also applicable to other types of noncrystalline materials including halide glass and cryopreseverants.[11,12]

4.4   SUMMARY OF OTHER RECENTLY DEVELOPED GFA CRITERIA/INDICATORS

The successful establishment of the γ parameter has stimulated new interests in quantifying GFA of BMGs. As a result, many new GFA indicators/criteria have been suggested lately. In general, these parameters/criteria originated from limited experimental data, often from one single alloy system reported by a single research group. Based on their definition, these parameters/criteria can be categorized into four groups.

(1) Based on thermodynamic calculations using Miedema's model. The work about GFA criteria/indicators in this category is certainly important, because it may provide a promising way to eventually predict glass formation in

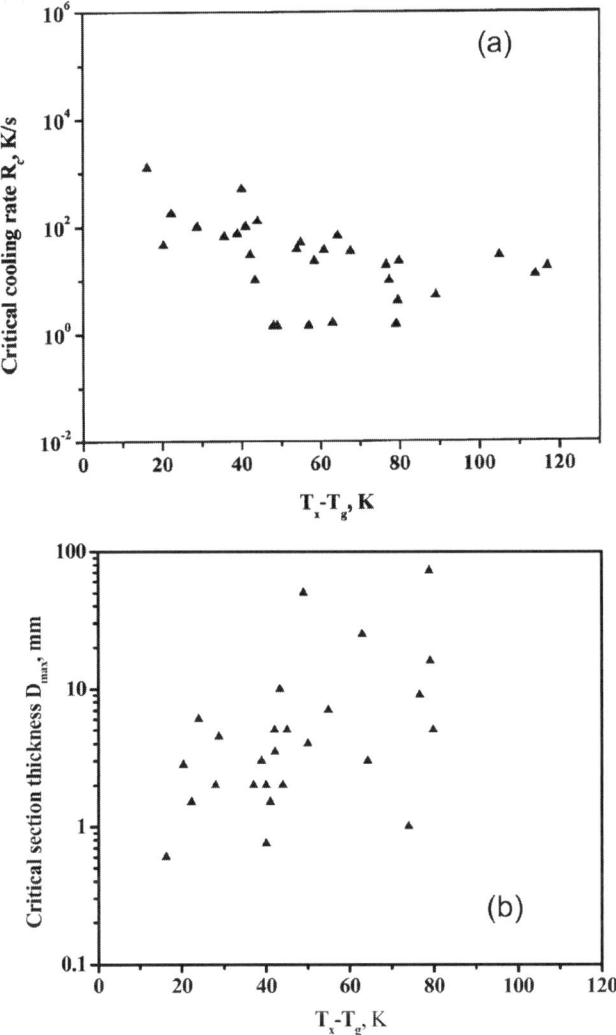

Fig. 4.5. The correlation between the supercooled liquid region ΔT_{xg} ($T_x - T_g$) and (**a**) the critical cooling rate as well as (**b**) the critical section thickness for typical bulk metallic glasses (reprinted from reference [10] with permission from Elsevier)

multicomponent glassy alloys. In the 1970s, Miedema[70,71] had first proposed an approach to calculating enthalpies in various binary systems for both the liquid and solid state. In the late 1980s, this approach was first used to predict composition range in binary transition-metal amorphous alloys.[72,73] In the 2000s, Inoue[74] and his coworkers had extended this approach to ternary amorphous alloys coupled with their empirical rules and calculated the mixing enthalpy (ΔH) and mismatch entropy (S_σ) for different glass-forming alloys. They found critical ΔH and S_σ values for obtaining high GFA in representative ternary metallic systems. According to the Miedema's model, the formation enthalpy of an amorphous phase (ΔH^{amor}), solid solutions (ΔH^{SS}), and intermetallic compounds (ΔH^{inter}) can be readily calculated[75]

$$\Delta H^{amor} = \Delta H^{chem}(amor) + \Delta H^{topo}, \tag{4.4}$$

$$\Delta H^{SS} = \Delta H^{chem}(SS) + \Delta H^{elastic} + \Delta H^{structure}, \tag{4.5}$$

$$\Delta H^{inter} = \Delta H^{chem}(inter), \tag{4.6}$$

where $\Delta H^{chem}(amor)$ is the chemical mixing enthalpy of amorphous state (glass), ΔH^{topo} is the topology enthalpy of glass, $\Delta H^{chem}(SS)$ is the chemical mixing enthalpy of solid solution, $\Delta H^{elastic}$ is the elastic enthalpy of solid solution calculated based on the continuous elastic model proposed by Eshelby and Friedel,[76–78] $\Delta H^{structure}$ is the structure enthalpy induced by the structural changes, and $\Delta H^{chem}(inter)$ is the chemical mixing enthalpy of intermetallic compound. ΔH^{inter} of the alloy with the composition between two adjacent intermetallic compounds was calculated using the level principle.

From the thermodynamic point of view, the formation of metastable amorphous state should also include two aspects (1) the driving force for the formation of glass, i.e., $-\Delta H^{amor}$ and (2) the resistance of glass formation against crystallization, i.e., the difference between the driving force for glass and intermetallic compound $\Delta H^{amor} - \Delta H^{inter}$. Figure 4.6 schematically shows effects of formation enthalpies of glass and intermetallic compounds on the GFA of different alloy compositions.[79] When two glass-forming alloys have the same ΔH^{amor} but different $\Delta H^{amor} - \Delta H^{inter}$ (alloys 1 and 2 in Fig. 4.6), their GFA is then dominated by $\Delta H^{amor} - \Delta H^{inter}$. The lower the value of $\Delta H^{amor} - \Delta H^{inter}$, the higher the GFA. Thus, the GFA of alloy 1 is better than alloy 2. On the other hand, when two glass-forming alloys have the same $\Delta H^{amor} - \Delta H^{inter}$ but different ΔH^{amor} (as alloys 2 and 3 in Fig. 4.6), their GFA is dominated by $-\Delta H^{amor}$. The higher the value of $-\Delta H^{amor}$, the higher the GFA. The correlations mentioned above can be expressed as follows:

$$GFA \propto \frac{-\Delta H^{amor}}{\Delta H^{amor} - \Delta H^{inter}}. \tag{4.7}$$

Hence Xia et al.[79] defined a parameter γ^* to evaluate GFA as below:

$$\gamma^* = \frac{-\Delta H^{amor}}{\Delta H^{amor} - \Delta H^{inter}}. \tag{4.8}$$

Note that the basic approaches of (4.1) and (4.8) are quite similar, although their final expressions are different. Equation (4.1) is formulated based on characteristic temperatures, whereas (4.8) is based on thermodynamic calculations of the formation enthalpies of phases.

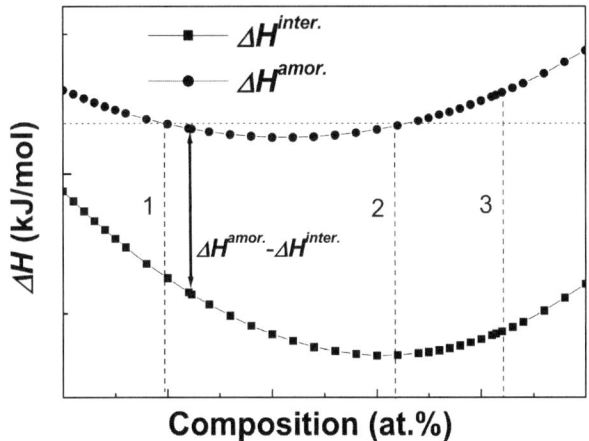

Fig. 4.6. The effects of formation enthalpies of glass and intermetallic compounds on the glass-forming ability in different alloys (reused with permission from reference [79]; Copyright 2006, American Institute of Physics)

Xia et al.[79] have applied this parameter to the binary Zr–Cu system, and Fig. 4.7a shows the corresponding dependence of the parameter γ^* on Zr concentration. The experimentally determined GFA (i.e., the maximum attainable diameter D_{max}) of the Zr–Cu binary alloys is shown in Fig. 4.7b.[80–83] The γ^* parameter and the attainable maximum diameter D_{max} show quite similar variation trend with the Zr concentration. Similar results were also observed in other binary systems including Cu–Hf[84] and Nb–Ni.[85] All of these experimental results validate the γ^* parameter, nevertheless, the challenge for this approach is to properly extend the current concept to ternary or even higher-order systems in which intermetallic phases are much complex and metastable compounds are likely to be formed.

(2) Based on characteristic temperatures of alloys. Similar to γ, these lately reported parameters are calculated from characteristic temperatures determined during either cooling or reheating of alloys, including the glass transition temperature T_g, onset crystallization temperature T_x, peak crystallization temperature T_p, onset melting point T_m, liquidus temperature T_l, and onset solidification temperature T_s. Their representatives include $T_g T_x / T_l T_m$, $T_x / (T_l - T_g)$, T_x / T_s, $(T_x - T_g) / (T_l - T_g)$, T_x / T_l, T_x / T_m, and so on.[60,66,86–89]

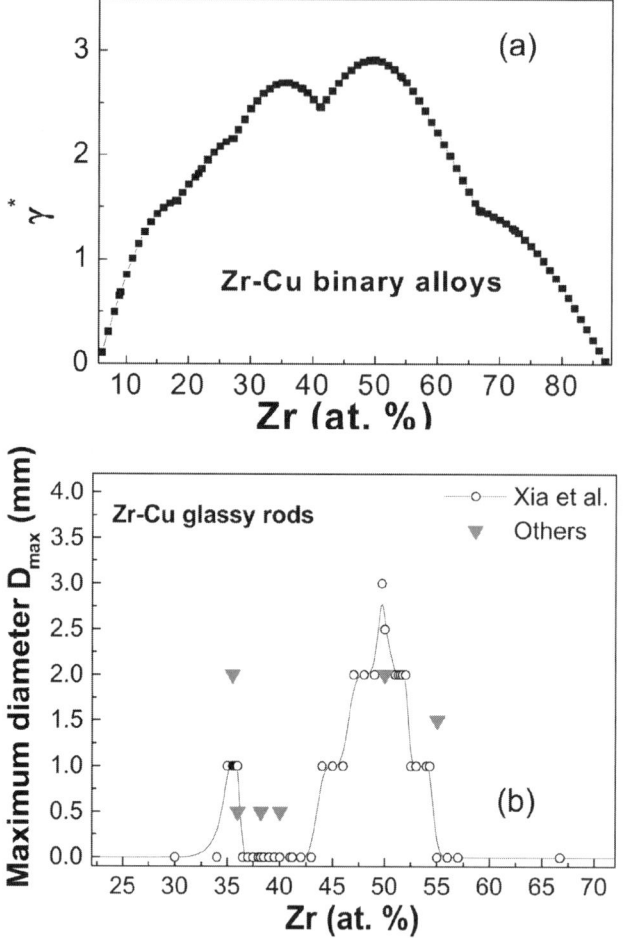

Fig. 4.7. Dependence of (**a**) the parameter γ^* and (**b**) the experimentally determined maximum diameter D_{max} on Zr concentration in the Cu–Zr binary system (reused with permission from reference [79]; Copyright 2006, American Institute of Physics)

(3) Based on fundamental properties of constituent elements.[67,90–95] In this group, GFA parameters are proposed based on the fundamental properties of constituent elements in metallic alloys. The related fundamental properties include atomic volume (or atomic size), atomic weight, density, heat of mixing, electronegativity, e/a (electrons per atom) ratio, electron structure, fusion enthalpy (ΔH_m) of each constituent element, mixing entropy (ΔS_{mix}), ionicity index, melting point of constituent elements, and elastic constant. A typical example in this category is the σ parameter which will be discussed in detail below.[94,95]

(4) Based on physical and thermal properties of alloys and/or their liquids.[68,96–99] Parameters in this category were suggested primarily with considerations of undercooled/superheat liquid behavior. These parameters are computed by measuring physical properties of alloys and their liquids, including viscosity, heat capacity, activation energy for glass formation and crystallization, melting point T_m, fusion enthalpy of resulting alloys, density, bulk modulus, etc. The most famous example in this group is the fragility parameter D, and a few others, such as superheated fragility M,[96] short-range ordering,[97] and Gibbs free energy minima,[98] were recently discussed.

4.5 LIMITATIONS OF ALL THE NEWLY DEVELOPED GFA PARAMETERS

Glass formation from a melt is essentially to retain liquid structure and simultaneously avoid crystallization as the temperature is lowered.[4] As mentioned earlier, two components, i.e., liquid phase stability and crystallization resistance, have to be taken into account as far as the GFA of a material is considered. Liquid stability also contains two aspects: the thermodynamic stability of the liquid/amorphous phase and the relative stability of the amorphous phase as compared to competing crystalline phases. Therefore, a reliable GFA indicator should consider all above aspects and can then be applied to different alloy systems.

The primary reason why γ is so effective is because it incorporates both liquid phase stability and crystalline resistance effects. Unlike the γ parameter, most of recently suggested indicators were proposed based only on a few compositions in one single system and solely concerned with either structure factors (e.g., atomic configuration, topological stability, electron stability, etc.) or liquid behavior (free energy minima, strong liquid behavior, etc.). Therefore, they might be useful in certain systems in which the concerned factors play predominant roles in glass formation. However, their universal applicability is definitely questionable. Next, we will use Fang's criteria and the σ parameter as examples to demonstrate thoroughly that these newly reported GFA indicators lack general applicability and cannot be utilized as a universal GFA gauge.

Fang et al.[67] have observed that the amorphous stability of Mg-based BMGs is related closely to two weighted parameters, the electronegativity difference Δx and the atomic size mismatch δ, and can be modeled by

$$\Delta x = \sqrt{\sum_{i=1}^{n} C_i \cdot (x_i - \overline{x})^2},$$ (4.9)

where n is the number of components of the alloy, x_i the Pauling electro-negativity of element i, C_i the atomic percentage of element i in the alloy, and $\bar{x}$ the mean value of the electronegativity for an alloy, which can be computed as follows:

$$\bar{x} = \sum_{i=1}^{n} C_i \cdot x_i. \qquad (4.10)$$

The atomic size parameter δ is expressed as

$$\delta = \sqrt{\sum_{i=1}^{n} C_i \cdot \left(1 - \frac{r_i}{\bar{r}}\right)^2}, \qquad (4.11)$$

where r_i is the covalent atomic radius of element i and $\bar{r}$ is the mean value of covalent atomic radius for a compound and can be calculated as follows:

$$\bar{r} = \sum_{i=1}^{n} C_i \cdot r_i. \qquad (4.12)$$

Interestingly, it has been demonstrated that the supercooled liquid region ΔT_{xg} has a strong correlation with both the atomic size difference δ and electronegativity difference Δx in Mg-based BMGs. More interestingly, a linear equation for identifying BMGs with expectable thermal stability has been deduced. On the other hand, we have confirmed that these two models do not have universal applicability; they cannot effectively reflect effects of size mismatch and bonding nature on thermal stability of BMGs in general.[100] For example, the correlation between the two parameters, i.e., the electro-negative difference Δx (Fig. 4.8a) and the atomic size parameter δ (Fig. 4.8b), and the supercooled liquid region ΔT for Zr-based BMGs is shown in Fig. 4.8.[100] As can be seen, the supercooled liquid region displays a "Christmas-tree"-like pattern as a function of Δx that ranges from 0.23 to 0.33 for 79 Zr-based BMGs, whereas there is no apparent trend between ΔT_{xg} and δ.

Another interesting example to be discussed is the σ parameter, which has been proposed solely for ternary and binary BMGs. Starting with Davies's concept that the liquid phase stability scales with the amount of melting temperature depression, a dimensionless melting temperature depression parameter ΔT^* to evaluate GFA of metallic glasses is introduced as follows[101]

$$\Delta T^* = \frac{T_m^{mix} - T_l}{T_m^{mix}}, \qquad (4.13)$$

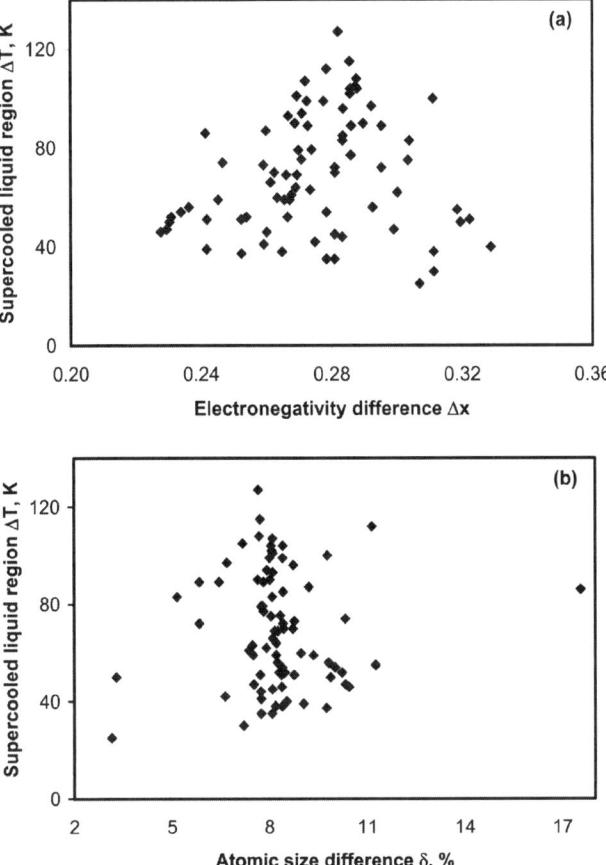

Fig. 4.8. Supercooled liquid region ΔT_{xg} as a function of (**a**) electronegativity difference Δx and (**b**) atomic size difference δ for Zr-based bulk metallic glasses (reprinted from reference [100] with permission from Elsevier)

where T_l is the liquidus temperature and $T_m^{mix} = \sum_i^n x_i \cdot T_m^i$, with x_i and T_m^i for the mole fraction and melting point, respectively, of the ith component in an n-component alloy system.

On the other hand, considering the atomic configuration of liquid phase, Egami and Waseda[6] have suggested the following composition criterion for the formation of the amorphous phase in binary alloy systems based on the atomic scale elasticity theory

$$x_B^{min} \left| \frac{(v_B - v_A)}{v_A} \right| = x_B^{min} \left| \left(\frac{r_B}{r_A} \right)^3 - 1 \right| \approx 0.1, \qquad (4.14)$$

where $x_B^{\min}$ is the minimum solute content, and v_i and r_i (i = A, B) are atomic volume and atomic radius, respectively. Above a certain level of atomic mismatch, the crystalline solid solution loses its stability and glass can form. The larger the atomic size difference is, the smaller the amount of solute is required to form an amorphous phase. However, the criterion estimates only the minimum solute content for the formation of amorphous phase. With the assumption that the extent of atomic mismatch can be considered as one of the parameters reflecting the GFA of BMGs, Park et al.[102] extended the overall effect of atomic size mismatch on GFA to ternary alloys by the following P' parameter:

$$P' = \frac{x_B}{x_B + x_C} \left| \frac{(v_B - v_A)}{v_A} \right| + \frac{x_C}{x_B + x_C} \left| \frac{(v_C - v_A)}{v_A} \right|, \qquad (4.15)$$

Combining the above two considerations of thermodynamic stability ΔT^* and atomic configuration P' of the liquid phase, Kim et al. have proposed the σ parameter using the fundamental properties of the constituents, which can be expressed as:[94]

$$\sigma = \Delta T^* \times P'. \qquad (4.16)$$

It is clear from the above definition that the σ parameter holds only for ternary or binary systems. To properly assess the validity and feasibility of this parameter, a relatively complete data set, including T_g, T_x, liquidus temperature T_l, the critical diameter $D_{\max}$, and calculated values of γ and σ, for most binary and ternary BMGs reported in literature was compiled in Table 4.3.[103] The correlation of the γ parameter (Fig. 4.9a) and σ parameter (Fig. 4.9b) with the critical diameter $D_{\max}$ for all the ternary and binary BMGs listed in Tables 4.1–4.3 is compared in Fig. 4.9. As seen clearly from the plot, the regression coefficient R^2 for the $\gamma - D_{\max}$ correlation is around 0.60, which is significantly higher than that for the $\sigma - D_{\max}$ correlation of 0.21. This observation apparently indicates that, even for binary and ternary BMGs, σ does not have a strong correlation with GFA as γ does. In addition, the statistical coefficient R^2 value of the $\gamma - D_{\max}$ regression for the binary and ternary BMGs is consistent with that reported previously for all types of multicomponent BMGs,[10,12] suggesting a universal validity of the γ indicator. As pointed out earlier, compared with the $\gamma - R_c$ (critical cooling rate) interrelationship as shown in Fig. 4.3a, the $\gamma - D_{\max}$ correlation always shows a lower R^2 value, which is due mainly to the fact that $D_{\max}$ is much more dependent on the fabrication process than R_c. For example, the $D_{\max}$ value of the alloy $Mg_{65}Cu_{25}Gd_{10}$ is 8 mm using a copper mold wedge casting technique[104] but decreased to 6 mm using a copper mold injection casting method.[63]

The main deficiency of the σ parameter is the lack of consideration of kinetic effects on GFA (i.e., the crystallization resistance). Normally, kinetic effects on GFA will be reflected by changes in T_x and T_g. As an example, for the alloy $Pd_{40}Ni_{40}P_{20}$, its D_{max} can be increased from 3 to 10 mm with the use of the flux melting technique. It was found that T_g and T_l of the flux-treated amorphous rods remain unchanged while T_x is increased by ~29 K.[105] As a result, the γ parameter increases from 0.415 to 0.432, effectively representing the GFA enhancement due to the use of the flux melting in this alloy. However, the σ indicator remains the same in both cases and cannot correctly reflect the resultant GFA enhancement. Similar phenomenon can also be seen for the effects of minor additions on GFA. It was reported that micro-alloying of proper elements can result in a dramatic increase in the GFA of various BMGs.[22,23,106] Such small amounts of additions will not affect T_l and T_g but increase T_x. Consequently, the γ parameter still works well in these cases but σ does not.

Generally speaking, numerous parameters with a large R^2 value can be figured out if the analysis is based only on limited data and lacks a statistical approach. As can be seen, different lines can be drawn in Fig. 4.9a based on limited data points selected and various resultant equations can then be induced. However, none of them would be able to represent the entire picture of GFA. Thus, it has to be ensured that the GFA parameter to be proposed should properly incorporate both kinetic and thermodynamic effects. Also, the mathematical operations during establishing the parameter must have physical meanings and the resulting parameter should hold across-the-board.

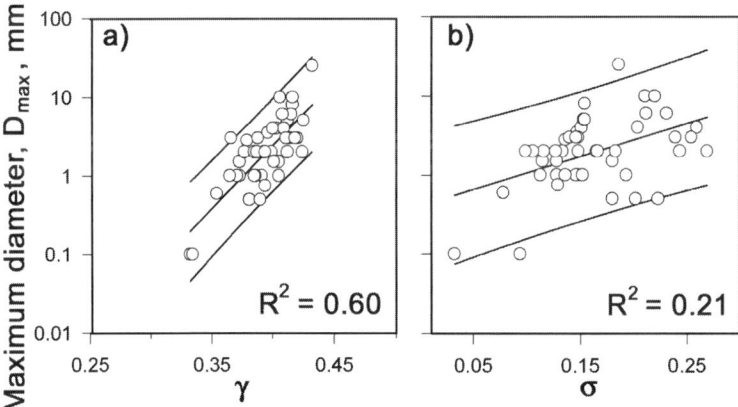

Fig. 4.9. Comparison of the correlations of the GFA parameters and the critical size D_{max} for ternary BMGs: (**a**) γ vs. D_{max} and (**b**) σ vs. D_{max} (reprinted from reference [103] with permission from Elsevier)

Table 4.3. Characteristic temperatures and critical size D_{max} as well as calculated γ and σ values for ternary BMGs[103]

Alloy	T_g (K)	T_x (K)	T_l (K)	σ	γ	D_{max} (mm)
$Mg_{77}Ni_{18}Nd_5$	429	437	887	0.094	0.332	0.1
$Mg_{90}Ni_5Nd_5$	426	449	919	0.032	0.334	0.1
$Mg_{80}Ni_{10}Nd_{10}$	454	471	878	0.078	0.353	0.6
$Mg_{65}Cu_{25}Gd_{10}$	423	484	740	0.154	0.416	8.0
$Mg_{65}Cu_{25}Tb_{10}$	414	487	733	0.153	0.425	5.0
$Mg_{65}Cu_{25}Sm_{10}$	418	470	723	0.154	0.412	5.0
$Mg_{65}Cu_{25}Dy_{10}$	422	492	750	0.147	0.420	3.0
$Mg_{65}Cu_{25}Pr_{10}$	413	446	784	0.128	0.373	1.0
$Mg_{65}Cu_{25}Nd_{10}$	423	456	744	0.147	0.391	1.0
$Mg_{65}Cu_{25}Ho_{10}$	417	473	751	0.146	0.405	1.0
$Cu_{60}Hf_{25}Ti_{15}$	730	790	1,177	0.184	0.414	4.0
$Cu_{50}Hf_{42.5}Al_{7.5}$	781	836	1,240	0.254	0.414	3.0
$Cu_{52.5}Hf_{40}Al_{7.5}$	779	833	1,250	0.239	0.411	3.0
$Cu_{50}Hf_{45}Al_5$	763	854	1,250	0.269	0.424	2.0
$Y_{56}Al_{24}Co_{20}$	636	690	1,078	0.182	0.403	1.5
$Zr_{54}Cu_{46}$	696	746	1,201	0.148	0.393	2.0
$Zr_{47}Cu_{46}Al_7$	705	781	1,163	0.146	0.418	3.0
$Ni_{60}Nb_{40}$	933	933	1,484	0.193	0.386	1.0
$Ni_{60}Nb_{30}Ta_{10}$	934	961	1,559	0.183	0.385	2
$Ca_{66.4}Al_{33.6}$	528	540	873	0.113	0.385	1.00
$Ca_{60}Al_{30}Ag_{10}$	483	531	868	0.125	0.393	2
$Ca_{63}Al_{32}Cu_5$	512	523	831	0.150	0.389	2
$Ca_{55}Mg_{15}Zn_{30}$	389	419	711	0.202	0.381	0.5
$Ca_{55}Mg_{20}Zn_{25}$	383	428	702	0.166	0.394	2.0
$Ca_{55}Mg_{25}Zn_{20}$	375	418	751	0.136	0.371	1.0
$Ca_{60}Mg_{10}Zn_{30}$	380	425	710	0.180	0.390	0.5
$Ca_{60}Mg_{15}Zn_{25}$	379	427	650	0.212	0.415	6.0
$Ca_{60}Mg_{17.5}Zn_{22.5}$	378	428	650	0.211	0.416	10.0
$Ca_{60}Mg_{20}Zn_{20}$	378	415	660	0.204	0.400	4.0
$Ca_{60}Mg_{25}Zn_{15}$	377	409	744	0.152	0.365	1.0
$Ca_{62.5}Mg_{17.5}Zn_{20}$	375	412	640	0.220	0.406	10.0
$Ca_{65}Mg_{15}Zn_{20}$	375	410	630	0.231	0.408	6.0
$Ca_{70}Mg_{10}Zn_{20}$	367	399	657	0.223	0.390	0.5

Melting point and atomic volume of pure metals are taken from http://web.mit.edu/3.091/ www/pt/pert12.html and http://web.mit.edu/3.091/www/pt/pert2.html, respectively. The critical size D_{max} is obtained by common copper mold casting.

4.6 PROSPECTIVE DIRECTIONS

The ultimate goal in this area is to establish a simple parameter for quantifying and eventually predicting GFA for alloy design of new BMGs with superior GFA. Parameters involving characteristic temperatures and/or properties of amorphous phase, such as the γ parameter, are powerful in measuring and understanding GFA. To use them to aid alloy design, however, samples with amorphous structure have to be prepared and properties of these glassy specimens have to be measured. A possible approach is to make glassy ribbons for different compositions via the melt-spinning technique and then use the GFA indicator to screen the compositions with potentially high GFA. Nevertheless, this approach is still time consuming and tedious. To quickly predict GFA in different systems, future efforts can be possibly focused on the following two aspects:

(1) Develop a practical criterion related only to fundamental properties. Ideally, the GFA parameter/criterion to be developed is related exclusively to fundamental properties of constituent elements (e.g., atomic size and electronegativity) and/or already known thermal and physical properties of lower-order systems, such as binary phase diagram, heat of mixing between binary atomic pairs, and so on. With these fundamental properties, quantitatively analyzing and predicting GFA in different systems would become possible. In this regard, development of the parameters as those from category III (e.g., the σ parameter as discussed earlier) is the right direction for future efforts in this area. However, glass formation is a complex phenomenon and so many factors can affect the GFA of the material. For example, we have demonstrated that both atomic size mismatch Δx and electronegativity difference δ as defined in (4.9) and (4.11), respectively, showed a weak correlation with BMGs. Experimental data indicate that both models Δx and δ need to be further refined and combined such that the overall effects of atomic size mismatch and electronegativity difference on GFA can be properly represented.[84] Therefore, the key challenge is to establish appropriate models to account for all the different factors and then weigh all contributions from each factor, eventually leading to reliable and applicable parameters.

(2) Computation approach to predicting GFA in different metallic alloys. As mentioned earlier, assessing the GFA via calculations of mixing enthalpy and mismatch entropy in binary and ternary alloys have been conducted based on Miedema's macroscopic model. Especially, the conceptual approach and derived parameter γ^* by Xia et al. are an important attempt in predicting GFA of metallic alloys. Nevertheless, the challenge in this area is to properly extend such an approach to high-order systems in which more complex and metastable crystalline phases compete.

Recently, there are also a few interesting reports about utilizing the Calphad approach to predict glass formation in several metallic systems.[107–111] A stimulating example is given by Shao et al.[108,109] who treated glass transition as a second-order phase transformation from the supercooled liquid phase and simulated thermodynamic stability of metallic glasses based on a full thermodynamic database of the given systems. They successfully demonstrated that glass transition temperature T_g, onset crystallization temperature T_x, and liquidus temperature T_l can be calculated. As a result, all temperature-related

GFA indicators can now be obtained with no need for characterizing glassy samples. Another exciting example was given by Chang's group[110,111] and they thermodynamically calculated liquidus projection surface of the multicomponent Zr–Ti–Ni–Cu–(Al) systems based on information built on lower-order constituent binaries and ternaries. Using the liquidus temperature as a guideline, they have successfully predicted several novel compositions with superior GFA. The most difficult part of such approaches is to establish reliable lower-order thermodynamic descriptions for multicomponent systems.

Additionally, there are other computation approaches attempting to predict glass formation and GFA from a topological and thermodynamic point of view.[112,113] For example, Miracle has proposed a structural model for metallic glasses based on the concept of dense packing of atomic clusters. The simulation results from this topological model are consistent with experimental observation reported in the literature.[113] Nevertheless, how to incorporate chemical effects on glass formation in such kind of models perhaps is necessary for future efforts.

ACKNOWLEDGMENT

This research was sponsored by the Division of Materials Sciences and Engineering, Office of Basic Energy Sciences, US Department of Energy under contract DE-AC05-00OR-22725 with UT-Battelle, LLC.

REFERENCES

1. D. E. Polk, Structure of glassy metallic alloys, *Acta Metall.* **20**, 485–491 (1972).
2. S. R. Nagel and J. Tauc, Nearly-free-electron approach to theory of metallic glass alloys, *Phys. Rev. Lett.* **35**, 380–383 (1975).
3. H. S. Chen and B. K. Park, Role of chemical bonding in metallic glasses, *Acta Metall.* **21**, 395–400 (1973).
4. K. S. Dubey and P. Ramachandrarao. Phase-diagram features and glass forming ability of liquid-quenched binary alloys, *Int. J. Rapid Solidif.* **5**, 127–135 (1990).
5. P. Ramachandrarao, On glass-formation in metal–metal systems, *Z. Metallkd.* **71**, 172–177 (1980).
6. T. Egami and Y. Waseda, Atomic size effect on the formability of metallic glasses, *J. Non-Cryst. Solids* **64**, 113–134 (1984).
7. S. H. Whang, New prediction of glass-forming ability in binary-alloys using a temperature composition map, *Mater. Sci. Eng.* **57**, 87–95 (1983).
8. Y. Li, S. C. Ng, C. K. Ong, H. H. Hng, and T. T. Goh, Glass forming ability of bulk glass forming alloys, *Scripta Mater.* **36**, 783–787 (1997).
9. M. L. F. Nascimento, L. A. Souza, E. B. Ferreira, and E. D. Zanotto, Can glass stability parameters infer glass forming ability? *J. Non-Cryst. Solids* **351**, 3296–3308 (2005).

10. Z. P. Lu and C. T. Liu, A new glass-forming ability criterion for bulk metallic glasses, *Acta Mater.* **50**, 3501–3512 (2002).

11. Z. P. Lu and C. T. Liu, A new approach to understanding and measuring glass formation in bulk amorphous materials, *Intermetallics* **12**, 1035–1043 (2004).

12. Z. P. Lu and C. T. Liu, Glass formation criterion for various glass-forming systems, *Phys. Rev. Lett.* **91**, 115505 (2003). http://link.aps.org/abstract/PRL/v91/e115505

13. A. Inoue, Stabilization of metallic supercooled liquid and bulk amorphous alloys, *Acta Mater.* **28**, 279–306 (2000).

14. W. L. Johnson, Bulk glass-forming metallic alloys: Science and technology, *MRS Bull.* **24**, 42–56 (1999).

15. A. L. Greer, Metallic glasses, *Science* **267**, 1947–1953 (1995).

16. K. F. Kelton, A new model for nucleation in bulk metallic glasses, *Philos. Mag. Lett.* **77**, 337–343 (1998).

17. D. R. Allen, J. C. Foley, and J. H. Perepezko, Nanocrystallization development during primary crystallization of amorphous alloys, *Acta Mater.* **46**, 431–440 (1998).

18. J. Schroers and W. L. Johnson, Ductile bulk metallic glass, *Phys. Rev. Lett.* **93**, 255506 (2004).

19. J. J. Lewandowski and A. L. Greer, Temperature rise at shear bands in metallic glasses, *Nat. Mater.* **5**, 15–18 (2006).

20. B. Yang, M. L. Morrison, P. K. Liaw, R. A. Buchanan, G. Y. Wang, C. T. Liu, and M. Denda, Dynamic evolution of nanoscale shear bands in a bulk-metallic glass, *Appl. Phys. Lett.* **86**, 141904 (2005).

21. J. Das, M. B. Tang, K. B. Kim, R. Theissmann, F. Baier, W. H. Wang, and J. Eckert, "Work-hardenable" ductile bulk metallic glass, *Phys. Rev. Lett.* **94**, 205501 (2005).

22. Z. P. Lu and C. T. Liu, Role of minor alloying additions in formation of bulk metallic glasses: A review, *J. Mater. Sci.* **39**, 3965–3974 (2004).

23. Z. P. Lu, C. T. Liu, J. R. Thompson, and W. D. Porter, Structural amorphous steels, *Phys. Rev. Lett.* **92**, 245503 (2004).

24. M. H. Cohen and D. Turnbull, Composition requirements for glass formation in metallic and ionic systems, *Nature* **189**, 131–132 (1961).

25. D. Turnbull, Under what conditions can a glass be formed? *Contemp. Phys.* **10**, 473–483 (1969).

26. Z. P. Lu, H. Tan, Y. Li, and S. C. Ng, The correlation between reduced glass transition temperature and glass forming ability of bulk metallic glasses, *Scripta Mater.* **42**, 667–673 (2000).

27. Z. P. Lu, Y. Li, and S. C. Ng, Reduced glass transition temperature and glass forming ability of bulk glass forming alloys, *J. Non-Cryst. Solids* **270**, 103–114 (2000).

28. A. Inoue, T. Zhang, and T. Masumoto, Zr–Al–Ni amorphous alloys with high glass transition temperature and significant supercooled liquid region, *Mater. Trans. JIM* **31**, 177–183 (1990).

29. T. A. Waniuk, J. Schroers, and W. L. Johnson, Critical cooling rate and thermal stability of Zr–Ti–Cu–Ni–Be alloys, *Appl. Phys. Lett.* **78**, 1213–1216 (2001).

30. A. Inoue, W. Zhang, T. Zhang, and K. Kurosaka, High-strength Cu-based bulk glassy alloys in Cu–Zr–Ti and Cu–Hf–Ti ternary systems, *Acta Mater.* **29**, 2645–2652 (2001).

31. A. Inoue, W. Zhang, T. Zhang, and K. Kurosaka, Formation and mechanical properties of Cu–Hf–Ti bulk glassy alloys, *J. Mater. Res.* **16**, 2836–2844 (2001).

32. T. D. Shen and R. B. Schwarz, Bulk amorphous Pd–Ni–Fe–P alloys: Preparation and characterization, *J. Mater. Res.* **14**, 2107–2115 (1999).

33. T. D. Shen and R. B. Schwarz, Bulk ferromagnetic glasses prepared by flux melting and water quenching, *Appl. Phys. Lett.* **75**, 49–52 (1999).

34. B. S. Murty and K. Hono, Formation of nanocrystalline particles in glassy matrix in melt-spun Mg–Cu–Y based alloys, *Mater. Trans. JIM* **41**, 1538–1544 (2000).

35. I. Schmidt, Glass forming ability of alloys on the basis of Fe–C, *Z. Metallkd.* **74**, 561–583 (1983).

36. D. R. Uhlmann, A kinetic treatment of glass formation, *J. Non-Cryst. Solids* **7**, 337–348 (1972).

37. H. A. Davies, The formation of metallic glasses, *Phys. Chem. Glasses* **17**, 159–173 (1976).

38. M. C. Weinberg, D. R. Uhlmann, and E. D. Zanotto, "Nose method" of calculating critical cooling rates for glass formation, *J. Am. Ceram. Soc.* **72**, 2054–2058 (1989).

39. N. Clavaguera, Non-equilibrium crystallization, critical cooling rates and transformation diagrams, *J. Non-Cryst. Solids* **162**, 40–50 (1993).

40. J. F. Loffler, J. Schroers, and W. L. Johnson, Time–temperature–transformation diagram and microstructures of bulk glass forming $Pd_{40}Cu_{30}Ni_{10}P_{20}$, *Appl. Phys. Lett.* **77**, 681–683 (2000).

41. D. D. Thornburg, Evaluation of glass formation tendency from rate dependent thermograms, *Mater. Res. Bull.* **9**, 1481–1485 (1974).

42. N. Clavaguera and M. T. Clavaguera-Mora, J. Casas-Vazquez, Crystallization kinetics of a Se–Ge–Sb alloy glass, *J. Non-Cryst. Solids* **22**, 23–27 (1976).

43. J. Schroers, J. F. Loffler, E. Pekarskaya, R. Busch, and W. L. Johnson, Crystallization of bulk glass forming Pd-based melts, *Mater. Sci. Forum* **360–362**, 79–84 (2001).

44. J. Schroers, R. Busch, S. Bossuyt, and W. L. Johnson, Crystallization behavior of the bulk metallic glass forming $Zr_{41}Ti_{14}Cu_{12}Ni_{10}Be_{23}$ liquid, *Mater. Sci. Eng. A* **304–306**, 287–291 (2001).

45. H. A. Davies, in *Rapidly Quenched Metals III*, Vol. 1, edited by B. Cantor (The Metals Society, London, 1978), pp. 1–21.

46. D. R. Uhlmann and H. Yinnon, in *Glasses Science and Technology*, Vol. 1, edited by D. R. Uhlmann and N. J. Kreidl (Academic, New York, 1983), pp. 1–47.

47. C. A. Angell, Relaxation in liquids, polymers and plastic crystals – Strong fragile patterns and problems, *J. Non-Cryst. Solids* **131**, 13–31 (1991).

48. C. A. Angell, Formation of glasses from liquids and biopolymers, *Science* **267**, 1924–1935 (1995).

49. M. C. Weinberg, Glass-forming ability and glass stability in simple systems, *J. Non-Cryst. Solids* **167**, 81–88 (1994).

50. W. B. Sheng, Correlation between critical section thickness and glass-forming ability criteria of Ti-based bulk amorphous alloys, *J. Non-Cryst. Solids* **351**, 3081–3086 (2005).

51. W. B. Sheng, Evaluation on the reliability of criteria for glass-forming ability of bulk metallic glasses, *J. Mater. Sci.* **40**, 5061–5066 (2005).

52. J. K. Lee, D. H. Bae, S. Yi, W. T. Kim, and D. H. Kim, Effects of Sn addition on the glass forming ability and crystallization behavior in Ni–Zr–Ti–Si alloys, *J. Non-Cryst. Solids* **333**, 212–220 (2004).

53. Q. S. Zhang, H. F. Zhang, Y. F. Deng, B. Z. Ding, and Z. Q. Hu, Bulk metallic glass formation of Cu–Zr–Ti–Sn alloys, *Scripta Mater.* **49**, 273–278 (2003).

54. E. S. Park and D. H. Kim, Formation of Ca–Mg–Zn bulk glassy alloy by casting into cone-shaped copper mold, *J. Mater. Res.* **19**, 685–688 (2004).

55. Y. C. Kim, W. T. Kim, and D. H. Kim, A development of Ti-based bulk metallic glass, *Mater. Sci. Eng. A* **375–377**, 127–135 (2004).

56. L. Xia, M. B. Tang, M. X. Pan, W. H. Wang, and Y. D. Dong, Glass forming ability and magnetic properties of $Nd_{48}Al_{20}Fe_{27}Co_5$ bulk metallic glass with distinct glass transition, *J. Phys. D: Appl. Phys.* **37**, 1706–1709 (2004).

57. J. H. Kim, J. S. Park, E. Fleury, W. T. Kim, and D. H. Kim, Effect of yttrium addition on thermal stability and glass forming ability in Fe–TM (Mn, Mo, Ni)–B ternary alloys, *Mater. Trans.* **45**, 2770–2775 (2004).

58. D. S. Sung, O. J. Kwon, E. Fleury, K. B. Kim, J. C. Lee, D. H. Kim, and Y. C. Kim, Enhancement of the glass forming ability of Cu–Zr–Al alloys by Ag addition, *Met. Mater. Int.* **10**, 575–579 (2004).

59. E. S. Park, D. H. Kim, T. Ohkubo, and K. Hono, Enhancement of glass forming ability and plasticity by addition of Nb in Cu–Ti–Zr–Ni–Si bulk metallic glasses, *J. Non-Cryst. Solids* **351**, 1232–1238 (2005).

60. O. N. Senkov and J. M. Scott, Glass forming ability and thermal stability of ternary Ca–Mg–Zn bulk metallic glasses, *J. Non-Cryst. Solids* **351**, 3087–3094 (2005).

61. Y. X. Wei, B. Zhang, R. J. Wang, M. X. Pan, D. Q. Zhao, and W. H. Wang, Erbium and cerium based bulk metallic glasses, *Scripta Mater.* **54**, 599–602 (2006).

62. K. F. Yao and R. Fang, Pd–Si binary bulk metallic glass prepared at low cooling rate, *Chin. Phys. Lett.* **22**, 1481–1483 (2005).

63. X. K. Xi, D. Q. Zhao, M. X. Pan, and W. H. Wang, On the criteria of bulk metallic glass formation in MgCu-based alloys, *Intermetallics* **13**, 638–641 (2005).

64. D. S. Song, J. H. Kim, E. Fleury, W. T. Kim, and D. H. Kim, Synthesis of ferromagnetic Fe-based bulk glassy alloys in the Fe–Nb–B–Y system, *J. Alloys Compd.* **389**, 159–164 (2005).

65. B. Zhang, M. X. Pan, D. Q. Zhao, and W. H. Wang, "Soft" bulk metallic glasses based on cerium, *Appl. Phys. Lett.* **85**, 61–63 (2004).

66. K. Mondal and B. S. Murty, On the parameters to assess the glass forming ability of liquids, *J. Non-Cryst. Solids* **351**, 1366–1371 (2005).

67. S. Fang, X. Xiao, L. Xia, Q. Wang, W. Li, and Y. D. Dong, Effects of bond parameters on the widths of supercooled liquid regions of ferrous BMGs, *Intermetallics* **12**, 1069–1072 (2004).

68. O. N. Senkov, D. B. Miracle, and J. M. Mullens, Topological criteria for amorphization based on a thermodynamic approach, *J. Appl. Phys.* **97**, 103502 (2005).

69. Z. P. Lu, C. T. Liu, C. A. Carmichael, W. D. Porter, and S. C. Deevi, Bulk glass formation in an Fe-based Fe–Y–Zr–M (M = Cr, Co, Al)–Mo–B system, *J. Mater. Res.* **19**, 921–929 (2004).

70. A. R. Miedema, R. Boom, and M. R. DeBoer, Heat of formation of solid alloys, *J. Less-Common Met.* **41**, 283–298 (1975).

71. R. Boom, M. R. DeBoer, and A. R. Miedema, Heat of mixing of liquid alloys-1, *J. Less-Common Met.* **45**, 237–245 (1976).

72. G. J. Van der Kolk, A. R. Miedema, and A. K. Niessen, On the composition range of amorphous binary transition-metal alloys, *J. Less-Common Met.* **145**, 1–17 (1988).

73. R. Coehoorn, G. J. Van der Kolk, J. J. Van den Broek, T. Minemura, and A. R. Miedema, Thermodynamics of the stability of amorphous alloys of 2 transition-metals, *J. Less-Common Met.* **140**, 307–396 (1988).

74. A. Takeuchi and A. Inoue, Calculations of mixing enthalpy and mismatch entropy for ternary amorphous alloys, *Mater. Trans. JIM* **41**, 1372–1378 (2000).

75. F. R. DeBoer, R. Boom, W. C. M. Mattens, A. R. Miedema, and A. K. Niessen, *Cohesion in Metals: Transition Metals Alloys* (North-Holland, Amsterdam, 1988).

76. J. Friedel, Electronic structure of primary solid solutions in metals, *Adv. Phys.* **3**, 446–507 (1954).

77. J. D. Eshelby, Distortion of a crystal by point imperfections, *J. Appl. Phys.* **25**, 255–261 (1954).

78. J. D. Eshelby, The continuum theory of lattice defects, *Solid State Phys.* **3**, 79–114 (1956).

79. L. Xia, S. S. Fang, Q. Wang, Y. D. Dong, and C. T. Liu, Thermodynamic modeling of glass formation in metallic glasses, *Appl. Phys. Lett.* **88**, 171905 (2006).

80. D. H. Xu, B. Lohwongwatana, B. G. Duan, W. L. Johnson, and C. Garland, Bulk metallic glass formation in binary Cu-rich alloy series – $Cu_{100-x}Zr_x$ ($x = 34$, 36, 38.2, 40 at.%) and mechanical properties of bulk $Cu_{64}Zr_{36}$ glass, *Acta Mater.* **52**, 2621–2624 (2004).

81. A. Inoue and W. Zhang, Formation, thermal stability and mechanical properties of Cu–Zr and Cu–Hf binary glassy alloy rods, *Mater. Trans.* **45**, 584–587 (2004).

82. D. Wang, Y. Li, B. B. Sun, M. L. Sui, K. Lu, and E. Ma, Bulk metallic glass formation in the binary Cu–Zr system, *Appl. Phys. Lett.* **84**, 4029–4032 (2004).

83. M. B. Tang, D. Q. Zhao, M. X. Pan, and W. H. Wang, Binary Cu–Zr bulk metallic glasses, *Chin. Phys. Lett.* **21**, 901–903 (2004).

84. L. Xia, D. Ding, S. T. Shan, and Y. D. Dong, The glass forming ability of Cu-rich Cu–Hf binary alloys, *J. Phys.: Condens. Mat.* **18**, 3543–3548 (2006).

85. L. Xia, W. H. Li, S. S. Fang, B. C. Wei, and Y. D. Dong, Binary Ni–Nb bulk metallic glasses, *J. Appl. Phys.* **99**, 026103 (2006).

86. X. S. Xiao, S. S. Fang, G. M. Wang, Q. Hua, and Y. D. Dong, Influence of beryllium on thermal stability and glass-forming ability of Zr–Al–Ni–Cu bulk amorphous alloys, *J. Alloys Compd.* **376**, 145–148 (2004).

87. N. Nishiyama and A. Inoue, Direct comparison between critical cooling rate and some quantitative parameters for evaluation of glass-forming ability in Pd–Cu–Ni–P alloys, *Mater. Trans.* **43**, 1913–1917 (2002).

88. J. H. Kim, J. S. Park, H. K. Lim, W. T. Kim, and D. H. Kim, Heating and cooling rate dependence of the parameters representing the glass forming ability in bulk metallic glasses, *J. Non-Cryst. Solids* **351**, 1433–1440 (2005).

89. Q. J. Chen, J. Shen, H. B. Fan, J. F. Sun, Y. J. Huang, and D. G. McCartney, Glass-forming ability of an iron-based alloy enhanced by Co addition and evaluated by a new criterion, *Chin. Phys. Lett.* **22**, 1736–1738 (2005).

90. X. F. Zhang, Y. M. Wang, J. B. Qiang, Q. Wang, D. H. Wang, D. J. Li, C. H. Shek, and C. Dong, Optimum Zr–Al–Co bulk metallic glass composition $Zr_{53}Al_{23.5}Co_{23.5}$, *Intermetallics* **12**, 1275–1278 (2004).

91. M. X. Xia, H. X. Zheng, J. Liu, C. L. Ma, and J. G. Li, Thermal stability and glass-forming ability of new Ti-based bulk metallic glasses, *J. Non-Cryst. Solids* **351**, 3747–3751 (2005).

92. A. H. Cai, G. X. Sun, and Y. Pan, Evaluation of the parameters related to glass-forming ability of bulk metallic glasses, *Mater. Des.* **27**, 479–488 (2006).

93. Y. M. Wang, X. F. Zhang, J. B. Qiang, Q. Wang, D. H. Wang, D. J. Li, C. H. Shek, and C. Dong, Composition optimization of the Al–Co–Zr bulk metallic glasses, *Scripta Mater.* **50**, 829–833 (2004).

94. E. S. Park, D. H. Kim, and W. T. Kim, Parameter for glass forming ability of ternary alloy systems, *Appl. Phys. Lett.* **86**, 061907 (2005).

95. E. S. Park and D. H. Kim, Effect of atomic configuration and liquid stability on the glass-forming ability of Ca-based metallic glasses, *Appl. Phys. Lett.* **86**, 201912 (2005).

96. Q. G. Meng, J. K. Zhou, H. X. Zheng, and J. G. Li, Fragility of superheated melts and glass-forming ability in Pr-based alloys, *Scripta Mater.* **54**, 777–781 (2006).

97. H. Tanaka, Relationship among glass-forming ability, fragility, and short-range bond ordering of liquids, *J. Non-Cryst. Solids* **351**, 678–690 (2005).

98. A. H. Cai, Y. Pan, and G. X. Sun, New thermodynamic parameter describing glass forming ability of bulk metallic glasses, *Mater. Sci. Technol.* **21**, 1222–1226 (2005).

99. X. F. Bian, B. A. Sun, L. N. Hu, and Y. B. Jia, Fragility of superheated melts and glass-forming ability in Al-based alloys, *Phys. Lett. A* **335**, 61–67 (2005).

100. Z. P. Lu, C. T. Liu, and Y. D. Dong, Effects of atomic bonding nature and size mismatch on thermal stability and glass-forming ability of bulk metallic glasses, *J. Non-Cryst. Solids* **341**, 93–100 (2004).

101. I. W. Donald and H. A. Davies, Prediction of glass-forming ability for metallic systems, *J. Non-Cryst. Solids* **30**, 77–85 (1978).

102. E. S. Park, W. T. Kim, and D. H. Kim, A simple model for determining alloy composition with large glass forming ability in ternary alloys, *Metall. Mater. Trans. A* **32**, 200–202 (2001).

103. Z. P. Lu, H. Bei, and C. T. Liu, Recent progress in quantifying glass-forming ability of bulk metallic glasses, *Intermetallics* **15**(5–6), 618–624 (2007).

104. J. Y. Lee, D. H. Bae, J. K. Lee, and D. H. Kim, Bulk glass formation in the Ni–Zr–Ti–Nb–Si–Sn alloy system, *J. Mater. Res.* **19**, 2221–2225 (2004).

105. A. Inoue and N. Nishiyama, Extremely low critical cooling rates of new Pd–Cu–P base amorphous alloys, *Mater. Sci. Eng. A* **226–228**, 401–405 (1997).

106. Z. P. Lu, C. T. Liu, and W. D. Porter, Role of yttrium in glass formation of Fe-based bulk metallic glasses, *Appl. Phys. Lett.* **83**, 2581–2583 (2003).

107. D. Kim, B. J. Lee, and N. J. Kim, Prediction of composition dependency of glass forming ability of Mg–Cu–Y alloys by thermodynamic approach, *Scripta Mater.* **52**, 969–972 (2005).

108. G. Shao, B. Lu, Y. Q. Liu, and P. Tsakiropoulos, Glass forming ability of multi-component metallic systems, *Intermetallics* **13**, 409–414 (2005).

109. G. Shao, Thermodynamic and kinetic aspects of intermetallic amorphous alloys, *Intermetallics* **11**, 313–324 (2003).

110. D. Ma, H. Cao, L. Ding, Y. A. Chang, K. C. Hsieh, and Y. Pan, Bulkier glass formability enhanced by minor alloying additions, *Appl. Phys. Lett.* **87**, 171914 (2005).

111. X. Y. Yan, Y. A. Chang, Y. Yang, F. Y. Xie, S. L. Chen, F. Zhang, S. Daniel, and M. H. He, A thermodynamic approach for predicting the tendency of multicomponent metallic alloys for glass formation, *Intermetallics* **9**, 535–538 (2001).

112. D. B. Miracle, A structural model for metallic glasses, *Nat. Mater.* **3**, 697–702 (2004).

113. P. Jalali and M. Li, Atomic size effect on critical cooling rate and glass formation, *Phys. Rev. B* **71**, 014206 (2005).

Chapter 5

MICROSTRUCTURE

M. K. Miller

Materials Science and Technology Division, Oak Ridge National Laboratory, Oak Ridge, TN 37831-6136, USA

5.1   INTRODUCTION

Microstructural characterizations are performed to investigate both the atomic structure and the stability of bulk metallic glasses (BMGs). Many characterizations with a variety of instruments are performed to identify the phases produced during preparation (i.e., quenching), annealing, and devitrification of BMG. The typical parameters that are quantified are the size, morphology, composition, crystal structure, and volume fraction of the phases formed. These parameters may be used to improve the alloy design process. Some techniques provide continuous monitoring of time-dependent processes, such as crystallization, as a function of time and temperature. High-resolution techniques are used to investigate the atomic structure of the glass including short- and medium-range order. Specialized characterizations may be performed to investigate the quantity and type of free-volume in the glass. Microstructural characterizations may also be performed in conjunction with mechanical property tests to investigate the type of failure, the source of crack initiation, and the interaction of shear bands that are produced under stress with microstructural features.

A simple visual inspection of cast ingots of BMGs can *quickly* reveal useful information. For example, a high reflectivity surface generally indicates that the ingot is amorphous, whereas a dull gray surface generally indicates that the ingot has devitrified or crystallized. Cracks, porosity, and coarse inclusions and precipitates may also be identified by visual inspection or examination in a light microscope at magnifications up to ~1,000 times.

However, the full characterization of a BMG requires instruments with higher resolution. In general, a complete characterization of the microstructure will require the use of multiple techniques.

The optimum microstructure for a BMG may not actually be a fully amorphous alloy but rather an alloy with a distribution of fine particles or precipitates[1] (sometimes called an in situ composite[2] or a precipitation-hardened alloy) or a true composite of a glass matrix and a second phase. Such obstacles can inhibit the propagation of shear bands in the glass, thereby improving the mechanical properties. These intrinsic or extrinsic (i.e., composite) microstructures may also be characterized. For example, the wetting of the BMG–particle interface, pullout behavior, particle deformation, and the propagation of the shear bands and the fracture path through the composite are of interest.

In this chapter, many of the techniques that have been used to characterize the microstructures of BMGs are reviewed. The techniques are divided broadly into the routine techniques, such as differential scanning calorimetry, X-ray diffraction, and scanning electron microscopy, that are used to evaluate parameters such as the glass-forming ability, the critical dimension, and examine the general microstructure, and the higher resolution, more specialized techniques, such as high-resolution and fluctuation electron microscopy, field ion microscopy, atom probe tomography, small angle scattering, and positron annihilation spectroscopy, that are used to characterize the nanometer-scale structure, phase composition, open volume, etc.

5.2   AMORPHOUS AND CRYSTALLINE STRUCTURES

All crystalline (or morphous) solids have structural order as the atoms are arranged on a well-defined three-dimensional lattice with a characteristic space group and a replicating unit cell of atoms. In some crystals, such as meteorites and galvanized zinc, the individual crystal grains are large enough to be visible to the naked eye. This long-range order of the atoms gives rise to sharp diffraction or Bragg peaks in scattering experiments. In some crystalline materials, the different elements prefer to reside on specific sites in the unit cell and thereby create ordered sublattices in the crystal.

Glasses, including BMGs, lack this long-range order and are said to be *amorphous*. Glasses have a topologically disordered distribution of atoms but properties of an isotropic solid. Historically, the structural model for glasses is that of dense random packing of hard spheres.[3,4] Hard sphere or "Bernal" models can satisfactorily describe monatomic systems (or alloys with constituent species having comparable atomic sizes and insignificant chemical short-range order), but fail to describe many binary metallic glasses and multielement BMGs, particularly metal–metalloid glasses, in which chemical short-range order is pronounced.

Glasses are not truly a random distribution of atoms because no two atoms can be closer than a typical bonding distance nor farther apart than a few nearest neighbors. This distribution of atoms gives rise to a diffuse intensity peak in X-ray, neutron and electron scattering experiments. However, glasses still exhibit some short-range order of the atoms. Structural information on short-range order is typically obtained from the pair correlation function.

Some amorphous materials are thought to exhibit order over a length scale that is slightly longer than that of short-range order but too short and nonrepetitive to be considered long-range order. This regime has been termed medium-range order (MRO). MRO can be described as local structural "units" such as polyhedra (or cages) that are connected by common edges, faces, or vertices, and arranged to fill three-dimensional space.[5–7] MRO should not be confused with the early stages of phase separation where embryos or small nanoclusters of a second crystalline or amorphous phase are formed. A full discussion of the concepts of order and disorder is beyond the scope of this chapter and has been discussed by Ossi.[8] Amorphous alloys do not contain the microstructural features that are observed in crystalline materials, such as grain boundaries, dislocations, stacking faults, etc.

Structural analysis of metallic liquids and glasses is complicated by their lack of long-range order and the length scale of interest – typically a few nearest neighbor distances. Therefore, specialized techniques are required to determine the structure. However, even modern advanced microstructural characterization techniques have difficulty in distinguishing between (or characterizing) amorphous, short- and medium-range order, and crystalline phases when their size is close to a few unit cell dimensions. In addition, a large fraction of the atoms is located at or near the surface of a nanometer-scale crystal. Therefore, surface relaxation and interfacial effects can distort the atomic positions and decrease the structural order.

In addition to the characterization of the structure of the amorphous state, several other microstructural features of BMGs are of interest. In particular, the decomposition path of the amorphous structure has been investigated in many studies, generally with the use of multiple techniques. Several different paths have been suggested, including precipitation and growth of one or more crystalline phase, spinodal decomposition, and phase separation followed by crystallization. It is generally assumed that the liquid from which the BMG is made is a single phase. However, it is possible for phase separation to occur in the liquid and for two immiscible liquid phases to form prior to quenching.

5.3 DIFFERENTIAL SCANNING CALORIMETRY

Differential scanning calorimetry (DSC) is typically used to determine the thermodynamics of phase transitions including crystallization (or devitrification) but does not provide any insight into the atomic rearrangements which occur during the transition or the identity and number of the phases. A differential scanning calorimeter measures the amount of energy (heat) absorbed or released by a sample as it is heated, cooled, or held at a constant temperature. In addition, DSC may be used to evaluate the temperature range and kinetics of the decomposition processes. Because the glass transition is a second-order transition, DSC may be used to measure the glass transition temperature T_g, as defined by the onset of the endothermic event, as well as the onset of crystallization temperature T_x. The glass-forming ability of the BMG can be estimated directly from these parameters, as discussed in Chap. 4.

A typical DSC plot for a $Cu_{47}Ti_{33}Zr_{11}Ni_8Si_1$ gas-atomized powder before consolidation into a BMG[9] is shown in Fig. 5.1. In this example, the endothermic glass transition temperature is followed by two exothermic events. These heating experiments are typically performed at a constant heating rate of 10–40 K min^{-1}. The data shown in Fig. 5.1 were obtained at a constant heating rate of 40 K min^{-1}. Since the incubation time for crystallization is a time- and temperature-dependent process, the estimated T_g and T_x temperatures are a function of the heating rate. Therefore, it is important to explicitly state

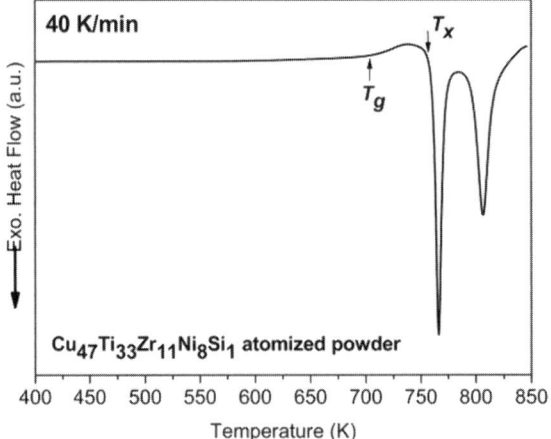

Fig. 5.1. Differential scanning calorimeter profile for a $Cu_{47}Ti_{33}Zr_{11}Ni_8Si_1$ gas-atomized powder before consolidation into a bulk metallic glass. The data were obtained at a constant heating rate of 40 K min^{-1}. The glass transition temperature T_g and the onset of crystallization temperature T_x are indicated (reprinted from reference [9] with permission of the American Physical Society)

the heating rate in all DSC experiments. The glass transition generally occurs at higher temperatures for faster cooling rates and is also influenced by strain. Each DSC experiment requires a fresh sample as the measurement alters the microstructure.

The start and finish times of each phase transition that are measured in a series of isothermal experiments at different temperatures may be used to construct time–temperature–transformation (TTT) diagrams, as shown in Fig. 4.1. Each resulting "C" curve provides a description of the kinetics of phase separation and may be used to design appropriate time and temperature regimes for more detailed microstructural characterization experiments. The critical cooling rate R_c may also be derived from the cooling rate that just avoids the nose of the first crystallization event in the TTT diagram.

5.4   X-RAY DIFFRACTION

X-ray diffraction (XRD) is frequently used to evaluate whether an as-prepared sample is fully amorphous from the characteristic diffuse intensity peak, and to detect and identify crystalline phases from their characteristic Bragg peaks in the diffraction patterns. In this technique, a monochromatic X-ray beam is passed through a thin sample and the intensity of the diffracted beam is measured as a function of diffraction angle, 2θ. To obtain constructive interference between the incident and the scattered waves in a crystalline phase, the path difference, $2d \sin \theta$, has to be a multiple of the wavelength, λ. Bragg's law gives the relationship between interplanar distance d and diffraction angle θ

$$2d \sin \theta = n\lambda. \tag{5.1}$$

The identity of one or more phases responsible for all the Bragg peaks is determined from the position and intensity of the peaks. The ternary and higher-order phase diagrams of these systems are usually not available due to the number of elements generally present in BMGs, and thus, it is difficult to predict the structures of all of the possible phases that may be present. Therefore, computer simulations of the positions and intensities of Bragg peaks for possible crystal structures are normally required.

A typical set of X-ray intensity plots[10] vs. diffraction angle 2θ for a $Zr_{49}Cu_{49-x}Al_x$ BMG with $x = 6, 8, 10$, and 12 at.% Al is shown in Fig. 5.2. As the aluminum content of the alloy was increased, the volume fraction, f, of a dendritic τ_3 ($Zr_{51}Cu_{28}Al_{21}$) phase increased from 0 to 7, 15, and 20% (Fig. 5.2), respectively, as shown in the scanning electron micrographs[10] in Fig. 5.3. The X-ray data show the characteristic diffuse intensity peak from the amorphous structure as well as several sharp Bragg peaks from the crystalline

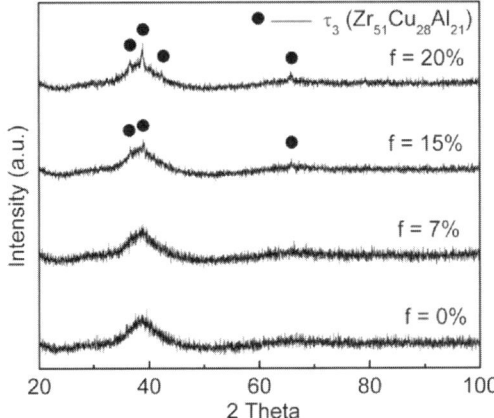

Fig. 5.2. Typical X-ray diffraction patterns for a series of $Zr_{49}Cu_{49-x}Al_x$ ($x = 6$, 8, 10, and 12%) BMG alloys. The diffuse intensity maximum arises from the amorphous phase in this alloy. Sharp Bragg peaks identified as a τ_3 ($Zr_{51}Cu_{28}Al_{21}$) phase are marked with a *bullet mark* and are evident in the 10 and 12% Al alloy that contain 15 and 20% volume fraction, f, of the dendritic τ_3 phase (reprinted from reference [10] with permission from Elsevier)

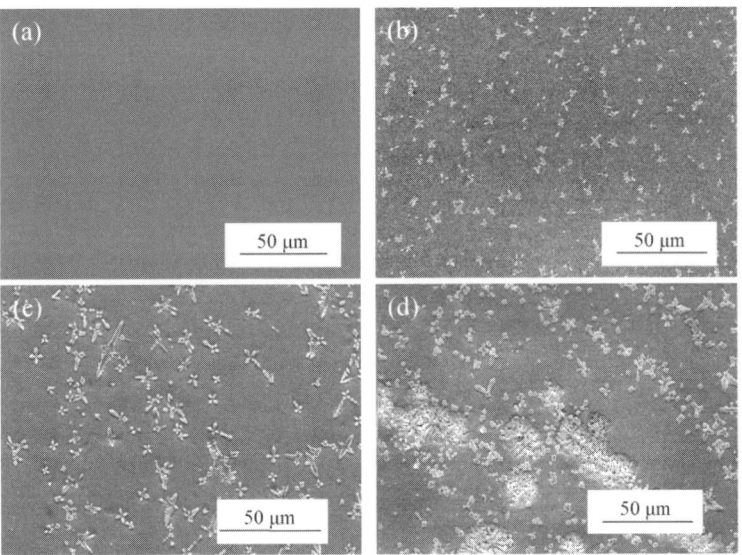

Fig. 5.3. Scanning electron micrographs from a series of $Zr_{49}Cu_{49-x}Al_x$ ($x = 6$, 8, 10, and 12%) BMG alloys. No second phase was evident in (**a**) the fully amorphous 6% Al alloy. The light dendritic phase in this in situ composite in the (**b**) 8% Al, (**c**) 10% Al, and (**d**) 12% Al alloys was identified as a τ_3 $Zr_{51}Cu_{28}Al_{21}$ phase from the sharp Bragg peaks in the X-ray diffraction patterns shown in Fig. 5.2 (reprinted from reference [10] with permission from Elsevier)

τ_3 phase in the high-aluminum content alloys. These results demonstrate that standard XRD is relatively insensitive to small volume fractions of crystalline phases, as no Bragg peaks are evident in the alloy with 7% of the τ_3 phase. Diffraction analysis, using either X-rays or neutrons, has proven effective in understanding the pairwise bonding that comprises the short-range order in these inherently disordered systems.[11–13] In general, these studies have only been a "snap shot" in time or temperature. One of the reasons for this was the long counting times necessary to obtain data with good signal-to-noise ratios out to the high reciprocal space necessary for accurate analysis. Third generation synchrotrons now have sufficient flux at high energies (>40 keV) so that transmission experiments, i.e., high-energy X-ray diffraction (HEXRD), similar to those at neutron facilities can be performed.[14,15] The high flux sources, in conjunction with new high efficiency area detectors, have reduced data collection times to a few minutes or even a few seconds.[16,17] These instruments provide a data collection rate comparable to thermal analysis at high reciprocal space that is sufficient for pair distribution function analysis.

For example, in situ devitrification studies were performed on a series of $Zr_2Cu_{1-x}Pd_x$ alloys at the Advanced Photon Source (APS) at Argonne National Laboratory in collaboration with the Midwest Universities Collaborative Access Team.[18–20] The HEXRD data were obtained at the 6ID-D beamline with an energy of 99.55 keV. Silicon double-crystal monochromators were used to select the wavelength of $\lambda = 0.012466$ nm. All millimeter–diameter samples were sealed under argon in thin-walled silica capillaries. Thermal analysis of the crystalline phases suggests that this system should behave as a simple solid solution as they share the same high-temperature C11b crystal structure. All the alloys in this system can be made amorphous by rapid solidification but they follow different devitrification paths. This suggests that, in the amorphous state, there are subtle differences in the short-range order that determine phase selection during devitrification. The amorphous Zr_2Cu devitrifies directly to the C11b structure whereas small amounts of Pd cause the system to first transform to a metastable, quasi-crystalline (i-phase) structure, as shown in Fig. 5.4.[18–20] Increasing the Pd level to $x = 0.5$ leads to the formation of a metastable C16 prior to the C11b phase. At higher Pd contents, the C16 phase is no longer observed. The overlay of the DSC traces shows that the phase transformations observed in the time-resolved HEXRD data perfectly match the thermal analysis. These results together with ab initio calculations suggest that electronic effects dominate the phase selection process and may dominate the short-range order in the as-quenched state.

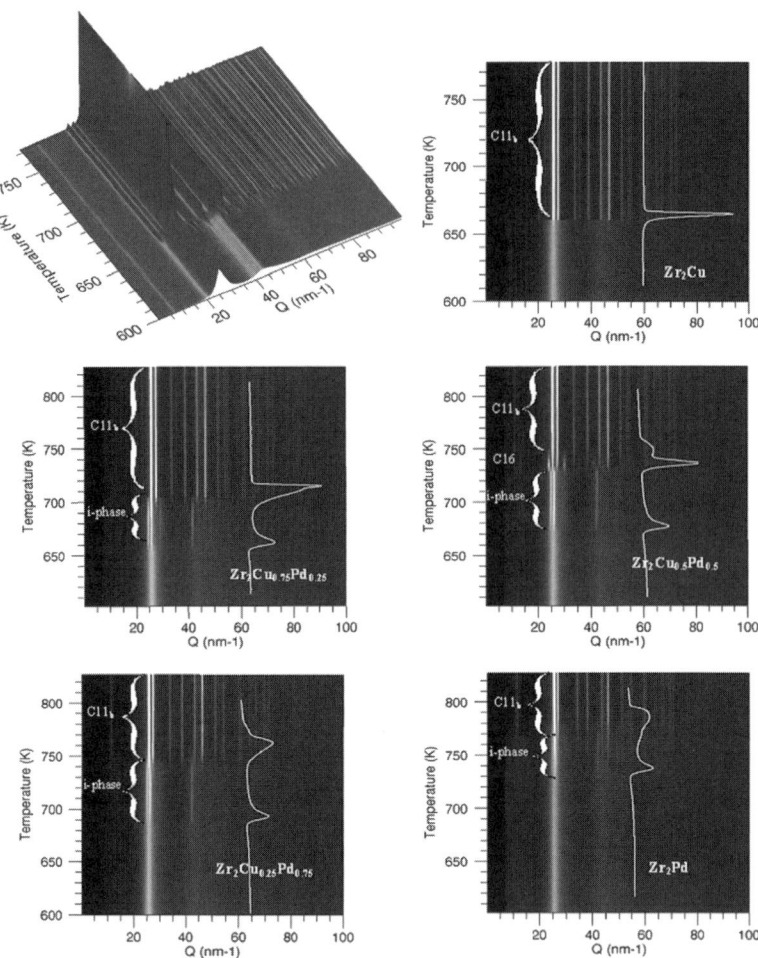

Fig. 5.4. *Top left*: a three-dimensional representation of the time-resolved XRD patterns obtained during devitrification of an amorphous Zr_2Cu alloy. The other images are two-dimensional representations of the XRD data (bright regions indicate stronger scattering intensity) of the five $Zr_2Cu_{1-x}Pd_x$ ($x = 0$, 0.25, 0.5, 0.75, and 1) alloys. The light lines are overlays of the DSC data obtained at the same 10 K min^{-1} heating rate. All structures are amorphous at 600 K (courtesy M. J. Kramer)

Extended X-ray absorption fine structure (EXAFS) measurements can also provide element-specific information on the local environment. Small angle X-ray scattering characterizations are discussed in Sect. 5.8.

5.5   ELECTRON MICROSCOPY

Electron microscopy is a cornerstone of the microstructural characterization of most materials as it has a wide variety of different techniques including scanning electron microscopy (SEM), transmission electron microscopy (TEM), and scanning transmission electron microscopy (STEM). Each of these techniques has many different modes of operation and types of signals that may be used to form images, determine the crystal structure through electron diffraction or high-resolution electron microscopy (HREM), and quantify the concentration variations through energy-dispersive X-ray spectroscopy (EDS), electron energy loss spectroscopy (EELS), etc.

5.5.1   Scanning electron and ion microscopy

SEM is typically performed on bulk specimens such as polished sections or fracture surfaces.[21] Images may be acquired with either the secondary or backscattered electrons that are produced from the interaction of the (1–30 keV) electron beam with the specimen. Amorphous BMGs generally exhibit uniform contrast in polished sections, as shown in Fig. 5.3a. The phase contrast that is produced with secondary electrons from a fully crystallized $Cu_{47}Ti_{33}Zr_{11}Ni_8Si_1$ BMG that was annealed for 24 h at 1,073 K is shown in Fig. 5.5a. The complex multiphase nature of crystallized BMGs is

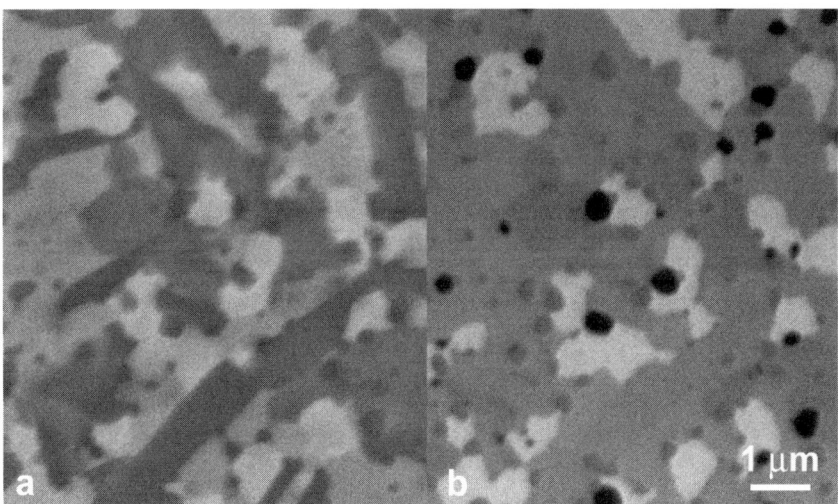

Fig. 5.5. (**a**) Secondary electron images obtained in a scanning electron microscope from a $Cu_{47}Ti_{33}Zr_{11}Ni_8Si_1$ BMG that was annealed for 24 h at 1,073 K. (**b**) Secondary electron image generated with a scanned ion beam. The different signals produce different contrast between the phases. At least five crystalline phases are evident after this heat treatment (material courtesy S. Venkataraman)

evident. In a dual-beam SEM/focused ion beam (FIB) instrument, secondary electron images may also be produced from the interaction of the (1–30 keV) scanning ion beam with the specimen, as shown in Fig. 5.5b. These images can provide different contrast to the secondary electron images produced from the electron beam. One advantage of the use of ions to produce the secondary electrons is that the interaction volume from a 30 keV gallium ion beam is typically only 10–20 nm compared to 1–2 μm from a 30 keV electron beam. The interaction volume in the low-voltage SEM may be reduced to the ~20 nm range by the use of low (1 keV) accelerating voltages.

The X-rays produced during the electron beam–specimen interaction may be used for compositional analysis in EDS. In modern low-voltage (<5 keV) field-emission scanning electron microscopes, the spatial resolution for X-ray microanalysis is typically 20–50 nm. Therefore, the predominant application of the SEM for BMG applications is to detect second phases and to assist in phase identification. In addition, SEM may be used to study fracture surfaces and crack propagation in tensile, compression and fatigue test specimens, and deformation and cracks in indentation specimens.

The dual-beam SEM/FIB instrument may be used to extract lift-out specimens from site-specific locations such as shear bands, underneath micro-hardness indents, phases, etc. for other microscopy techniques.[22] The size and morphology of the crystalline regions or fine scale (<50 μm) composites can also be reconstructed from serial sections produced by slice and view techniques. In this method, thin slices of known thickness are milled from the surface with the gallium ion beam and the newly exposed surface of the specimen is then imaged with either the electron or ion beam. These images may then be combined in a computer to produce the size and morphology of the phases present.

SEM has also been used to estimate the temperature rise at shear bands, as direct measurements are hampered by the short time and length scales of the shear. Lewandowski and Greer[23] used a novel experimental method based on coating the surface of a BMG sample with a 50-nm-thick layer of tin and then deforming the specimen under four-point bend loading until failure. SEM examination revealed that the fusible tin coating had formed spherical beads in the vicinity of the shear bands. From the measurement of the half-width of these beads, the true half-width of the hot zone around the shear band was estimated to be approximately 200–1,000 nm. This value is significantly larger than the accepted 10–20 nm thickness of the shear band.[24,25] In addition, the maximum actual temperature was estimated to be between 3,400 and 8,600 K over a few nanoseconds. Although in practice, the actual temperature rise may be significantly lower. This study indicated that shear band operation cannot be fully adiabatic and the temperature rise cannot be the factor controlling the thickness of the shear band.

5.5.2 Transmission electron microscopy

TEM requires thin foil or FIB-thinned lift-out specimens and accelerating voltages (typically $\geq$100 keV) sufficient for the electron beam to penetrate the specimen. Suitable thicknesses for the electron transparent region of a TEM specimen depend primarily on the material, accelerating voltage, and type of analysis. The thickness is typically between 50 and 200 nm. These thin specimens are fabricated by a combination of mechanical grinding, electropolishing, or ion milling techniques or from lift-out techniques in a dual-beam SEM/FIB instrument.

In suitably equipped transmission electron microscopes, phase identification can be made from selected area electron diffraction patterns combined with EDS and EELS. Amorphous materials generate diffuse rings in the electron diffraction patterns. Crystalline regions may be detected from the contrast in the images and from the presence of a regular array of spots in the diffraction pattern when the specimen is tilted to an orientation at which there is constructive interference from atomic planes. The lattice (planar) spacings and hence the crystal structure of the crystalline phase can be determined from the positions (in reciprocal space) and intensities of these spots in a series of diffraction patterns taken at different crystal orientations. By tilting the transmitted electron beam such that the diffracted beam is along the optic axis and is selecting the diffracted beam with an objective aperture (blocking the transmitted beam) in the column, dark field TEM images, where only the properly oriented crystalline phase(s) are illuminated, can be obtained. Other microscopic features, such as shear bands or voids,[26–30] can be detected from changes in contrast in the images. High-angle annular dark field (HAADF) imaging may also be used to investigate local compositional fluctuations provided the atomic number differences between the partitioning elements are great enough.

5.5.3 High-resolution electron microscopy

HREM may be used to detect nanoscale crystalline regions, investigate MRO, and identify the crystalline phases.[31] The success of this technique depends on the preparation of extremely thin (<50 nm) high-quality (surface-clean) specimens.[32]

A typical HREM image of an as-cast $Pd_{40}Ni_{40}P_{20}$ BMG alloy and the diffuse electron diffraction ring characteristic of an amorphous structure are shown in Fig. 5.6a. Small 2–4 nm diameter crystalline regions are evident from the regular lattice fringes in the image shown in Fig. 5.6b. In this example, the crystallized microstructure was produced in situ by the interaction of the electron beam with the specimen during examination (i.e., beam heating).

The interpretation of these HREM images is not intuitive as they can change dramatically with under- and overfocus conditions. Therefore, interpretation of HREM images requires extensive computer simulations of the possible structures.

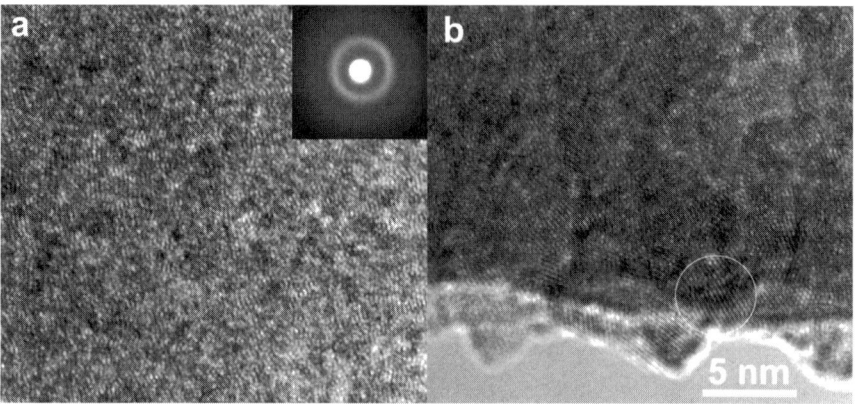

Fig. 5.6. (a) High-resolution electron microscopy images and electron diffraction pattern of an as-cast $Pd_{40}Ni_{40}P_{20}$ bulk metallic glass. (b) Some crystallization occurred during examination in the HREM near the edge of the thin region of the foil (micrographs courtesy K. L. More and material courtesy R. B. Schwartz)

5.5.4 Fluctuation electron microscopy

Fluctuation electron microscopy (FEM) is a relatively new electron microscopy technique proposed by Treacy and Gibson[33] in 1996 that is uniquely sensitive to MRO on the 1–2 nm length scale in disordered or amorphous materials. FEM probes MRO by measuring the point-to-point variation in scattered intensity. This measurement can be performed with tilted dark field or hollow-cone dark field imaging in the TEM (or by collecting a large number of scattering patterns from different nanoscale regions in the STEM).[33–35] The scattered intensity for all regions in a disordered (amorphous) material will be approximately the same. Therefore, a dark field image in the TEM will exhibit little variation in contrast intensity. In materials exhibiting MRO, some of these small, ordered regions will be oriented so that they scatter incident electrons strongly and other regions will not. Therefore, significant point-to-point variations (salt and pepper speckle) will be observed in the intensity of the dark field TEM image. More intensity variations or speckle in the dark field image implies more MRO. The normalized variance $V(k)$ can be used to quantify the degree of speckle

$$V(k,Q) = \frac{\langle I^2(\mathbf{r},k,Q) \rangle}{\langle I(\mathbf{r},k,Q) \rangle^2} - 1, \qquad (5.2)$$

where $I(\mathbf{r}, k, Q)$ is the image intensity as a function of position $\mathbf{r}$ in the image, the dark field scattering vector magnitude k, and the objective aperture size Q. The brackets indicate averaging over the image. Peaks in the variance provide information as a function of either k (called "variable coherence microscopy") or Q ("variable resolution microscopy") about the type, length scale, and degree of MRO.

Few FEM studies have been performed on BMGs to date. The suitability of this technique for the characterization of MRO in BMGs was demonstrated by Hufnagel et al.[36–38] in $Zr_{57}Ti_5Cu_{20}Ni_8Al_{10}$ and $Zr_{57}Ta_5Cu_{20}Ni_8Al_{10}$ alloys. The peak in the variance is related to structural order on the 1–2 nm length scale, as shown in Fig. 5.7. The Ta-containing alloy exhibited stronger MRO than the Ti-containing alloy. It was also demonstrated that the method used to fabricate the TEM specimen influenced the order, i.e., ion milling introduced more structural order than electropolishing.[36,37] This study also revealed that small changes in the alloy composition can have a significant effect on the MRO. FEM has also been performed on melt-spun $Al_{92}Sm_8$ amorphous alloys.[39–41] This $Al_{92}Sm_8$ alloy is not considered to be a true BMG as the size of the smallest dimension is less than 1 mm that is considered to be the critical cutoff value for bulk designation. It has been suggested that MRO influences the nucleation of the crystalline phases.

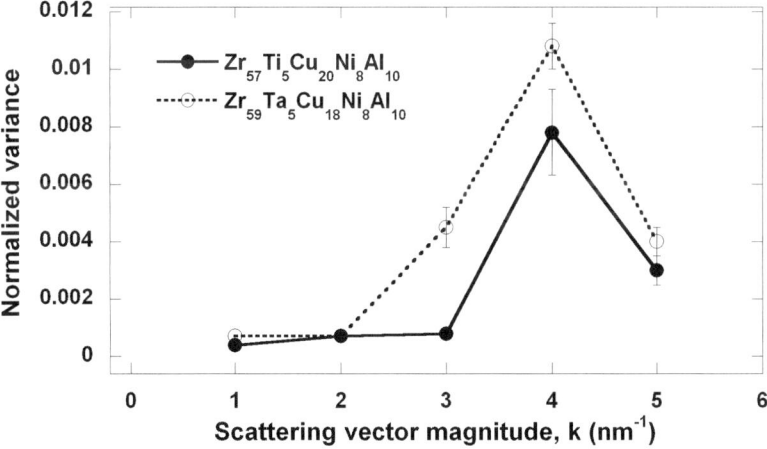

Fig. 5.7. Normalized variance for two amorphous alloys obtained from fluctuation electron microscopy. The peak in the variance is related to structural order on the 1–2 nm length scale. The Ta-containing alloy has stronger medium-range order than the Ti-containing alloy (reprinted from reference [37] with permission from Elsevier)

FEM measurements suggested that 1–2 nm diameter regions with Al crystal-like order are associated with primary crystallization and the results are consistent with a quenched-in cluster model of primary crystallization.[39–41]

5.6   FIELD ION MICROSCOPY

Several groups have used field ion microscopy (FIM)[42,43] to study BMGs due to its ability to image individual atoms. In fact, FIM was the first technique to image individual atoms over 50 years ago. This technique requires that a needle-shaped specimen with an end radius of less than 50 nm be fabricated from bulk materials. These needles are usually fabricated by either electropolishing or ion milling in a dual-beam SEM/FIB. The specimen is inserted into an ultrahigh vacuum chamber and cooled to cryogenic temperatures. The specimen is positioned approximately 50 mm in front of a microchannel plate and phosphor screen detector. A trace of image gas, typically 10^{-5} mbar of neon or helium, is introduced into the vacuum system. A high positive voltage V is then gradually applied to the specimen. The image gas atoms in close proximity to the apex of the specimen are polarized by the electric field F and then attracted to the specimen. At a certain voltage depending on the specimen radius r and the material, the image gas atoms on the surface of the specimen are field ionized and the resulting ions are then repelled from the specimen toward the detector. The field required for ionization depends on the ionization potentials, sublimation energy, and work function of the material, and is typically between 19 and 45 V nm^{-1} and is given by

$$F = \frac{V}{kr},\qquad(5.3)$$

where k is a numerical constant between 2 and 5. The microchannel plate converts each ion into approximately a million electrons. The resulting electrons are drawn across the gap to the phosphor screen by a small accelerating voltage where they produce an image. The dots in the field ion micrographs are individual atoms, as shown in Fig. 5.8. The typical magnification of the field ion microscope η is given to a first approximation by $\eta = d/\xi r$, where d is the specimen to detector distance and ξ is a projection parameter known as the *image compression factor*.

A field ion micrograph of an amorphous material exhibits a random distribution of atoms, as shown in Fig. 5.8a for an as-quenched $Pd_{40}Ni_{40}P_{20}$ BMG. Surface reconstruction limits the accuracy of reconstructing the atom positions. Therefore, field ion images cannot be used to reconstruct the lattice of crystal structures or investigate MRO. The presence of concentric rings in a field ion image is due to the atomic terraces of a low Miller index region of a crystalline phase, as shown in Fig. 5.8b for the same alloy that was annealed above the onset of crystallization for 1 h at 410°C. However, the absence of these rings does not necessarily indicate that the specimen is amorphous as high index regions from high solute concentration alloys generally do not exhibit well-defined ring structures. Different phases generally exhibit characteristic contrast due to their different compositions and evaporation fields, as shown in Fig. 5.8b. Individual atoms or phases cannot be identified solely by FIM. Atom probe tomography is required to identify phases from their compositions.

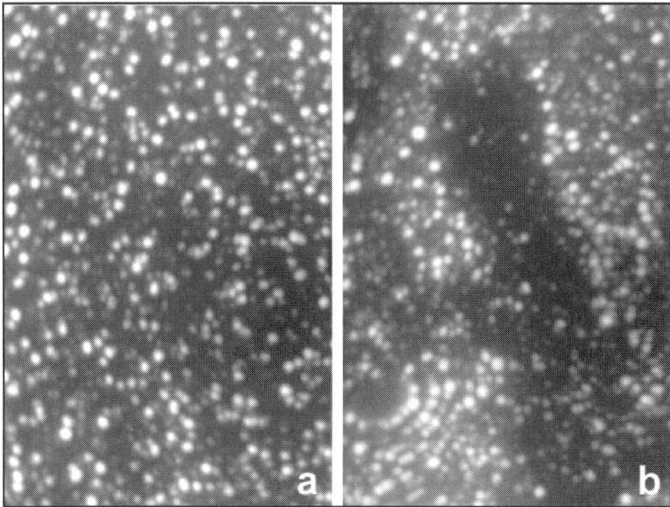

Fig. 5.8. Field ion micrographs of a $Pd_{40}Ni_{40}P_{20}$ bulk metallic glass in (**a**) the fully amorphous as-quenched state and (**b**) after annealing for 1 h at 410°C above the onset of crystallization. The brightly imaging atoms are randomly distributed in the amorphous condition. Partial rings due to atomic terraces and brightly and darkly imaging regions due to different crystalline phases are evident in the crystallized condition

The surface of the specimen may be carefully removed by raising the voltage on the specimen. This process is known as *field evaporation*. Field evaporation enables the atomic scale three-dimensional morphology of the phases present to be visualized and reconstructed from a series of field ion micrographs. These sequences of images may be used to reconstruct the size and shape of small voids or precipitates in an alloy either from film[44,45] or digitally,[46] assuming that all atom species present in the BMG contribute to the field ion image. However, these types of reconstructions are more easily performed from atom probe data.[47] Field evaporation is also used at the start of the experiment to remove surface contamination, corrosion, and any damage or surface irregularities produced during specimen preparation.

5.7   ATOM PROBE TOMOGRAPHY

The three-dimensional atom probe (3DAP) is a derivative of the field ion microscope that features a time-of-flight mass spectrometer with single atom sensitivity. The identity of each atom is determined from its flight time in the mass spectrometer.[47] A schematic diagram of a modern local electrode atom probe (LEAP®) is shown in Fig. 5.9.

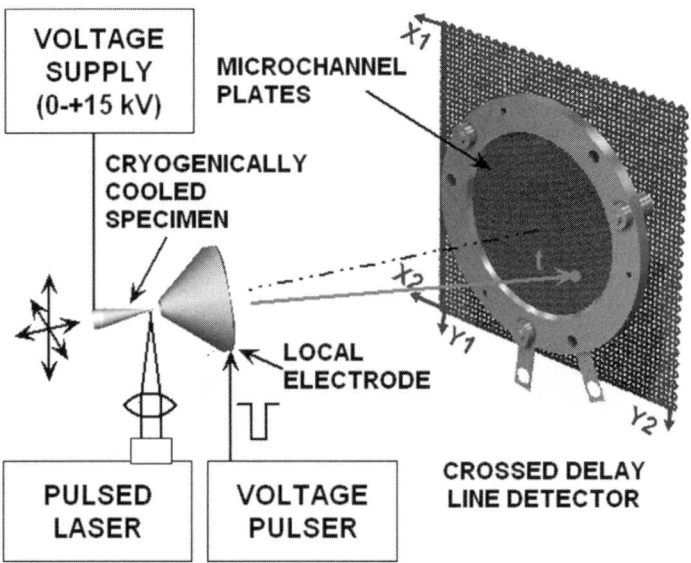

Fig. 5.9. A local electrode atom probe (LEAP) equipped with both a high-voltage pulse generator and a pulsed laser to field evaporate atoms from the needle-shaped specimen. The time of flight and the impact position of each atom are determined from the position-sensitive single atom, crossed delay line detector

A state-of-the-art instrument may use two alternative methods to field evaporate atoms from the needle-shaped cryogenically cooled specimen. The traditional method is to use short (1–10 ns) high-voltage pulses. However, this method is limited to materials with good electrical conductivity due to the attenuation of the high-voltage pulse as it travels down the length of the specimen to the apex region in high resistivity materials. Voltage pulsing is problematic for zirconium-based alloys due to a combination of their poor electrical conductivity at cryogenic temperatures and their brittle nature. However, several atom probe studies of zirconium-based[48–54] and other types[55–65] of BMGs have been performed. The alternative method is to use a 0.1–20 ps duration laser pulse focused on the apex of the needle-shaped specimen. Laser pulsing is also beneficial for brittle materials, such as devitrified BMGs, as it applies less stress to the specimen. Both these types of pulses may be applied to the specimen at rates up to 250 kHz.

The LEAP features a wide field of view, position-sensitive, crossed delay line detector that provides the x and y positions of the impacts of the individual atoms field evaporated from the specimen. As the instrument is a simple point projection microscope, these coordinates are, to a first approximation, linearly related to the position of the atom in the specimen just prior to field evaporation. The third coordinate is determined from the order in which the atom was field evaporated. The use of a pair of microchannel plates in the position-sensitive detector limits the detection efficiency of the atom probe to 50–60%. The atom probe has unlimited mass range and is equally sensitive to all elements. The resulting data are the x, y, z, and mass-to-charge state ratio m/n of the atoms. These data may be reconstructed with a variety of different methods.[47]

The atom probe is used to visualize the size, shape, and composition of small regions. Phase identity is derived from the local composition by counting the number of atoms of each element in the volume, i.e., the concentration of solute i is $c_i = n_i/n$, where n_i is the number of atoms of

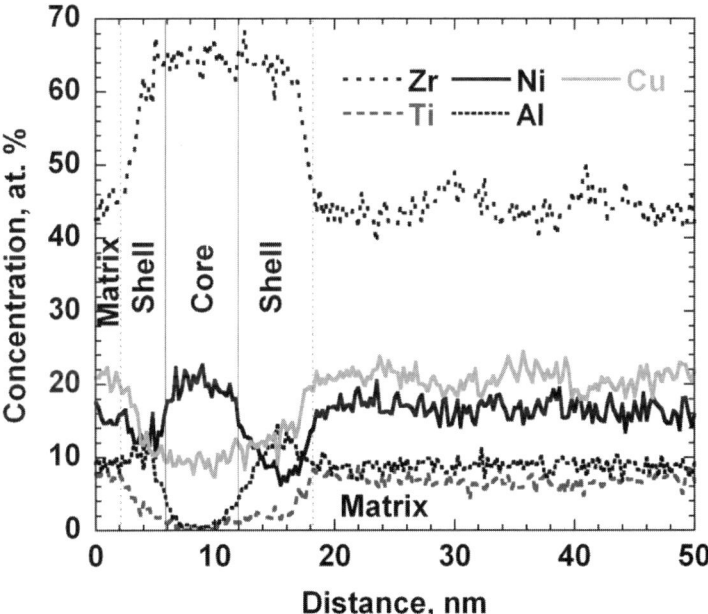

Fig. 5.10. A composition profile obtained in a local electrode atom probe from a $Zr_{52.5}Cu_{17.9}Ni_{14.6}Al_{10}Ti_5$ bulk metallic glass that was annealed for 15 h at 663 K. Both the shell and the core can be considered as the Zr_2M phase with different levels of Ni, Cu, Al, and Ti. The Zr_2M shell phase and matrix show opposite solute partitioning trends for all elements

solute i and n is the total number of atoms in the sample. Composition variations may be determined by constructing one-dimensional concentration profiles by subsampling the volume of analysis, as shown in Fig. 5.10 for a $Zr_{52.5}Cu_{17.9}Ni_{14.6}Al_{10}Ti_5$ BMG that was annealed for 15 h at 663 K. This profile was positioned to intersect both a precipitate and the surrounding matrix. The precipitate exhibited two distinct regions. Both the shell and the core can be considered as the Zr_2M phase with different levels of Ni, Cu, Al, and Ti. The Zr_2M shell phase and matrix show opposite solute partitioning trends for all elements. Two-dimensional concentration maps and isoconcentration surfaces may also be constructed from the atom-by-atom data. Even with its inherent high spatial resolution, investigation of short local order is hampered by the 50–60% detection efficiency of the technique. Therefore, the method can only be applied to follow trends rather than absolute measurements.

5.8   SCATTERING TECHNIQUES

Either X-rays or neutrons may be used to characterize BMGs. X-rays are scattered by the electrons surrounding the nucleus of an atom. Therefore, heavy elements with more electrons scatter X-rays more efficiently than light atoms. Neutrons interact with the nucleus and have cross-sections that vary with element and isotope. Small angle neutron scattering (SANS) may be used to estimate the size (radius of gyration), number density, and volume fraction of phases present in the microstructure. SANS studies have been performed on a number of different BMGs.[66–87] Radial distribution functions can also be determined from X-ray or neutron diffraction data to investigate short- and medium-range order. An example of time-resolved HEXRD data was presented in Sect. 5.4.

Although small angle scattering is a powerful tool to study phase separation over the nanometer to millimeter length scales, it does not provide any direct phase information. Diffraction is good for probing atomic ordering over the 0.1–1 nm length scale but provides little information unless the local atomic structure changes. Therefore, simultaneous measurements of small angle scattering and wide angle diffraction are particularly useful in revealing microstructural behavior over multiple length scales. High-energy X-rays and neutrons are highly penetrating so the measurement is representative of the bulk material. The typical configuration for the two types of detectors used in a small angle X-ray scattering experiment is shown in Fig. 5.11. The typical BMG sample size is 10–20 mm in diameter by 0.1–2 mm thick. The sample may be surrounded by a furnace for in situ measurements at different temperatures. The short times required to acquire data from high-energy sources enable statistically significant measurements to be obtained as a function of time at temperature and thus, the decomposition and devitrification processes can be monitored in real time.

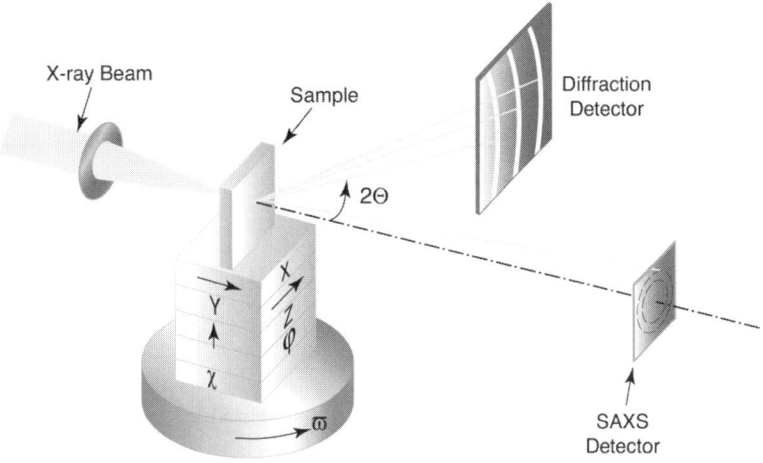

Fig. 5.11. Small angle and wide angle X-ray scattering. High-energy X-rays can easily penetrate the sample, ensuring that the measurements are representative of the bulk (courtesy X.-L. Wang)

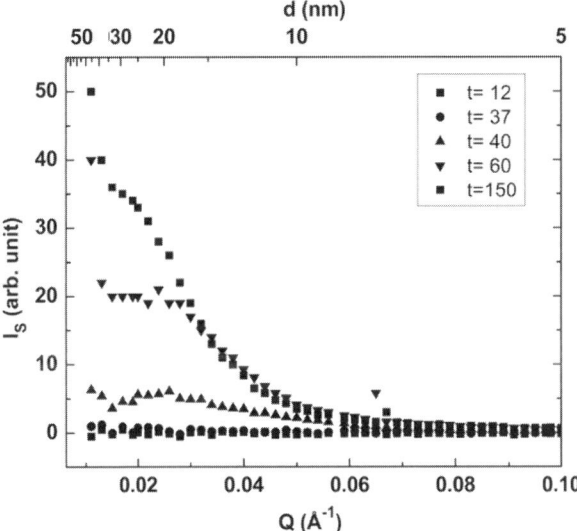

Fig. 5.12. Evolution of small angle scattering intensity during crystallization of a $Zr_{52.5}Cu_{17.9}Ni_{14.6}Al_{10}Ti_5$ bulk metallic glass. The scattering data were collected in situ with the use of high-energy synchrotron ($E = 77$ keV) X-rays at the Advanced Photon Source, Argonne National Laboratory while the specimen was being heated at a rate of 10 K min^{-1} until the specimen reached 696 K at $t = 53$ min and then held at that temperature. A broad peak at $Q \sim 0.25$ Å^{-1} develops in late stages of crystallization. The top scale gives the corresponding length scale (data courtesy X.-L. Wang[87])

The evolution of small angle scattering intensity during crystallization of an 8-mm-diameter, 1-mm-thick disk of $Zr_{52.5}Cu_{17.9}Ni_{14.6}Al_{10}Ti_5$ BMG[88] is shown in Fig. 5.12. The scattering data were collected in situ with the high-energy synchrotron ($E = 77$ keV) X-ray facility at the Advanced Photon Source, Argonne National Laboratory. A broad peak at $Q \sim 0.25$ Å^{-1} develops during the late stages of crystallization. The top scale gives the corresponding length scale ($d = 2\pi/Q$) that is being probed.

5.9   POSITRON ANNIHILATION SPECTROSCOPY

Positron annihilation spectroscopy (PAS) is a nondestructive method for studying the defects in solids.[89] The PAS method has been applied to the characterization of the free-volume in BMGs by measuring positron lifetimes.[90–104] In this method, positrons are injected into the sample from a radioisotope source, such as ^{22}Na. The typical specimen configuration consists of two ~1 cm^2 by ~1 mm thick plates sandwiching the positron source. A polished surface is required on each plate. The energetic positrons slow down and become thermalized within the solid in a few picoseconds, and then diffuse within the sample. After some time, the position is annihilated with an electron from the sample resulting in the emission of two γ-rays ($e^+ + e^- \rightarrow \gamma_1 + \gamma_2$). The γ-rays can escape the system without interacting and they contain information about the electronic environment around the annihiation site. Two principle groups of measurements may be made, as shown in Fig. 5.13. Positron lifetime measurements examine the sensitivity of positrons to the electron density at the annihilation site. Doppler broadening spectroscopy (DOBS) and angular correlation of the annihilation radiation (ACAR) examine electron momentum distribution in the sample.

Positron lifetime measurements are capable of distinguishing different types of open-volume defects as the positron lifetime is directly related to the size of the defects. In a lifetime experiment, the time is measured between when the positron is implanted into the specimen and an almost simultaneous birth γ-ray (at 1,274 keV) is generated and when one of the annihilation γ-rays (at 511 keV) is produced, as shown in Fig. 5.13. In a homogeneous defect-free metal, positrons are annihilated at the same characteristic lifetime, typically 100–200 ps, for a given material. During diffusion, the positron may become trapped into a localized state at open volumes in the material, which yields a characteristic increase in the lifetime due to the reduced local electron density in open volumes.

Doppler broadening measurements are based on the conservation of momentum and are sensitive to the chemical environment of the open volumes. The momentum p_L is measured from the amount of Doppler shift,

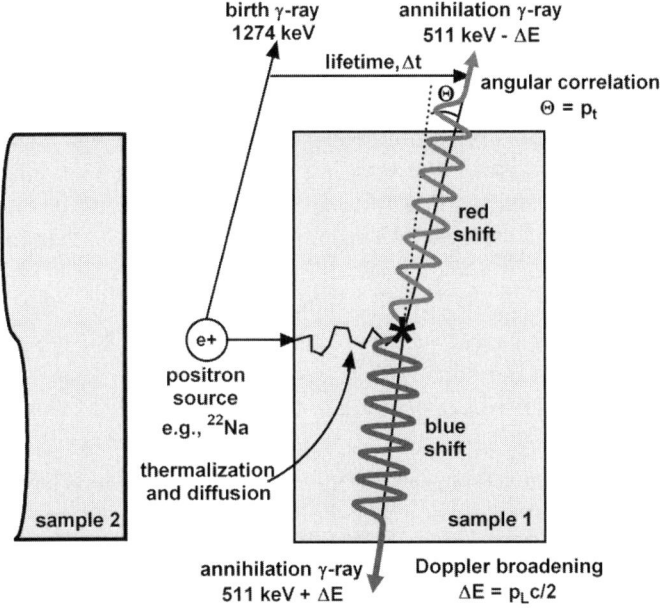

Fig. 5.13. Positron annihilation spectroscopy showing lifetime measurement, Doppler broadning, and angular correlation

$\Delta E = p_{\mathrm{L}} c / 2$, of the emitted γ-rays, where c is the speed of light. As the emission of the γ-rays is isotropic, the detector measures both the upshifted and downshifted rays and this produces a broadening of the annihilation peak. In solids, annihilation predominately occurs with the conduction of valence electrons and a low probability of interactions with the core electrons. The low momentum annihilation with valence electrons gives information on the atomic binding energy or open-volume character. The high momentum annihilation with core electrons gives information on the elemental or chemical character. The S shape parameter is determined from the relative area of the central (low momentum) part of the γ-ray energy spectrum and the W shape parameter from the relative contributions of the peak tails (high momentum) to the total peak area. High S:W ratios indicate many or large open-volume defects and low ratios indicate few or small defects. Modern two-detector coincidence systems are capable of measuring Doppler broadning of annihilation radiation (2D-DBAR) spectra with backgrounds that are ~500 times smaller than the conventional one-detector method. The diagonal deviation from the central 511 keV energy is a measure of Doppler broadning which is generally generated from annihilation with the core electrons.

Several studies have indicated that the open volume in Zr-based and $Pd_{40}Cu_{30}Ni_{10}P_{20}$ BMGs decreased when the sample was annealed below the glass transition temperature. This behavior is consistent with structural relaxation of the disordered atomic structure. The open volume has also been studied after heavy deformation. Inhomogeneous deformation was found to increase the free-volume in both Zr- and Cu-based BMGs. This result is consistent with the generation of free-volume during flow. For example, the results of a lifetime experiment on a 3-mm-thick plate of $Zr_{58.5}Nb_{2.8}Cu_{15.6}Ni_{12.8}Al_{10.3}$ that was annealed for 1 h at 573 K and then cold rolled between 10 and 50% indicated three lifetime components (defined below) from fits to the positron lifetime spectrum. The changes in these lifetimes with reductions in thickness are shown in Fig. 5.14. The first lifetime component τ_1, due to annihilation in the bulk, decreased from 0.15 to 0.12 ns as the thickness of the sheet was reduced. This result indicated that the size of the Bernal holes decreased as the large holes became flow defects and the relative concentration decreased as larger defects formed.

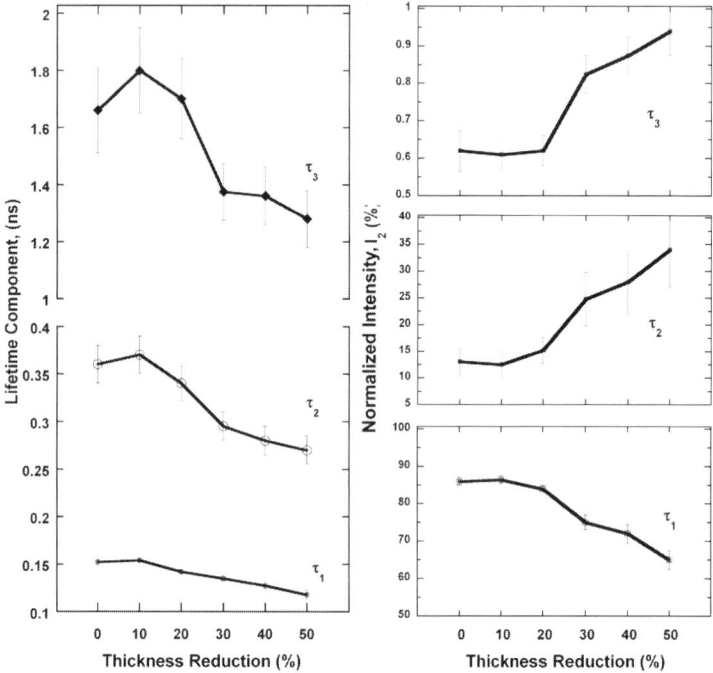

Fig. 5.14. Positron annihilation spectroscopy lifetime measurements from a 3-mm-thick plate of $Zr_{58.5}Nb_{2.8}Cu_{15.6}Ni_{12.8}Al_{10.3}$ BMG showing the change in the three different lifetime components during reduction of thickness from 10 to 50%. The first lifetime component τ_1 was due to annihilation in the bulk and indicated that the size of the Bernal holes decreased as the large holes became flow defects and the relative concentration decreased as larger defects formed. The second lifetime component τ_2 is due to annihilation in flow defects. The third lifetime component τ_3 is due to annihilation in nanoscale voids (data courtesy K. M. Flores, E. Sherer, H. Chen, and Y. C. Jean)

The second lifetime component τ_2 decreased from 0.38 to 0.26 ns and was due to annihilation in flow defects. This change indicated that the larger defects broke up and redistributed as flow progressed. The third lifetime component τ_3 decreased from 1.8 to 1.3 ns and was due to annihilation in nanoscale voids. The radius of the nanovoids was estimated to be between 0.2 and 0.26 nm and their concentration increased sharply between 20 and 30% reduction in thickness.

5.10   SUMMARY

A variety of techniques can be applied to study the atomic structure and bulk microstructure of BMGs. DSC is a thermoanalytical technique for detecting enthalpic processes that occur during the glass transition, crystallization, and other phase transitions, and may be used to evaluate the temperature range and kinetics of the BMG decomposition processes, but it does not provide any phase identification. XRD and, in particular, the time-resolved variants provide information on the presence and identity of crystalline phases and the devitrification process. Electron microscopy (SEM, TEM, and HREM), field ion microscopy, atom probe tomography, and small angle scattering techniques also enable crystalline phases to be identified, and their size, number density, volume fraction, morphology, and distribution to be determined. Electron microscopy also enables the width and extent of shear bands to be characterized.

The nature of atomic structure of glasses is difficult to accurately characterize due to the short length scale and the lack of long-range order. Even with advanced characterization techniques, it is challenging to make a distinction between truly amorphous solids and crystalline solids in which the size of the crystals is less than unit cell dimensions. FEM and radial distribution functions from SANS data enable short- and medium-range order to be investigated. PAS, reconstructed sequences of field ion images and atom probe tomography permit the size and shape of small open volumes in the BMG to be characterized either globally or at a site-specific location. It is generally advantageous to use multiple techniques to provide a complete microstructural characterization over a variety of length scales.

ACKNOWLEDGMENTS

The author would like to thank K. F. Russell, S. C. Glade, K. M. Flores, T. C. Hufnagel, M. J. Kramer, B. D. Wirth, X.-L. Wang, and Jun Xun for their assistance. Research at the Oak Ridge National Laboratory SHaRE User Facility was sponsored by Basic Energy Sciences, US Department of Energy. LEAP® is a registered trademark of Imago Scientific Instruments.

REFERENCES

1. A. Leonhard, L. Q. Xing, M. Heilmaier, A. Gebert, J. Eckert, and L. Schutz, Effect of crystalline precipitation on the mechanical behavior of bulk glass forming Zr-based alloys, *Nanostruct. Mater.* **10**, 805–817 (1998).

2. H. Choi-Yim, R. D. Conner, and W. L. Johnson, In situ composite formation in the Ni–(Cu)–Ti–Zr–Si system, *Scripta Mater.* **53**, 1467–1470 (2005).

3. J. D. Bernal, Geometry of the structure of monatomic liquids, *Nature* **185**, 68–70 (1960).

4. J. D. Bernal, The structure of liquids, *Proc. R. Soc. Lond. A* **280**, 299–322 (1964).

5. D. B. Miracle, A structural model for metallic glasses, *Nat. Mater.* **3**(10), 697–702 (2004).

6. D. B. Miracle, W. S. Sanders, and O. N. Senkov, The influence of efficient atomic packing on the constitution of metallic glasses, *Philos. Mag. A* **83**(20), 2409–2428 (2003).

7. H. W. Sheng, W. K. Luo, F. M. Alamgir, J. M. Bai, and E. Ma, Atomic packing and short-to-medium-range order in metallic glasses, *Nature* **439**(7075), 419–425 (2006).

8. P. M. Ossi, *Disordered Materials: An Introduction* (Springer, Berlin Heidelberg New York, 2003).

9. S. Venkataraman, H. Hermann, C. Mickel, L. Schultz, D. J. Sordelet, and J. Eckert, Calorimetric study of the crystallization kinetics of $Cu_{47}Ti_{33}Zr_{11}Ni_8Si_1$ metallic glass, *Phys. Rev. B* **75**, 104206 (2007).

10. X. L. Fu, Y. Li, and C. A. Schuh, Temperature, strain rate and reinforcement volume fraction dependence on plastic deformation in metallic glass composites, *Acta Mater.* (2007). doi:10.1016/j.actamat.2007.01.009

11. Y. Waseda, *The Structure of Non-Crystalline Materials: Liquids and Amorphous Solids* (McGraw-Hill, New York, 1980), p. 326.

12. T. Egami and S. J. L. Billinge, in *Underneath the Bragg Peaks: Structural Analysis of Complex Materials*, edited by R. W. Cahn, *Pergamon Materials Series* (Elsevier, Oxford, 2003).

13. T. Proffen, S. J. L. Billinge, T. Egami, and D. Louca, Structural analysis of complex materials using the atomic pair distribution function – A practical guide, *Z. Kristallogr.* **218**(2), 132–143 (2003).

14. L. Hennet, S. Krishnan, A. Bytchkov, T. Key, D. Thiaudiere, P. Melin, I. Pozdnyakova, M. L. Saboungi, and D. L. Price, X-ray diffraction on high-temperature liquids: Evolution towards time-resolved studies, *Int. J. Thermophys.* **26**(4), 1127–1136 (2005).

15. L. Margulies, M. J. Kramer, R. W. McCallum, S. Kycia, D. R. Haeffner, J. C. Lang, and A. I. Goldman, New high temperature furnace for structure refinement by powder diffraction in controlled atmospheres using synchrotron radiation, *Rev. Sci. Instrum.* **70**(9), 3554–3561 (1999).

16. P. J. Chupas, X. Qiu, J. C. Hanson, P. L. Lee, C. P. Grey, and S. J. L. Billinge, Rapid acquisition pair distribution function (RA-PDF) analysis, *J. Appl. Crystallogr.* **36**(6), 1342–1347 (2003).

17. A. K. Gangopadhyay, G. W. Lee, K. F. Kelton, J. R. Rogers, A. I. Goldman, D. S. Robinson, T. J. Rathz, and R. W. Hyers, Beamline electrostatic levitator for in situ high energy X-ray diffraction studies of levitated solids and liquids, *Rev. Sci. Instrum.* **76**(7), 073901 (2005).

18. B. S. Murty, D. H. Ping, and K. Hono, Nanoquasicrystallization of binary Zr–Pd metallic glasses, *Appl. Phys. Lett.* **77**(8), 1102–1104 (2000).

19. D. J. Sordelet, E. Rozhkova, M. F. Besser, and M. J. Kramer, Synthesis route-dependent formation of quasicrystals in $Zr_{70}Pd_{30}$ and $Zr_{70}Pd_{20}Cu_{10}$ amorphous alloys, *Appl. Phys. Lett.* **80**(25), 4735–4737 (2002).
20. M. J. Kramer, M. F. Besser, N. Yang, E. Rozhkova, D. J. Sordelet, Y. Zhang, and P. L. Lee, Devitrification studies of Zr–Pd and Zr–Pd–Cu metallic glasses, *J. Non-Cryst. Solids* **317**(1–2), 62–70 (2003).
21. J. Goldstein, D. E. Newbury, D. C. Joy, C. E. Lyman, P. Echlin, E. Lifshin, L. C. Sawyer, and J. R. Michael, *Scanning Electron Microscopy and X-Ray Microanalysis*, 3rd ed. (Springer, Berlin Heidelberg New York, 2003).
22. L. A. Giannuzzi and F. A. Stevie, editors, *Introduction to Focused Ion Beams: Instrumentation, Theory, Techniques and Practice* (Springer, Berlin Heidelberg New York, 2005).
23. J. J. Lewandowski and A. L. Greer, Temperature rise at shear bands in metallic glasses, *Nat. Mater.* **5**(1), 15–18 (2006).
24. P. E. Donovan and W. M. Stobbs, The structure of shear bands in metallic glasses, *Acta Metall.* **29**(8), 1419–1436 (1981).
25. E. Pekarskaya, C. P. Kim, and W. L. Johnson, In-situ transmission electron microscopy studies of shear bands in a bulk metallic glass based composite, *J. Mater. Res.* **16**(9), 2513–2518 (2001).
26. H. Chen, Y. He, G. J. Shiflet, and S. J. Poon, Deformation-induced nanocrystal formation in shear bands of amorphous alloys, *Nature* **367**, 541–543 (1994).
27. J.-J. Kim, Y. Choi, S. Suresh, and A. S. Argon, Nanocrystallization during nanoindentation of a bulk amorphous metal alloy at room temperature, *Science* **295**, 654–657 (2002).
28. J. Li, Z. L. Wang, and T. C. Hufnagel, Characterization of nanometer-scale defects in metallic glasses by quantitative high-resolution transmission electron microscopy, *Phys. Rev. B* **65**, 144201 (2002).
29. W. H. Jiang, F. E. Pinkerton, and M. Atzmon, Mechanical behavior of shear bands and the effect of their relaxation in a rolled amorphous Al-based alloy, *Acta Mater.* **53**, 3469–3477 (2005).
30. T. Yano, Y. Yorikado, Y. Akeno, F. Hori, Y. Yokoyama, A. Iwase, A. Inoue, and T. J. Konno, Relaxation and crystallization behavior of the $Zr_{50}Cu_{40}Al_{10}$ metallic glass, *Mater. Trans.* **46**(12), 2886–2892 (2005).
31. J. C. H. Spence, *High-Resolution Electron Microscopy*, 3rd ed. (Oxford University Press, New York, 2003).
32. H. J. Chang, E. S. Park, Y. C. Kim, and D. H. Kim, Observation of artifact-free amorphous structure in Cu–Zn-based alloy using transmission electron microscopy, *Mater. Sci. Eng. A* **406**, 1190124 (2005).
33. M. M. J. Treacy and J. M. Gibson, Variable coherence microscopy: A rich source of structural information from disordered materials, *Acta Cryst. A* **52**, 212–220 (1996).
34. P. M. Voyles, J. M. Gibson, and M. M. J. Treacy, Fluctuation microscopy: A probe of atomic correlations in disordered materials, *J. Electron Microsc.* **49**(2), 259–266 (2000).
35. M. M. J. Treacy, J. M. Gibson, L Fan, D. J. Paterson, and I. McNulty, Fluctuation microscopy: A probe of medium-range order, *Rep. Prog. Phys.* **68**(12), 2899–2944 (2005).
36. J. Li, X. Gu and T. C. Hufnagel, Medium-range order in metallic glasses studied by fluctuation microscopy, *Microsc. Microanal.* **7**(Suppl. 2), 1260–1261 (2001).
37. T. C. Hufnagel, C. Fana, R. T. Ott, J. Li, and S. Brennan, Controlling shear band behavior metallic glasses through microstructural design, *Intermetallics* **10**(11–12), 1163–1166 (2002).
38. J. Li, X. Gu and T. C. Hufnagel, Using fluctuation microscopy to characterize structural order in metallic glasses, *Microsc. Microanal.* **9**, 509 (2003).

39. W. G. Stratton, J. Hamann, J. H. Perepezko, and P. M. Voyles, Medium-range order in high Al-content amorphous alloys measured by fluctuation electron microscopy, in *Amorphous and Nanocrystalline Metals, MRS Symposium Proceedings*, Vol. 806, MM9.4.1 (2004).

40. W. G. Stratton, P. M. Voyles, J. Hamann, and J. H. Perepezko, Medium-range order in high Al-content amorphous alloys measured by fluctuation electron microscopy, *Microsc. Microanal.* **10**(Suppl. 2), 788–789 (2004).

41. W. G. Stratton, J. Hamann, J. H. Perepezko, P. M. Voyles, S. V. Khare, and X. Mao, Aluminum nanoscale order in amorphous $Al_{92}Sm_8$ measured by fluctuation electron microscopy, *Appl. Phys. Lett.* **86**(14), 141910 (2005).

42. E. W. Müller and T. T. Tsong, *Field Ion Microscopy* (Elsevier, New York, 1969).

43. K. M. Bowkett and D. A. Smith, *Field Ion Microscopy* (North Holland, Amsterdam, 1970).

44. L. A. Beavan, R. M. Scanlan, and D. N. Seidman, The defect structure of depleted zones in irradiated tungsten, *Acta Metall.* **19**, 1339–1350 (1971).

45. D. N. Seidman, The direct observation of point defects in irradiated or quenched metals by quantitative field-ion microscopy, *J. Phys. F: Met. Phys.* **3**, 393–421 (1973).

46. F. Vurpillot, M. Gilbert, and B. Deconihout, Towards the three-dimensional field ion microscope, *Surf. Interface Anal.* **39**, 273–277 (2007). doi:10.1002/sia.2490

47. M. K. Miller, *Atom Probe Tomography* (Springer, Berlin Heidelberg New York, 2000).

48. R. Busch, S. Schneider, A. Peker, and W. L. Johnson, Decomposition and primary crystallization in undercooled $Zr_{41.2}Ti_{13.8}Cu_{12.5}Ni_{10.0}Be_{22.5}$ melts, *Appl. Phys. Lett.* **67**, 1544–1546 (1995).

49. R. Busch, Y. J. Kim, S. Schneider, and W. L. Johnson, Atom probe field ion microscope and levitation studies of the decomposition and crystallization of undercooled Zr–Ti–Cu–Ni–Be melts, *Mater. Sci. Forum* **225–227**, 77–82 (1996).

50. R. Busch, E. Bakke, and W. L. Johnson, On the glass forming ability of bulk metallic glasses, *Mater. Sci. Forum* **235–238**, 327–335 (1997).

51. M. K. Miller, K. F. Russell, P. M. Martin, R. Busch, and W. L. Johnson, Characterization of bulk metallic glasses with the atom probe, *J. Phys. IV* **6**(C5), 217–222 (1996).

52. D. H. Ping, K. Hono, and A. Inoue, Oxygen distribution in $Zr_{65}Cu_{15}Al_{10}Pd_{10}$ nanocrystalline alloys, *Mater. Sci. Forum* **307**, 31–38 (1999).

53. N. Wanderka, U. Czubayko, P. Schubert-Bischoff, and M.-P. Macht, Primary crystals in ZrTiCuNiBe bulk glasses, *Mater. Sci. Eng. A* **270**, 44–47 (1999).

54. S. Scudino, U. Kuhn, L. Schultz, D. Nagahama, K. Hono, and J. Eckert, Microstructure evolution upon devitrification and crystallization kinetics of $Zr_{57}Ti_8Nb_{2.5}Cu_{13.9}Ni_{11.1}Al_{7.5}$ melt-spun glassy ribbon, *J. Appl. Phys.* **95**, 3397–3403 (2004).

55. M. K. Miller, D. J. Larson, R. B. Schwarz, and Y. He, Decomposition in $Pd_{40}Ni_{40}P_{20}$ metallic glass, *Mater. Sci. Eng. A* **250**, 141–145 (1998).

56. S. C. Glade, J. F. Loffler, S. Bossuyt, W. L. Johnson, and M. K. Miller, Crystallization of amorphous $Cu_{47}Ti_{34}Zr_{11}Ni_8$, *J. Appl. Phys.* **89**, 1573–1579 (2001).

57. H. G. Read, K. Hono, A. P. Tsai, and A. Inoue, Preliminary atom probe studies of PdNi(Cu)P supercooled liquids, *Mater. Sci. Eng. A* **226–228**, 453–457 (1997).

58. M. K. Miller, T. D. Shen, and R. B. Schwarz, Atom probe studies of metallic glasses, *J. Non-Cryst. Solids* **317**, 10–16 (2003).

59. K. B. Kim, Y. Zhang, P. J. Warren, and B. Cantor, Crystallization behaviour in a new multicomponent $Ti_{16.6}Zr_{16.6}Hf_{16.6}Ni_{20}Cu_{20}Al_{10}$ metallic glass developed by the equiatomic substitution technique, *Philos. Mag.* **83**, 2371–2381 (2003).

60. M. K. Miller, T. D. Shen, and R. B. Schwarz, Atom probe tomography study of the decomposition of a bulk metallic glass, *Intermetallics* **10**, 1047–1052 (2002).

61. M. K. Miller, S. C. Glade, and W. L. Johnson, Phase separation in $Cu_{47}Ti_{33}Zr_{11}Ni_8Si_1$, *Surf. Interface Anal.* **36**, 598–600 (2004).

62. J. C. Oh, T. Ohkubo, Y. C. Kim, E. Fleury, and K. Hono, Phase separation in $Cu_{43}Zr_{43}Al_7Ag_7$ bulk metallic glass, *Scripta Mater.* **53**, 165–169 (2005).

63. E. S. Park, D. H. Kim, T. Ohkubo, and K. Hono, Enhancement of glass forming ability and plasticity by addition of Nb in Cu–Ti–Zr–Ni–Si bulk metallic glasses, *J. Non-Cryst. Solids* **351**, 1232–1238 (2005).

64. Y. Hirotsu, T. G. Nieh, A. Hirata, T. Ohkubo, and N. Tanaka, Local atomic ordering and nanoscale phase separation in a Pd–Ni–P bulk metallic glass, *Phys. Rev. B* **73**, 012205 (2006).

65. M. K. Miller, C. T. Liu, J. A. Wright, W. Tang, and K. Hildal, APT characterization of some iron-based bulk metallic glasses, *Intermetallics* **14**, 1019–1026 (2006).

66. S. Schneider, P. Thiyagarajan, and W. L. Johnson, Formation of nanocrystals based on decomposition in the amorphous $Zr_{41.2}Ti_{13.8}Cu_{12.5}Ni_{10}Be_{22.5}$ alloy, *Appl. Phys. Lett.* **68**(4), 493–495 (1996).

67. A. Wiedenmann and J. M. Liu, Dynamic scaling phenomena in phase separation of amorphous $Cu_{12.5}Ni_{10}Zr_{41}Ti_{14}Be_{22.5}$, as probed by small-angle neutron scattering, *Solid State Commun.* **100**(5), 331–335 (1996).

68. S. Schneider, U. Geyer, P. Thiyagarajan, R. Busch, R. Schulz, K. Samwer, and W. L. Johnson, Phase separation and crystallization in the bulk amorphous $Zr_{41.2}Ti_{13.8}Cu_{12.5}Ni_{10}Be_{22.5}$ alloy, *Mater. Sci. Forum* **225–227**, 59–64 (1996).

69. A. Wiedenmann, U. Keiderling, M. P. Macht, and H. Wollenberger, Decomposition and crystallization of bulk amorphous $Zr_{41}Ti_{14}Cu_{12.5}Ni_{10}Be_{22.5}$ alloys studied by small angle neutron scattering, *Mater. Sci. Forum* **225–227**, 71–76 (1996).

70. P. Uebele, A. Wiedenmann, H. Hermann, and K. Wetzig, Decomposition kinetics of bulk amorphous $Zr_{41}Ti_{14}Cu_{12.5}Ni_{10}Be_{22.5}$ alloys studied by computer simulation and small-angle neutron scattering, *J. Appl. Crystallogr.* **30**, 613–617 (1997).

71. A. Wiedenmann, Small-angle neutron scattering investigations of nanoscaled microtructures, *J. Appl. Crystallogr.* **30**, 580–585 (1997).

72. U. Gerold, A. Wiedenmann, U. Keiderling, and H. J. Fecht, Decomposition and crystalliation of the bulk amorphous $Zr_{11}Ti_{34}Cu_{47}Ni_8$ alloy studied by SANS, *Physica B* **234–236**, 995–996 (1997).

73. H. Hermann, A. Wiedenmann, and P. Uebele, SANS study and a model for partially ordering during decomposition in bulk amorphous $Zr_{41}Ti_{14}Cu_{12.5}Ni_{10}Be_{22.5}$ amorphous alloys, *Physica B* **241–243**, 352–354 (1997).

74. S. Schneider, P. Thiyagarajan, U. Geyer, and W. L. Johnson, SANS of bulk metallic ZrTiCuNiBe glasses, *Physica B* **241–243**, 918–920 (1997).

75. P. Uebele, H. Hermann, A. Wiedenmann, and K. Wetzig, Analysis of SANS-data with models of stochastic geometry, *Physica B* **234–236**, 426–427 (1997).

76. K. Shibata, T. Higuchi, A. P. Tsai, M. Imai, and K. Suzuki, Isothermal evolution of long-range order in bulk metallic glass $Ni_{15}Pt_{60}P_{25}$ near glass transition: In-situ SANS measurement, *Prog. Theor. Phys. Suppl.* **126**, 75–78 (1997).

77. S. Schneider, U. Geyer, P. Thiyagarajan, and W. L. Johnson, Time and temperature dependence of decomposition and crystallization in a multicomponent bulk metallic glass forming alloy, *Mater. Sci. Forum* **235–238**, 337–342 (1997).

78. J.-M. Liu, A. Wiedenmann, U. Gerold, and H. Wollenberger, Crystallization and phase separations in amorphous $Cu_{12.5}Ni_{10}Zr_{41}Ti_{14}Be_{22.5}$ alloy, as investigated by SANS, *Mater. Sci. Forum* **235–238**, 523–528 (1997).

79. J. M. Liu, Demonstration of spinodal decomposition in amorphous $Cu_{12.5}Ni_{10}Zr_{41}Ti_{14}Be_{22.5}$ alloy, *Solid State Commun.* **105**(1), 71–75 (1998).

80. N. Wanderka, Q. Wei, R. Doole, M. Jenkins, S. Friedrich, M.-P. Macht, and H. Wollenberger, Crystallization of the supercooled liquid $Zr_{41}Ti_{14}Cu_{12.5}Ni_{10}Be_{22.5}$, *Mater. Sci. Forum* **269–272**, 773–778 (1998).

81. J. F. Loffler, S. Bossuyt, S. C. Glade, W. L. Johnson, W. Wagner, and P. Thiyagarajan, Crystallization of bulk amorphous Zr–Ti(Nb)–Cu–Ni–Al, *Appl. Phys. Lett.* **77**(4), 525–527 (2000).

82. S. E. H. Abaidia and A. Wiedenmann, Thermal stability of the bulk metallic glass $Zr_{46.75}Ti_{8.25}Cu_{7.5}Ni_{10}Be_{27.5}$ studied by SANS, *Physica B* **276–278**, 454–455 (2000).

83. J. F. Loffler, W. L. Johnson, W. Wagner, and P. Thiyagarajan, Comparison of the decomposition and crystallization behavior of Zr and Pd based bulk amorphous alloys, *Mater. Sci. Forum* **343–346**, 179–184 (2000).

84. J. F. Loffler and W. L. Johnson, Model for decomposition and crystallization of Zr-based bulk amorphous alloys near the glass transition, *Mater. Sci. Eng. A* **304–306**, 670–673 (2001).

85. A. Hoell, F. Bley, A. Wiedenmann, J. P. Simon, A. Mazuelas, and P. Boesecke, Composition fluctuations in the demixed supercooled liquid state of $Zr_{41}Ti_{14}Cu_{12.5}Ni_{10}Be_{22.5}$: A combined ASAXS and SANS study, *Scripta Mater.* **44**(8–9), 2335–2339 (2001).

86. E. Pekarskaya, J. F. Loffler, and W. L. Johnson, Microstructural studies of crystallization of a Zr-based bulk metallic glass, *Acta Mater.* **51**(14), 4045–4057 (2003).

87. X. P. Tang, J. F. Loffler, W. L. Johnson, and Y. Wu, Devitrification of the $Zr_{41.2}Ti_{13.8}Cu_{12.5}Ni_{10.0}Be_{22.5}$ bulk metallic glass studied by XRD, SANS, and NMR, *J. Non-Cryst. Solids* **317**(1–2), 118–122 (2003).

88. X.-L. Wang, APS Science 2003, The Annual Report of the Advanced Photon Source at Argonne National Laboratory, pp. 19–20, ANL-04/07, Argonne National Laboratory, USA.

89. I. Prochazka, Positron annihilation spectroscopy, *Mater. Struct.* **8**, 55–60 (2001).

90. C. Nagel, K. Ratzke, E. Schmidtke, J. Wolff, U. Geyer, and F. Faupel, Free-volume changes in the bulk metallic glass $Zr_{46.7}Ti_{8.3}Cu_{7.5}Ni_{10}Be_{27.5}$ and the undercooled liquid, *Phys. Rev. B* **57**(17), 10224–10227 (1998).

91. C. Nagel, K. Ratzke, E. Schmidtke, F. Faupel, and W. Ulfert, Positron-annihilation studies of free-volume changes in the bulk metallic glass $Zr_{65}Al_{7.5}Ni_{10}Cu_{17.5}$ during structural relaxation and at the glass transition, *Phys. Rev. B* **60**(13), 9212–9215 (1999).

92. P. Asoka-Kumar, J. Hartley, R. Howell, P. A. Sterne, and T. G. Nieh, Chemical ordering around open-volume regions in bulk metallic glass $Zr_{52.5}Ti_5Al_{10}Cu_{17.9}Ni_{14.6}$, *Appl. Phys. Lett.* **77**(13), 1973–1975 (2000).

93. K. M. Flores, D. Suh, R. Howell, P. Asoka-Kumar, P. A. Sterne, and R. H. Dauskardt, Flow and fracture of bulk metallic glass alloys and their composites, *Mater. Trans.* **42**(4), 619–622 (2001).

94. K. M. Flores, D. Suh, R. H. Dauskardt, P. Asoka-Kumar, P. A. Sterne, and R. H. Howell, Characterization of free volume in a bulk metallic glass using positron annihilation spectroscopy, *J. Mater. Res.* **17**(5), 1153–1161 (2002).

95. P. Asoka-Kumar, R. Howell, T. G. Nieh, P. A. Sterne, B. D. Wirth, R. H. Dauskardt, K. M. Flores, D. Suh, and G. R. Odette, Opportunities for materials characterization using high-energy positron beams, *Appl. Surf. Sci.* **194**(1–4), 160–167 (2002).

96. D. Suh, R. H. Dauskardt, P. Asoka-Kumar, P. A. Sterne, and R. H. Howell, Temperature dependence of positron annihilation in a Zr–Ti–Ni–Cu–Be bulk metallic glass, *J. Mater. Res.* **18**(9), 2021–2024 (2003).

97. D. Suh and R. H. Dauskardt, Effects of open-volume regions on relaxation time-scales and fracture behavior of a Zr–Ti–Ni–Cu–Be bulk metallic glass, *J. Non-Cryst. Solids* **317**(1–2), 181–186 (2003).

98. B. P. Kanungo, S. C. Glade, P. Asoka-Kumar, and K. M. Flores, Characterization of free volume changes associated with shear band formation in Zr- and Cu-based bulk metallic glasses, *Intermetallics* **12**(10–11), 1073–1080 (2004).

99. L. C. Damonte, Nuclear techniques characterization of short-range order in Zr–TM–Cu–Al–Ni (TM = Hf, Ti, Fe) bulk metallic glasses, *Ann. Chim. Sci. Mat.* **27**, 61–67 (2002).

100. A. Rehmet, K. Gunther-Schade, K. Ratzke, U. Geyer, and F. Faupel, Quenching rate dependence of free volume in a Zr–Cu–Ni–Ti–Be glass as probed by positron annihilation lifetime spectroscopy, *Phys. Status Solidi A* **201**(3), 467–470 (2004).

101. K. M. Flores, Structural changes and stress state effects during inhomogeneous flow of metallic glasses, *Scripta Mater.* **54**(3), 327–332 (2006).

102. Y. Shang, Study of structural free volumes in the supercooled liquid state of bulk amorphous ZrTiCuNiBe, *Mater. Lett.* **59**(24–25), 3177–3180 (2005).

103. J. F. Wang, L. Liu, J. Z. Xiao, T. Zhang, B. Y. Wang, C. L. Zhou, and W. Long, Ageing behaviour of $Pd_{40}Cu_{30}Ni_{10}P_{20}$ bulk metallic glass during long-time isothermal annealing, *J. Phys. D* **38**, 946–949 (2005).

104. D. Suh, P. Asoka-Kumar, and R. H. Dauskardt, The effects of hydrogen on viscoelastic relaxation in Zr–Ti–Ni–Cu–Be bulk metallic glasses: Implications for hydrogen embrittlement, *Acta Mater.* **50**(3), 537–551 (2002).

Chapter 6

DEFORMATION BEHAVIOR

T. G. Nieh

Department of Materials Science and Engineering, The University of Tennessee, Knoxville, TN 37996-2200, USA

6.1   INTRODUCTION

Although metallic glasses synthesized by rapid quenching from the melt were first discovered in 1960 by Duwez and coworkers at Caltech, the study of the mechanical behavior of metallic glasses only started in the early 1970s. Metallic glasses were found to deform elastically and exhibit negligible plasticity in uniaxial tension. Despite a limited macroscopic tensile plastic strain (<0.5%), exceptionally high strain (~100) was observed to take place within localized shear bands. One of the scientific questions naturally arises: how does a shear band nucleate and propagate in a medium presumably consisting of randomly packed atoms? Several theories, including the free-volume model and the dislocation model, were subsequently proposed to address the shear band formation and propagation. It was impossible to carry out irrevocable experiments to prove these theories at that time due to the limited size of the samples, and thus a poorly defined stress state during mechanical testing, and the lack of advanced analytical tools. However, the situation changed after the successful development of bulk metallic glasses (BMGs) in many alloy systems, e.g., Zr-, Mg-, Pd-, La-, Cu-, Ti-, and Fe-based. As a result of this development, research of BMGs has been very active, especially in the area of mechanical deformation.

6.2   INHOMOGENEOUS DEFORMATION

The mechanical behavior of metallic glasses is characterized by either inhomogeneous or homogeneous deformation. Inhomogeneous deformation usually occurs when a metallic glass is deformed at room temperature (i.e., low temperature) and is characterized by the formation of localized shear bands, followed by the rapid propagation of these bands, and sudden fracture. Macroscopically, a typical stress–strain curve of a BMG in tension consists of only elastic region with a fracture strain of about 2%.[1] In compression, the stress–strain curve consists of elasticity followed by yielding and perfect plasticity, i.e., the absence of work hardening. However, serrated flow usually takes place at an early stage of compressive deformation, prior to the occurrence of macroscopic yielding.[2] In fact, the stress at the onset of microyielding is about 0.9 of the fracture strength. Despite a limited macroscopic plasticity, local strain within a shear band can be sometimes quite significant (~10).[3] A shear band is typically 10–20 nm in width.[4] The generally accepted view on inhomogeneous deformation in metallic glasses is the free-volume theory.[5,6] In Sect. 6.3, the macroscopic observations and their microscopic explanation with a particular emphasis on the shear banding will be presented.

6.3   YIELDING CRITERION: VON MISES VS. MOHR-COULOMB

An important characteristic of the yield criteria for polycrystalline solids is their symmetry, predicting yield stresses of equal magnitude in either tension or compression. In contrast, metallic glasses have displayed asymmetric yield behavior in several experimental studies.[2,7] As a result of strength asymmetry, the shear-off angles of a test sample under tension and compression are also expected to be different, as demonstrated in Fig. 6.1 and Table 6.1. Donovan[8] studied a $Pd_{40}Ni_{40}P_{20}$ alloy and argued that the yielding of metallic glasses obeys the Mohr-Coulomb criterion instead of the von Mises criterion.

The Mohr-Coulomb criterion depends not only on the applied shear stress τ but also on the stress normal to the shear displacement σ_n

$$\tau_y = \tau_0 + \alpha \sigma_n. \qquad (6.1)$$

Here, τ_y is the effective shear yield stress, τ_0 a constant, and α a system-specific coefficient that controls the strength of the normal stress effect. Schuh and Lund[9] carried out atomistic calculation and the results confirmed that metallic glasses exhibit strength asymmetry in tension and compression, suggesting the Mohr-Coulomb behavior. Using the value of $\alpha = 0.123 \pm 0.004$ found in their analyses, (6.1) predicts a compressive shear angle of $\theta = 41.5 \pm 0.15°$. This value is in agreement with previous experiments with a range of $\theta = 39.5–43.7°$ in a variety of metallic glass compositions.

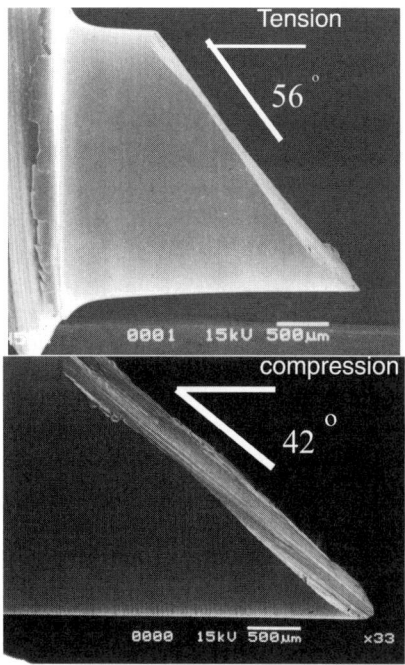

Fig. 6.1. Shear-off angles of a Zr-based BMG upon fracture are different in (**a**) tension (56°) and (**b**) compression (42°) (reprinted from reference [13] with permission from Elsevier)

Table 6.1. Summary of mechanical data for several bulk metallic glasses

Alloy	Strain rate (s⁻¹) (tension or compression)	Yield strength (MPa)	Shear-off angles	References
$Zr_{41.25}Ti_{13.5}Cu_{12.5}Ni_{10}Be_{22.5}$	10^2–10^4 (C)		45°	[10]
	10^{-3} (C)		42°	[11]
	10^{-3} (C)	1,911	42°	[12]
	10^{-3} (T)	1,910	90°	[12]
	10^{-3} (T)	1,978	57°ᵃ	[12]
$Pd_{40}Ni_{40}P_{20}$	10^{-3} (T)	1,600	56°	[13]
	10^{-3} (C)	1,600	42°	[2]
$Zr_{52.5}Al_{10}Ti_5Cu_{17.9}Ni_{14.6}$	10^{-3} (T)		56°	[1]

ᵃWith an applied hydrostatic pressure of 50–575 MPa.

6.4 DYNAMIC DEFORMATION AND STRAIN RATE EFFECT

Limited data have been reported for the strain rate effect on the macroscopic strength of BMG. Maddin and Masumoto[14] showed that the fracture stress of $Pd_{80}Si_{20}$ filaments decreased with increasing strain rate at 273 K. Kawamura et al.[15] also reported that the room temperature tensile strength of a rapidly solidified $Zr_{65}Al_{10}Ni_{10}Cu_{15}$ ribbon specimen decreased with increasing strain rate in the strain rate ranging from 10^{-4} to 10^0 s^{-1}. In contrast, Bruck et al.[10] reported that the compressive strength of a bulk $Zr_{41.25}Ti_{13.5}Cu_{12.5}Ni_{10}Be_{22.5}$ metallic glass was independent of strain rate. Mukai et al.[2,13] also studied the mechanical behavior of $Pd_{40}Ni_{40}P_{20}$ over a wide strain rate range (1×10^{-3} to 2.8×10^3 s^{-1}) in both tension and compression, and observed that the fracture strength of the material is virtually independent of strain rate, except at an extremely high strain rate of above 10^3 s^{-1}, at which the fracture strength is reduced. The reduced fracture strength is a result of the fact that at dynamic strain rate the rate of shear band emission is not sufficiently fast to accommodate the applied strain rate and thus causes an early fracture of the test sample. Data of fracture strength as a function of strain rate are summarized in Fig. 6.2. It is apparent that the strain rate dependence of the strength in amorphous alloys varies with the specimen size. However, compared to a simple shear-off surface observed at the strain rate of 10^{-3} s^{-1}, the fracture surface at the dynamic strain rate of 2.8×10^3 s^{-1} was noted to be extremely rough, suggesting higher fracture energy at a higher strain rate. The rough surface was caused by simultaneous operation of multiple shear bands.

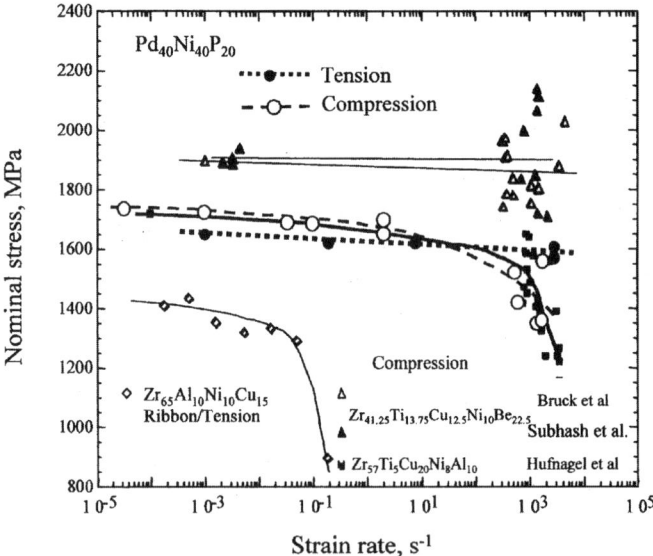

Fig. 6.2. Variation of the fracture strength as a function of strain rate for $Pd_{40}Ni_{40}P_{20}$ (reprinted from reference [2] with permission from Elsevier)

6.5 STRAIN HARDENING

As pointed out previously, a typical stress–strain curve in compression does not exhibit any strain hardening. The absence of strain hardening was rationalized by the absence of dislocations in an amorphous material. However, there have been some recent experiments suggesting this may not be the case.

For example, Das et al.[16] recently reported the observation of significant increase in the flow stress during compression (i.e., strain hardening) in a high-strength bulk $Cu_{47.5}Zr_{47.5}Al_5$ metallic glass. As a result of work hardening, the alloy exhibits a large ductility of 18%. Based upon the fracture surface observations, they argued that the large ductility was attributed to special microstructural features at the atomic scale, which enable easy and homogeneous nucleation of the shear bands and continuous multiplication during deformation. However, the special atomic scale microstructural features were not described in detail.

Yang et al.[17] also recently reported the observation of strain hardening in a $Zr_{52.5}Cu_{17.9}Ni_{14.6}Al_{10.0}Ti_{5.0}$ BMG. Using a controlled cyclic instrumented nanoindentation technique, they found the hardness of the BMG increased each time when the sample was reloaded immediately after unloading and then gradually reduced to a stable value before next unloading. The observed hardening–recovery phenomenon and hardness variations were explained using the Spaepen's model[5] which involves free-volume accumulation and annihilation in amorphous structure. They also found the phenomenon to be independent of the loading rate (or strain rate) within loading rates between 200 and 20,000 $\mu N\ s^{-1}$. This rate independence resulted from the relatively fast propagation rate of shear bands ($>10^{-5}$ s).[18]

In contrast, Bei et al.[19] recently reported softening of shear bands. They conducted compression experiments with a Zr-based BMG to various strains and subsequently performed hardness measurements as a function of compressive strain. The hardness was found to decrease with increasing deformation strain (or shear band density), and a simple composite model was used to describe the hardness variation as a function of shear band density. Their result suggested that a sheared region is weaker than its surrounding undeformed amorphous region. However, the structure of the sheared region was not characterized, and the reason why a shear region was weak was also not given. In summary, there is no general agreement on whether work hardening or softening actually occurs in BMGs.

6.6   SHEAR BAND NUCLEATION AND PROPAGATION

BMGs normally fracture in a catastrophic manner, indicating that shear bands propagate with an extremely fast speed. Specifically, the duration of shear band propagating in BMGs is typically about 10^{-5} s.[18] Thus, the rate controlling process for shear banding is the nucleation. Several studies[11,20] have reported the observation of stress serration during the compression of BMGs and attributed the phenomenon to the emission of shear bands. Recently, Schuh et al.[21–23] used nanoindentation to probe mechanical properties of different BMG alloy systems and observed stepped load–displacement curve punctuated by discrete bursts of plasticity. These discrete "pop-in" events correspond to the activation of individual shear bands, and the character of serrations is strongly dependent on the indentation loading rate; slower indentation rates promote more conspicuous serrations, and rapid indentations suppress serrated flow. An example of $Pd_{40}Ni_{40}P_{20}$ BMG is given in Fig. 6.3 in which pop-in events are evident. Analysis of the experimental data reveals a critical applied strain rate, above which serrated flow is completely suppressed. Furthermore, careful separation of the plastic and elastic contributions to deformation reveals that, at sufficiently low indentation rates, plastic deformation occurs entirely in discrete events of

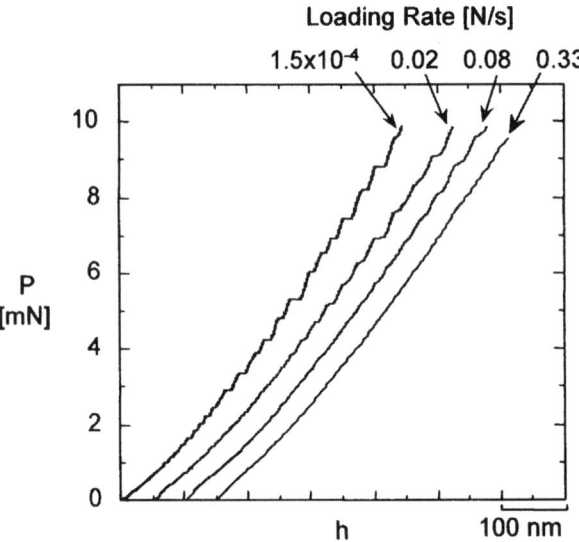

Fig. 6.3. Load (P) vs. indentation depth (h) for the loading segment of four indents on $Pd_{40}Ni_{40}P_{20}$ with different loading rates (reprinted from reference [21] with permission from the Materials Research Society)

isolated shear banding, whereas at the highest rates, deformation is continuous, without any evidence of discrete events at any size scale. All the present results are consistent with a kinetic limitation for the nucleation of shear bands, where at high rates, a single shear band cannot accommodate the imposed strain rapidly enough, and consequently multiple shear bands must operate simultaneously. Most recently, Schuh et al.[24] further extended their room temperature nanoindentation results to high temperatures, and proposed the presence of a second homogeneous deformation regime. The second homogeneous regime occurs at high deformation rates even well below the glass transition, and arises when deformation rates exceed the characteristic rate for shear band nucleation, kinetically forcing strain distribution. By extending an existing Argon model[6] for glass deformation to explore shear band nucleation kinetics, they found that the second regime could be quantitatively rationalized and the natural frequency of $6 \times 10^4 \text{ s}^{-1}$ for shear band nucleation was extracted from the data. Also, from this analysis the critical radius of a shear band as it transitions from nucleation to propagation was estimated to be in the submicron range. From the indentation data obtained at different temperatures, Schuh et al.[24] were able to construct a new deformation map for $Pd_{40}Ni_{40}P_{20}$ and $Mg_{65}Cu_{25}Gd_{10}$ metallic glasses, as shown in Fig. 6.4. In contrast to the deformation map proposed previously by Spaepen,[5] the new one delineates a temperature-dependent regime at high strain rates where glass plasticity becomes more homogeneous.

6.7 SHEAR BAND TEMPERATURES

At ambient temperature, the plastic flow of metallic glasses is localized into shear bands.[1,2] This localization and the vein patterns seen on fracture surfaces are consistent with shear softening in the bands. The extent to which this softening results from local heating has remained controversial, with estimates of the local temperature rise ranging from less than 0.1 to a few thousand kelvin.[3–8,10–12] This large discrepancy in temperature evaluation is a result of great difficulty of directly measuring temperature in extremely small distances (~10 nm in shear band width[12,13]) and short time scales (~10^{-5} s for shear band propagation).[9] Lewandowski and Greer[25] conducted bending experiments with a Zr-based BMG coated with indium (melting point ~157°C) and observed melting of the coating near shear bands after bending. They concluded that there was a temperature rise in the band but shear band operation was not fully adiabatic according to their heat conduction analysis.

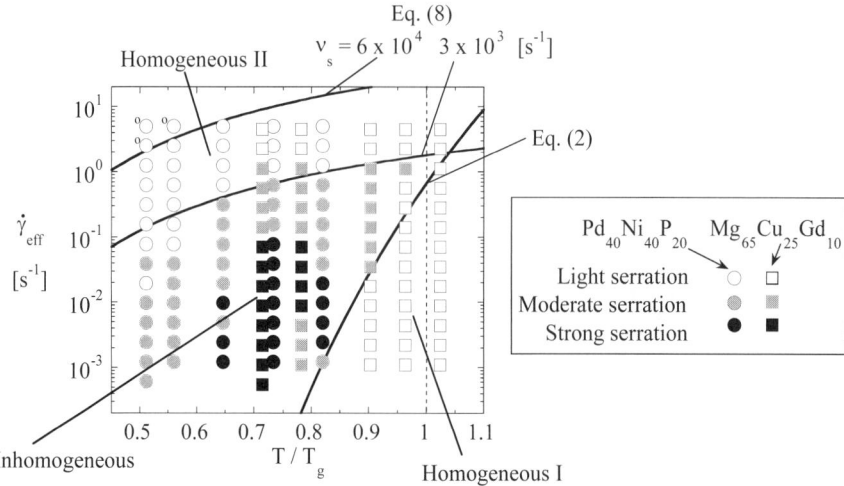

Fig. 6.4. The deformation map of $Pd_{40}Ni_{40}P_{20}$ and $Mg_{65}Cu_{25}Gd_{10}$ metallic glasses, plotting the effective strain rate during indentation vs. the homologous test temperature. The symbols are shaded according to the degree of flow serration, and the transitions from inhomogeneous to homogeneous flow are described in reference [24] (reprinted from reference [24] with permission from Elsevier)

It is known[26] that viscosity drops rapidly (Fig. 6.5) and significant softening occurs once the temperature is near the glass transition temperature. This can lead to rapid shear band propagation and catastrophic failure in BMGs. Thus, it is conceivable that the temperature rise within shear bands during deformation may have a close correlation with the glass transition temperature of a BMG.

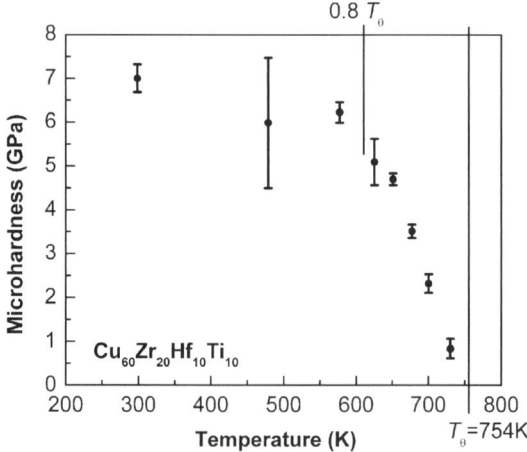

Fig. 6.5. Hardness as a function of temperature for amorphous Cu alloy. Hardness value decreases rapidly at $\sim 0.8 T_g$ ($T_g = 754$ K) (reprinted from reference [25] with permission from Elsevier)

Recently, with a shear-transformation-zone (STZ) model and by balancing the mechanical work and heat generation within a STZ unit, Yang et al.[27] showed that substantial temperature increase inside the shear band can occur by collective STZ deformation. The calculated temperature rise within a shear band ΔT_s was

$$\Delta T_s = \frac{\alpha \sigma_f}{2 \rho C_p},\qquad (6.2)$$

where ρ is the density, C_p is the heat capacity, σ_f is the nominal fracture strength, and $\alpha \approx 0.9$ is the ratio of plastic work converted to heat. Yang et al. also conducted in situ measurements on a Zr-based BMG with a high-speed infrared thermographic camera to capture the heat generated as a result of shear band emission. From the heat measurements they were able to calculate the temperature rise within a shear band and found it in good agreement with the theoretical predictions. Based upon the theoretical analysis, Yang et al. further computed the shear band temperature of a variety of BMGs with (6.2). The shear band temperatures calculated at the fracture strength for nine BMGs with six different alloy systems are shown in Fig. 6.6. It is evident that all calculated shear band temperatures are remarkably close to the glass transition temperature T_g. This interesting discovery reveals that catastrophic failure of BMG is caused by the sudden drop in viscosity inside the shear band as a result of local heating to the temperature close to the glass transition.

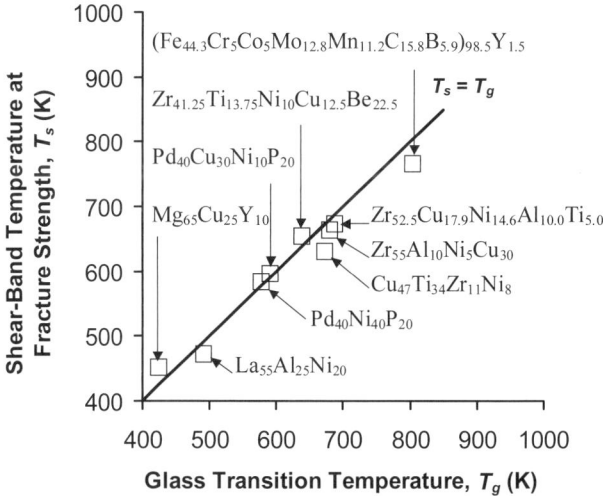

Fig. 6.6. The shear band temperatures calculated at the fracture for nine BMGs with six different alloy systems (reprinted from reference [27] with permission from the Materials Research Society)

6.8 STRENGTH

Two mechanical properties of BMG with major concern are the strength and ductility. Experimentally, the elastic limit of a BMG is typically 2%. Thus, the shear strength τ_y is typically about 0.02μ, where μ is the shear modulus. Johnson and Samwer[28] recently used the classical Frenkel's analysis of the shear strength of solids to offer a physical explanation for the linear dependence. The strength of a BMG is expected to be closely related to the physical parameters associated with atomic cohesive energy, such as glass transition temperatures T_g, elastic modulus,[28,29] and thermal expansion co-efficients.[30] The relationship between strength and T_g is of particular interest because T_g is also a key parameter governing the glass-forming ability.[31]

Recently, Yang et al.[32] noted that there are remarkable similarities between the physical processes of plastic deformation and glass transition of glassy materials – both phenomena are facilitated by collective atomic motions which require sufficient energy to overcome the bonding between atoms and result in an increase of the free-volume.[5,6] Whereas the thermal energy for glass transition spreads around the entire body of the solids during heating, the mechanical energy of shearing is highly confined in the localized shear bands in BMGs. Given the same atomic bonding force to overcome and a similar end product – reshuffled amorphous structure with similar amount of defects (free-volume) increased – the energy density – a variable that only depends on the initial and final states – can be expected to be similar for the two processes. By equating the two energy densities, one obtains

$$\tau_y \gamma_0 \approx \int_{T_0}^{T_g} \rho C_p \, dT, \tag{6.3}$$

where τ_y is the maximum shear stress upon yielding, $\gamma_0 = 1$ is the shear strain of the basic shear unit (STZ) in shear bands, ρ is the density of the material, C_p is the heat capacity at a constant pressure, and T_g and T_0 are the glass transition and ambient temperatures, respectively. Substituting different variables and with further mathematical manipulation, a simplified equation is obtained as

$$\sigma_y \approx \frac{6\rho_0 Nk}{M\gamma_0}(T_g - T_0) = \frac{6Nk}{\gamma_0}\frac{T_g - T_0}{V} = 50\frac{T_g - T_0}{V} = 50\frac{\Delta T_g}{V}, \tag{6.4}$$

where σ_y is the yield strength, N is Avogadro's number, k is the Boltzmann's constant, ρ_0 is the density at the ambient temperature, V is the molar volume, and M is the molar mass. With an additional assumption that the fracture strength σ_f is about 1.1 times of σ_y,[32] the equation is reduced to

$$\sigma_f = 1.1\sigma_y = 55\frac{\Delta T_g}{V} = 55\frac{\rho_0}{M}\Delta T_g. \tag{6.5}$$

The calculated fracture strength from (6.5) vs. the published experimental values of 27 BMGs from 11 different BMG systems is shown in Fig. 6.7. Evidently, a remarkable agreement is observed between the calculated and measured strength values. Moreover, the slope (ratio of strength over $\Delta T_g/V$) in Fig. 6.7 coincides exactly with the theoretically predicted value of $6Nk/r\gamma_0$. The above model offers a physical understanding for the strength of BMGs based on their unique deformation mode of localized shear banding (i.e., inhomogeneous deformation). Also, a simple unified equation is derived to allow the calculation of the low-temperature strength of a BMG directly from their T_g and other material constants that can be physically determined.

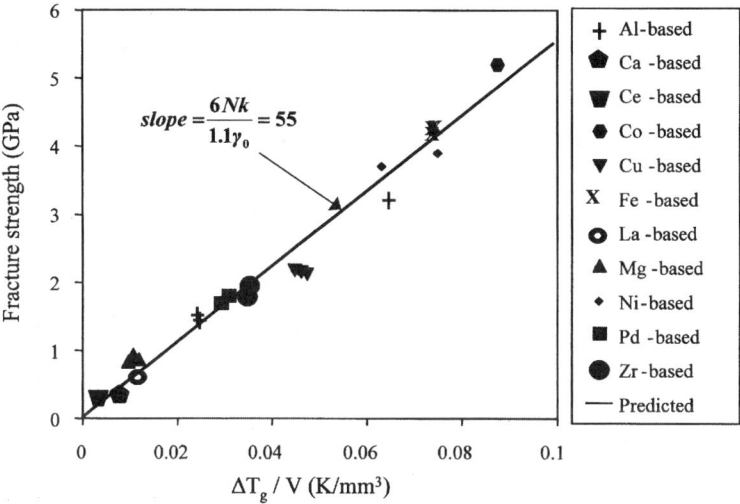

Fig. 6.7. The calculated fracture strength from a free-volume model (6.5) vs. the published experimental values of 27 BMGs from 11 different BMG systems (reprinted from reference [32] with permission from the American Institute of Physics)

6.9 STRUCTURE OF SHEAR BANDS

Chen et al.[33] were first to report direct transmission electron microscopy observations of crystallization within the shear bands of an $Al_{90}Fe_5Gd_5$ amorphous alloy induced by bending. The crystals were face-centered cubic Al, 7–10 nm in diameter, and seemed to form as a consequence of local atomic rearrangements in regions of high plastic strain. To further investigate the mechanism for the formation of nanocrystallites in this alloy, Jiang et al.[34,35] carried out bending experiments at a temperature of –40°C, and subsequently examined the microstructure by transmission electron microscopy. A high density of nanocrystals was observed within the shear bands, and it was argued that deformation-assisted atomic transport leads to nanocrystallization.

The formation of nanocrystals has also been observed during nano-indentation of a Zr-based BMG at room temperature.[36] Using atomic force microscopy and transmission electron microscopy, Kim et al. found that nanocrystallites nucleate in and around shear bands produced near indents, and that they are the same as crystallites formed during annealing without deformation at 783 K. The authors argued that it was a consequence of flow dilatation inside the bands and of the attendant, radically enhanced, atomic diffusional mobility inside the shear bands. Chen et al.[37] also reported, in a Zr–Cu glassy alloy, that strain softening caused by localized shearing was effectively prevented by nanocrystallization, which was formed in situ by plastic flow within the shear bands, leading to large plasticity and strain hardening.

Li et al.[38,39] used quantitative HRTEM and fluctuation microscopy to compare the structure of shear bands in a metallic glass with that of undeformed regions. They observed a large number of voids (density of one in 100 nm^3) of approximately 1 nm diameter in the shear bands, and argued that these voids are the result of coalescence of excess free-volume upon cessation of plastic flow in the shear band. Structural changes, particularly the generation of nanovoids and nanocrystals in shear bands, have been studied by positron annihilation spectroscopy (PAS).[40,41] Most recently, Hirotsu et al.[42] employed a spherical-aberration-corrected (Cs-corrected) high-resolution TEM to examine the local structure in shear bands in a $Pd_{40}Ni_{40}P_{20}$ BMG. TEM specimens were prepared from cast rod mechanically thinned down to about 20 μm in flowing water, followed by ion thinning. To minimize possible artifacts, ion thinning was carried out with low-voltage Ar-beam at 200 V with a glancing angle of 10°. Under appropriate imaging condition, the authors observed the presence of fcc Pd–Ni type, phosphide compound-like nanoclusters with sizes of 1–2 nm embedded in a dense-randomly packed amorphous matrix, as shown in Fig. 6.8. The density of nanoclusters was very high ($>10^{22}$ m^{-3}) and the interface between the nanocluster and the amorphous matrix was structurally diffuse. These nanoclusters were not stable crystalline phases per se but rather, they were more like medium-range-ordered clusters. The exact mechanism causing the formation of these nanoclusters is still unknown.

As discussed earlier, there is a local heating within a shear band, and the temperature rise in the shear band is close to the glass transition temperature. This temperature rise will obviously induce a structural change. Depending upon the thermal stability of each BMG system, nanocrystallites or nanoclusters can both possibly form as a result of local heating. On the other hand, deformation-enhanced mass transport is also possible. It is particularly noted that there are several recent reports showing that grain growth can take place during deformation of a nanocrystalline solid even at temperatures below the room temperature.[43,44] The structure in an amorphous solid is expected to be even more unstable than that in a nanocrystalline solid.

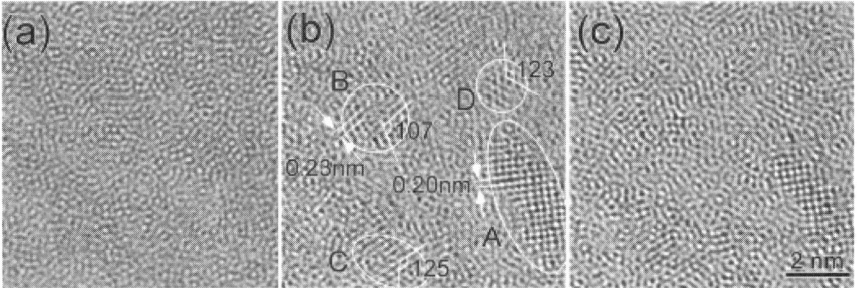

Fig. 6.8. High-resolution images of a $Pd_{40}Ni_{40}P_{20}$ BMG taken by a Cs-corrected TEM at Cs = 2 μm and $\Delta f \sim$ 1 nm (**a**), 5 nm (**b**), and 9 nm (**c**). The presence of medium-range-ordered nanoclusters is evident (reprinted from reference [42] with permission from the American Physical Society)

6.10 PLASTICITY: POISSON'S RATIO AND SOLIDITY INDEX

The plasticity or ductility is a more complicated issue than the strength. Traditionally, one way of distinguishing a ductile solid from a brittle one is by means of a solidity index S, which is defined as the ratio of the shear modulus to the bulk modulus

$$S = \frac{\mu}{B}, \tag{6.6}$$

where μ and B are the shear and bulk moduli, respectively. This index is zero for liquid and reaches its maximum value of 1.3 for diamond. This was originally suggested by Pugh in 1954,[45] and developed by Gilman et al.[46] Cottrell[47] has suggested that a dividing point of $S = 0.3$ is about right, brittleness being associated with $S > 0.3$.

Novikov and Sokolov[48] recently indicated that the fragility (brittleness) of a glass-forming liquid is intimately linked to a very basic property of the corresponding glass phase: the relative strength of shear and bulk moduli. Thus, Lewandowski et al.[49] extended the solidity parameter from crystalline solids to metallic glasses and found that a similar correlation, with the critical value of μ/B for metallic glasses (0.41–0.43) more sharply defined than for crystalline metals, i.e., metallic glasses with a μ/B value over 0.41

should be ductile. This critical value applies also for annealing-induced embrittlement of metallic glasses.

In view of the fact that for isotropic materials, which metallic glasses are, the solidity index can also be expressed as

$$S = \frac{\mu}{B} = \frac{3(1-2\nu)}{2(1+\nu)},$$ (6.7)

where ν is the Poisson's ratio. It is readily seen from the equation that a low solidity index corresponds to a high Poisson's ratio. In other words, BMGs with a high Poisson's ratio should be ductile. Schroers and Johnson[50] first demonstrated that this is indeed the case in a Pt-based BMG. The alloy exhibited a 20% plastic compression strain prior to fracture. They argued that a high Poisson's ratio can promote multiple shear band formation, suppress crack initiation and, thus, improve plasticity of a BMG. This was further confirmed in a bulk amorphous steel system.[51] Gu et al. recently investigated a $Fe_{65}Mo_{14}C_{15}B_6$ bulk amorphous steel doped with lanthanides to provide systematic variations of the elastic moduli, and found that an onset of plasticity is observed as Poisson's ratio approaches 0.32 from below. The findings are in support of the idea that there is a universal critical Poisson's ratio for plasticity in metallic glasses.

6.11 BMG COMPOSITES: DUCTILIZATION AND TOUGHENING

There are two composite approaches used to improve the plasticity and toughness of BMGs – extrinsic and intrinsic.[52] The extrinsic approach is to artificially introduce second phases into a glass matrix system, whereas the intrinsic approach, in contrast, is to design a chemical composition such that second phases are formed in situ during processing. The composite concept is borrowed from the intensive development of glass/ceramic matrix composite in the 1980s. In fact, it is similar to the strength and toughness improvement in silicon carbide fiber-reinforced glass matrix (lithium-alumino-silica) composites.[53]

In the case of extrinsic composites, Conner et al.[54] demonstrated that fracture strain as well as the toughness of a $Zr_{41.25}Ti_{13.75}Cu_{12.5}Ni_{10}Be_{22.5}$ glass can be significantly improved by reinforcing the glass matrix with either continuous tungsten fibers or 1080-steel wires. The increase in compressive toughness was suggested to come from the fibers restricting shear band propagation, promoting the generation of multiple shear bands and additional delamination of interfaces between fiber and matrix. The extrinsic approach has also been extended to Mg-based BMGs.

In the case of intrinsic composites, Hays et al.[55] cleverly designed a ductile metal-reinforced BMG matrix composite based on glass-forming compositions in the Zr–Ti–Cu–Ni–Be system. Primary dendrite growth and solute partitioning in the molten state yielded a microstructure consisting of a ductile crystalline Ti–Zr–Nb phase with bcc structure, in a Zr–Ti–Nb–Cu–Ni–Be BMG matrix. They further demonstrated that, under unconstrained mechanical loading, organized shear band patterns develop throughout the sample. This resulted in a dramatic increase in the plastic strain to failure, impact resistance, and toughness of the metallic glass. Several intrinsic composites have also been developed, including a 5 vol.% Ta-reinforced Zr-based BMG composite[56] and a Ti-based BMG composite.[57]

It is now generally accepted that composite is the appropriate approach to improve the mechanical properties of brittle BMGs. It is noted, however, from the composite point of view that the mechanical performance of a composite is dependent upon the relative strength, shape, size, and distribution of the second phases. Designing a composite with optimum performance would require not only sorting out of the appropriate chemical composition but also developing a well-controlled subsequent thermal processing.

6.12 HOMOGENEOUS DEFORMATION

Homogeneous deformation in metallic glasses usually takes place at high temperatures ($>0.70T_g$),[6] as indicated in Fig. 6.5, and the material can often exhibit a significant plasticity. The transition temperature from inhomogeneous to homogeneous deformation (or brittle-to-ductile transition) is strongly dependent upon strain rate,[58,59] indicating that homogeneous deformation is associated with a rate (or diffusional relaxation) process.

Metallic glasses, when deformed in the supercooled liquid region, exhibit extraordinarily large plasticity. For example, a tensile elongation of 20,000% has been reported in a $La_{55}Al_{25}Ni_{20}$ alloy in the supercooled liquid region. It is generally observed in the study of homogeneous deformation that BMG behaves like a Newtonian fluid at low strain rates, but becomes non-Newtonian at high strain rates. Such a transition has been observed in many BMG systems.[15,60–65] A typical example showing the Newtonian to non-Newtonian transition is given in Fig. 6.9.

The specific strain rate at which the transition occurred depends upon the testing temperature, and specifically the transition takes place at increasingly high strain rates as the testing temperature is reduced. Kawamura et al. argued that the non-Newtonian behavior is associated with stress overshoot at high strain rate (or high stress), and the stress overshoot was caused by a change in atomic mobility because of a rapid, deformation-induced, change of free-volume.

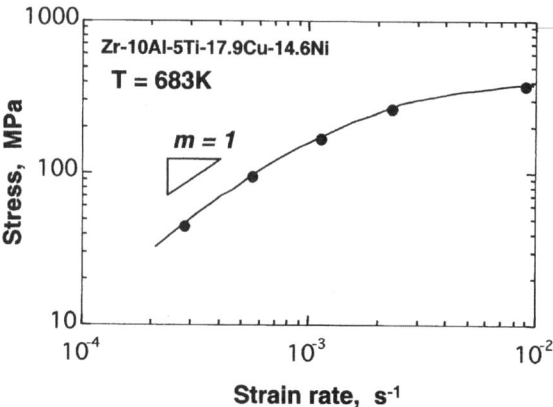

Fig. 6.9. Stress–strain rate relation for a Zr–10Al–5Ti–17.9Cu–14.6Ni glassy alloy shows Newtonian flow at low strain rates but non-Newtonian at high strain rates (reprinted from reference [68] with permission from Elsevier)

Both Reger-Leonhard et al.[62] and Bletry et al.[61] invoked the free-volume theory[5] to explain the departure from Newtonian flow. They argued that at higher strain rates inhomogeneous flow occurs concurrently with the homogeneous flow ($m = 1$) causing a deviation from the linearity in the strain rate–stress relation. The non-Newtonian data were well correlated with Newtonian ones with the equation

$$\dot{\varepsilon} = \dot{\varepsilon}_\mathrm{o} \sinh \frac{\sigma \gamma_\mathrm{o} \Omega_\mathrm{f}}{M k_\mathrm{B} T}, \qquad (6.8)$$

where $\dot{\varepsilon}_\mathrm{o}$ is the reference strain rate, γ_o is the local strain produced by the shear site of volume Ω_f, $M = \sqrt{3}$, and k_B and T have their usual meanings. However, structural evolution, in particular nanocrystal formation, during deformation was not considered in the model. Spaepen et al.[66] recently applied the free-volume theory to analyze compressive creep data on bulk $Pd_{41}Ni_{10}Cu_{29}P_{20}$ samples and found an overall activation volume of 106 $Å^3$, which is approximately eight atomic volumes. This result indicates that multiatom defects, not single-atom mechanism, are involved in the homogeneous deformation processes.

Conversely, Nieh et al.[60] proposed that the non-Newtonian behavior is associated with structural instability of the BMG alloys during high-temperature deformation. They subsequently conducted microbeam X-ray diffraction[67] and high-resolution TEM experiments[68] on deformed samples to demonstrate that, even though tests were carried out in the supercooled liquid region, nanocrystallization still occurred. The in situ nanocrystallization in BMGs deformed in the supercooled liquid region has been widely recognized.[61–65,69–71]

Mechanistically, when nanocrystallization occurs and a material contains nanocrystals dispersed in an amorphous structure, the total deformation rate can be expressed to a first approximation by

$$\dot{\gamma}_{\text{total}} = (1 - f_{\text{v}})\dot{\gamma}_{\text{am}} + f_{\text{v}}\dot{\gamma}_{\text{cry}}, \quad (6.9)$$

where $\dot{\gamma}_{\text{total}}$ is the total strain rate, $\dot{\gamma}_{\text{am}}$ and $\dot{\gamma}_{\text{cry}}$ are the strain rates resulting from the amorphous and crystalline phases, respectively, and f_{v} is the volume fraction of the crystalline phase. The plastic flow of the pure amorphous matrix is described by $\dot{\gamma}_{\text{am}} = A\tau$, but the plastic flow of the nanocrystalline phase is described by another power-law function with a nonlinear power-law dependence. Then (6.9) can be rewritten as

$$\dot{\gamma}_{\text{total}} = (1 - f_{\text{v}})A\tau + f_{\text{v}}B\tau^{\text{n}}, \quad (6.10)$$

where A and B are material constants. It is evident from the equation that, in the presence of nanocrystals, the resultant strain rate sensitivity is no longer unity – the value for ideal Newtonian viscous flow.

Although there is a difference in the flow behavior, i.e., Newtonian vs. non-Newtonian, a large ductility can generally be obtained when a BMG is deformed homogeneously in the supercooled liquid. Advantage of this high ductility has been taken to fabricate sophisticated structural components.[72–74] A microgear fabricated superplastically from a Pd-based BMG is shown in Fig. 6.10.[72] The excellent formability is readily seen. Due to the size limitation of BMG that can be synthesized, the application would probably be some miniature components but with high payoff value, such as MEMS devices or biomedical structures.

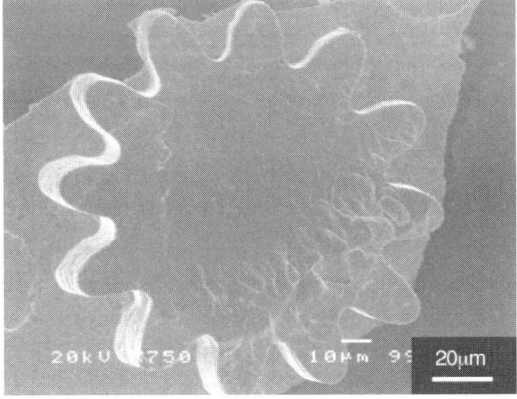

Fig. 6.10. A microgear fabricated superplastically from a Pd-based BMG

ACKNOWLEDGMENT

This work was supported by the Division of Materials Sciences and Engineering, Office of Basic Energy Sciences, US Department of Energy under contract DE-FG02-06ER46338 with the University of Tennessee.

REFERENCES

1. C. T. Liu, L. Heatherly, D. S. Easton, C. A. Carmichael, J. H. Schneibel, C. H. Chen, J. L. Wright, M. H. Yoo, J. A. Horton, and A. Inoue, Test environment and mechanical properties of Zr-base bulk amorphous alloys, *Metall. Mater. Trans. A* **29**(7), 1811–1820 (1998).
2. T. Mukai, T. G. Nieh, Y. Kawamura, A. Inoue, and K. Higashi, Influence of strain rate on compressive mechanical behavior of $Pd_{40}Ni_{40}P_{20}$ bulk metallic glass, *Intermetallics* **10**(11–12), 1071–1077 (2002).
3. T. Masumoto and R. Maddin, Structural stability and mechanical properties of amorphous metals, *Mater. Sci. Eng.* **19**(1), 1–24 (1975).
4. E. Pekarskaya, C. P. Kim, and W. L. Johnson, In situ transmission electron microscopy studies of shear bands in a bulk metallic glass based composite, *J. Mater. Res.* **16**(9), 2513–2518 (2001).
5. F. Spaepen, A microscopic mechanism for steady state inhomogeneous flow in metallic glasses, *Acta Metall.* **25**(4), 407–415 (1977).
6. A. S. Argon, Plastic deformation in metallic glasses, *Acta Metall.* **27**, 47–58 (1979).
7. Z. F. Zhang, J. Eckert, and L. Schultz, Difference in compressive and tensile fracture mechanisms of $Zr_{59}Cu_{20}Al_{10}Ni_8Ti_3$ bulk metallic glass, *Acta Mater.* **51**(4), 1167–1179 (2003).
8. P. E. Donovan, A yield criterion for $Pd_{40}Ni_{40}P_{20}$ metallic glass, *Acta Mater.* **37**(2), 445–456 (1989).
9. C. A. Schuh and A. C. Lund, Atomistic basis for the plastic yield criterion of metallic glass, *Nat. Mater.* **2**(7), 449–452 (2003).
10. H. A. Bruck, A. J. Rosakis, and W. L. Johnson, The dynamic compressive behavior of beryllium bearing bulk metallic glasses, *J. Mater. Res.* **11**(2), 503–511 (1996).
11. W. J. Wright, R. Saha, and W. D. Nix, Deformation mechanisms of the $Zr_{40}Ti_{14}Ni_{10}Cu_{12}Be_{24}$ bulk metallic glass, *Mater. Trans. JIM* **42**(4), 642–649 (2001).
12. P. Lowhaphandu, S. L. Montgomery, and J. J. Lewandowski, Effects of superimposed hydrostatic pressure on flow and fracture of a Zr–Ti–Ni–Cu–Be bulk amorphous alloy, *Scripta Mater.* **41**(1), 19–24 (1999).
13. T. Mukai, T. G. Nieh, Y. Kawamura, A. Inoue, and K. Higashi, Dynamic response of a $Pd_{40}Ni_{40}P_{20}$ bulk metallic glass in tension, *Scripta Mater.* **46**(1), 43–47 (2002).
14. R. Maddin and T. Masumoto, Deformation of amorphous palladium-20 at% silicon, *Mater. Sci. Eng.* **9**, 153–162 (1972).
15. Y. Kawamura, T. Shibata, A. Inoue, and T. Masumoto, Deformation behavior of $Zr_{65}Al_{10}Ni_{10}Cu_{15}$ glassy alloy with wide supercooled liquid region, *Appl. Phys. Lett.* **69**(9), 1208–1210 (1996).
16. J. Das, M. B. Tang, K. B. Kim, R. Theissmann, F. Baier, W. H. Wang, and J. Eckert, Work-hardenable ductile bulk metallic glass, *Phys. Rev. Lett.* **94**, 205501 (2005).

17. B. Yang, L. Riester, and T. G. Nieh, Strain hardening in a bulk metallic glass under nanoindentation, *Scripta Mater.* **54**, 1277–1280 (2006).
18. T. C. Hufnagel, T. Jiao, Y. Li, L. Q. Xing, and K. T. Ramesh, Deformation and failure of $Zr_{57}Ti_5Cu_{20}Ni_8Al_{10}$ bulk metallic glass under quasi-static and dynamic compression, *J. Mater. Res.* **17**(6), 1441–1445 (2002).
19. H. Bei, S. Xie, and E. P. George, Softening caused by profuse shear banding in a bulk metallic glass, *Phys. Rev. Lett.* **96**, 105503 (2006).
20. H. Kimura and T. Masumoto, A model of the mechanics of serrated flow in an amorphous alloy, *Acta Metall.* **31**(2), 231–240 (1983).
21. C. A. Schuh, T. G. Nieh, and Y. Kawamura, Rate dependence of serrated flow during nanoindentation of a bulk metallic glass, *J. Mater. Res.* **17**(7), 1651–1654 (2002).
22. C. A. Schuh and T. G. Nieh, A nanoindentation study of serrated flow in bulk metallic glasses, *Acta Mater.* **51**(1), 87–99 (2003).
23. C. A. Schuh, A. S. Argon, T. G. Nieh, and J. Wadsworth, The transition from localized to homogeneous plasticity during nanoindentation of an amorphous metal, *Philos. Mag. A* **83**(22), 2585–2597 (2003).
24. C. A. Schuh, A. C. Lund, and T. G. Nieh, New regime of homogeneous flow in the deformation map of metallic glasses: Elevated temperature nanoindentation experiments and mechanistic modeling, *Acta Mater.* **52**(20), 5879–5891 (2004).
25. P. Wesseling, T. G. Nieh, W. H. Wang, and J. J. Lewandowski, Preliminary assessment of flow, notch toughness, and high temperature behavior of $Cu_{60}Zr_{20}Hf_{10}Ti_{10}$ bulk metallic glass, *Scripta Mater.* **51**(2), 151–154 (2004).
26. J. J. Lewandowski, M. Shazly, and A. S. Nouri, Intrinsic and extrinsic toughening of metallic glasses, *Scripta Mater.* **54**(3), 337–341 (2006).
27. B. Yang, C. T. Liu, T. G. Nieh, M. Morrison, P. K. Liaw, and R. A. Buchanan, Localized heating and fracture criterion for bulk metallic glasses, *J. Mater. Res.* 21(4), 915–922 (2006).
28. W. L. Johnson and K. Samwer, A universal criterion for plastic yielding of metallic glasses with a $(T/T_g)^{2/3}$ temperature dependence, *Phys. Rev. Lett.* **95**, 195501 (2006).
29. A. Inoue, B. L. Shen, H. Koshiba, H. Kato, and A. R. Yavari, Ultra-high strength above 5000 MPa and soft magnetic properties of Co–Fe–Ta–B bulk glassy alloys, *Acta Mater.* **52**, 1631–1637 (2004).
30. H. S. Chen, Glassy metals, *Rep. Prog. Phys.* **43**, 353–432 (1980).
31. Z. P. Lu and C. T. Liu, A new approach to understanding and measuring glass formation in bulk amorphous materials, *Intermetallics* **12**, 1035–1043 (2004).
32. B. Yang, C. T. Liu, and T. G. Nieh, Unified equation for the strength of bulk metallic glasses, *Appl. Phys. Lett.* **88**(22), 221911 (2006).
33. H. Chen, Y. He, G. J. Shiflet, and S. J. Poon, Deformation-induced nanocrystal formation in shear bands of amorphous alloys, *Nature* **367**, 541–543 (1994).
34. W. H. Jiang, F. E. Pinkerton, and M. Atzmon, Deformation-induced nanocrystallization in an Al-based amorphous alloy at a subambient temperature, *Scripta Mater.* **48**(8), 1195–1200 (2003).
35. W. H. Jiang and M. Atzmon, Mechanically-assisted nanocrystallization and defects in amorphous alloys: A high-resolution transmission electron microscopy study, *Scripta Mater.* **54**, 333–336 (2006).
36. J. J. Kim, Y. Choi, S. Suresh, and A. S. Argon, Nanocrystallization during nanoindentation of a bulk amorphous metal alloy at room temperature, *Science* **295**, 654–657 (2002).
37. M. Chen, A. Inoue, W. Zhang, and T. Sakurai, Extraordinary plasticity of ductile bulk metallic glasses, *Phys. Rev. Lett.* **96**, 245502 (2006).

38. J. Li, X. Gu, and T. C. Hufnagel, Using fluctuation microscopy to characterize structural order in metallic glasses, *Microsc. Microanal.* **6**, 509–515 (2003).

39. J. Li, F. Spaepen, and T. C. Hufnagel, Nanometre-scale defects in shear bands in a metallic glass, *Philos. Mag. A* **82**(13), 2623–2630 (2002).

40. K. M. Flores, Structural changes and stress state effects during inhomogeneous flow of metallic glasses, *Scripta Mater.* **54**, 327–332 (2006).

41. B. P. Kanungo, S. C. Glade, P. Asoka-Kumar, and K. M. Flores, Characterization of free volume changes associated with shear band formation in Zr- and Cu-based bulk metallic glasses, *Intermetallics* **12**, 1073–1080 (2004).

42. Y. Hirotsu, T. G. Nieh, A. Hirata, T. Ohkubo, and N. Tanaka, Local atomic ordering and nanoscale phase separation in a Pd–Ni–P bulk metallic glass, Phys. Rev. B 73, 012205 (2006). http://link.aps.org/abstract/PRB/v73/e012205

43. K. Zhang, J. R. Weertman, and J. A. Eastman, Rapid stress-driven grain coarsening in nanocrystalline Cu at ambient and cryogenic temperatures, *Appl. Phys. Lett.* **87**, 061921 (2005).

44. D. Pan, T. G. Nieh, and M. W. Chen, Strengthening and softening of nanocrystalline nickel during multi-step nanoindentation, *Appl. Phys. Lett.* **88**(16), 161922 (2006).

45. S. F. Pugh, Relations between the elastic moduli and plastic properties of polycrystalline pure metals, *Philos. Mag.* **45**, 823–843 (1954).

46. J. J. Gilman, B. J. Cunningham, and A. C. Holt, Method for monitoring the mechanical state of a material, *Mater. Sci. Eng. A* **125**, 39–42 (1990).

47. A. H. Cottrell, The art of simplification in materials science, *MRS Bull.* **22**(5), 15 (1997).

48. V. N. Novikov and A. P. Sokolov, Poissons ratio and the fragility of glass-forming liquids, *Nature* **431**, 961–963 (2004).

49. J. J. Lewandowski, W. H. Wang, and A. L. Greer, Intrinsic plasticity or brittleness of metallic glasses, *Philos. Mag. Lett.* **85**(2), 77–87 (2005).

50. J. Schroers and W. L. Johnson, Ductile bulk metallic glass, *Phys. Rev. Lett.* **93**, 255506 (2004).

51. X. J. Gu, A. G. McDermott, and S. J. Poon, Critical Poisson's ratio for plasticity in Fe–Mo–C–B–Ln bulk amorphous steel, *Appl. Phys. Lett.* **88**, 211905 (2006).

52. J. J. Lewandowski, M. Shazly, and A. S. Nouri, Intrinsic and extrinsic toughening of metallic glasses, *Scripta Mater.* **54**(3), 337–341 (2006).

53. J. J. Brennan and K. M. Prewo, Silicon carbide fiber reinforced glass–ceramic matrix composites exhibiting high strength and toughness, J. Mater. Sci. 17, 2371–2383 (1982).

54. R. D. Conner, R. B. Dandliker, and W. L. Johnson, Mechanical properties of tungsten and steel fiber reinforced $Zr_{41.25}Ti_{13.75}Cu_{12.5}Ni_{10}Be_{22.5}$ metallic glass matrix composites, *Acta Mater.* **46**(17), 6089–6102 (1998).

55. C. C. Hays, C. P. Kim, and W. L. Johnson, Microstructure controlled shear band pattern formation and enhanced plasticity of bulk metallic glasses containing in situ formed ductile phase dendrite dispersions, *Phys. Rev. Lett.* **84**(13), 2901–2904 (2000).

56. C. Fan, H. Li, L. J. Kecskes, K. Tao, H. Choo, P. K. Liaw, and C. T. Liu, Mechanical behavior of bulk amorphous alloys reinforced by ductile particles at cryogenic temperatures, *Phys. Rev. Lett.* 96, 145506 (2006).

57. G. He, J. Eckert, and W. Loser, Stability, phase transformation and deformation behavior of Ti-base metallic glass and composite, *Acta Mater.* **51**, 1621–1631 (2003).

58. Y. Kawamura, T. Nakamura, and A. Inoue, Superplasticity in $Pd_{40}Ni_{40}P_{20}$ metallic glass, *Scripta Mater.* **39**(3), 301–306 (1998).

59. A. L. Mulder, R. J. A. Derksen, J. W. Drijver, and S. Radelaar, Presented at the *Proceedings of 4th International Conference on Rapidly Quenched Metals*, Sendai, 1982 (unpublished).

60. T. G. Nieh, T. Mukai, C. T. Liu, and J. Wadsworth, Superplastic behavior of a Zr–10Al–5Ti–17.9Cu–14.6Ni metallic glass in the supercooled liquid region, *Scripta Mater.* **40**(9), 1021–1027 (1999).

61. M. Bletry, P. Guyot, Y. Brechet, J. J. Blandin, and J. L. Soubeyroux, Homogeneous deformation of bulk metallic glasses in the super-cooled liquid state, *Mater. Sci. Eng. A* **387–389**, 1005–1011 (2004).

62. A. Reger-Leonhard, M. Heilmaier, and J. Eckert, Newtonian flow of $Zr_{55}Cu_{30}Al_{10}Ni_5$ bulk metallic glassy alloys, *Scripta Mater.* **43**, 459–464 (2000).

63. J. P. Chu, C. L. Chiang, T. Mahalingam, and T. G. Nieh, Plastic flow and tensile ductility of a bulk amorphous $Zr_{55}Al_{10}Cu_{30}Ni_5$ alloy at 700 K, *Scripta Mater.* **49**(5), 435–440 (2003).

64. C. L. Chiang, J. P. Chu, C. T. Lo, Z. X. Wang, W. H. Wang, J. G. Wang, and T. G. Nieh, Homogeneous plastic deformation in bulk amorphous $Cu_{60}Zr_{20}Hf_{10}Ti_{10}$ alloy, *Intermetallics* **12**, 1057–1061 (2004).

65. D. H. Bae, J. M. Park, J. H. Na, D. H. Kim, Y. C. Kim, and J. K. Lee, Deformation behavior of Ti–Zr–Ni–Cu–Be metallic glass and composite in the supercooled liquid region, *J. Mater. Res.* **19**(3), 937–942 (2004).

66. F. Spaepen, Homogeneous flow of metallic glasses: A free volume perspective, *Scripta Mater.* **54**(3), 363–367 (2006).

67. T. G. Nieh, J. Wadsworth, C. T. Liu, G. E. Ice, and K.-S. Chung, Extended plasticity in the supercooled liquid region of bulk metallic glasses, *Mater. Trans. JIM* **42**(4), 613–618 (2001).

68. T. G. Nieh, J. Wadsworth, C. T. Liu, Y. Ohkubo, and Y. Hirotsu, Plasticity and structure instability in a bulk metallic glass deformed in the supercooled liquid region, *Acta Mater.* **49**(15), 2887–2896 (2001).

69. J. P. Chu, C. L. Chiang, T. G. Nieh, and Y. Kawamura, Superplasticity in a bulk amorphous Pd–40Ni–20P alloy: A compression study, *Intermetallics* **10**(11–12), 1191–1195 (2002).

70. W. J. Kim, D. S. Ma, and H. G. Jeong, Superplastic flow in a $Zr_{65}Al_{10}Ni_{10}Cu_{15}$ metallic glass crystallized during deformation in a supercooled liquid region, *Scripta Mater.* **49**(11), 1067–1073 (2003).

71. G. Wang, J. Shen, J. F. Sun, Y. J. Huang, J. Zou, Z. P. Lu, Z. H. Stachurski, and B. D. Zhou, Superplasticity and superplastic forming ability of a Zr–Ti–Ni–Cu–Be bulk metallic glass in the supercooled liquid region, *J. Non-Cryst. Solids* **351**(3), 209–217 (2005).

72. Y. Saotome, K. Itoh, T. Zhang, and A. Inoue, Superplastic nanoforming of Pd-based amorphous alloy, *Scripta Mater.* **44**(8–9), 1541–1545 (2001).

73. Y. Saotome, T. Hatori, T. Zhang, and A. Inoue, Superplastic micro/nano-formability of $La_{60}Al_{20}Ni_{10}Co_5Cu_5$ amorphous alloy in supercooled liquid state, *Mater. Sci. Eng. A* **304**, 716–720 (2001).

74. J. P. Chu, C. L. Chiang, H. Wijaya, R. T. Huang, C. W. Wu, B. Zhang, W. H. Wang, and T. G. Nieh, Compressive deformation of a bulk Ce-based metallic glass, *Scripta Mater.* **55**, 227–230 (2006).

Chapter 7

FATIGUE AND FRACTURE BEHAVIOR

Gongyao Wang and Peter K. Liaw

Department of Materials Science and Engineering, University of Tennessee, Knoxville, TN 37996, USA

7.1 INTRODUCTION

Structural components are frequently subjected to repeated or cyclic loading. The resulting cyclic stresses, which may be far below the ultimate tensile strength of materials, can result in a microscopic physical damage to the material. The microscopic damage can accumulate with continued cyclic loading until it develops into a crack that could lead to the catastrophic failure. This process of damage and failure due to cyclic loading is called fatigue.[1]

Fatigue is a dynamic phenomenon that accounts for ~90% of all service failures from mechanical causes. In general, the fatigue life involves the number of loading cycles to initiate and propagate a crack to a critical size. Fatigue failure occurs in three stages: crack initiation, stable crack growth, and fast fracture. The main factors that contribute to fatigue failures include the number of load cycles, stress range, mean stress, and local stress concentrations. It is necessary to take these factors into account in the design of materials for structural components.[2]

7.1.1 Definitions

Fatigue is a progressive, localized, and permanent structural-damage process that occurs when a material is subjected to cyclic or fluctuating stresses that have maximum values less (often much less) than the ultimate tensile strength of the material.

If plastic deformations are small and localized in the vicinity of the crack tip, the main part of the component deforms elastically. This is called the high-cycle fatigue (HCF).[3] Historically, most attention has focused on situations that require more than 10^4 cycles to failure where stresses are low, and the deformation is primarily elastic. Low-cycle fatigue (LCF) occurs when the cyclic loading is accompanied by elastoplastic deformations in the bulk of the component.[3] In this case, the stress is high enough for the plastic deformation to occur. To describe this process in terms of the stress is less useful. Therefore, the strain in the material offers a simpler description.

7.1.2 Stress–life (*S–N*) curve

Stress (σ) is usually defined as the force per unit area. A nominal stress (S) is calculated from the load or moment or their combination as a matter of convenience and is only equal to σ in certain situations.[1] Stresses can be applied in three ways: torsion, axial, and bending.

The elastic stress-concentration factor, K_t, is the ratio between the peak stress at the root of the notch and the nominal stress, $K_t = \sigma/S$, which would be present if a stress concentration did not occur.[1] The stress range, $\Delta\sigma$, is the difference between the maximum (σ_{max}) and minimum (σ_{min}) values, i.e., $\Delta\sigma = \sigma_{max} - \sigma_{min}$. The mean stress, σ_m, is the average of the maximum and minimum values. The stress amplitude, σ_a, is half the stress range, $\Delta\sigma/2$. The stress (load) ratio, R, is the ratio of the minimum stress and the maximum stress, $R = \sigma_{min}/\sigma_{max}$, as demonstrated in Fig. 7.1.

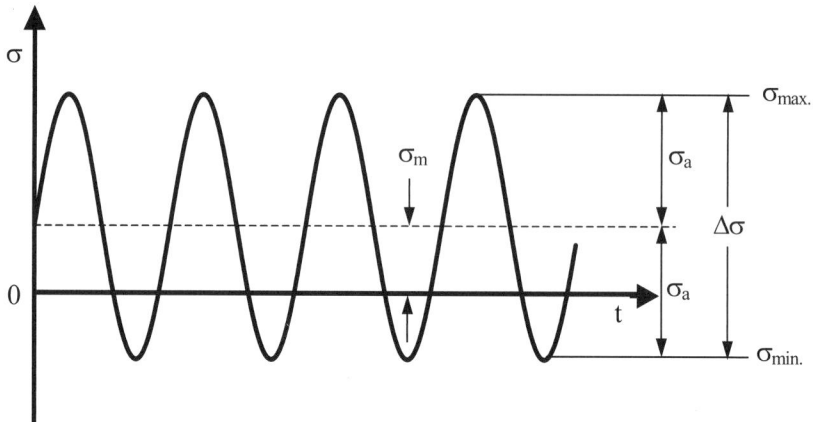

Fig. 7.1. Definitions for cyclic stresses

The amplitude of the stress, stress range, or maximum stress is commonly plotted vs. the number of cycles to failure, N. This curve is called a stress–life or *S–N* curve.[2] To represent conveniently both long and short endurances on the one diagram, a logarithmic scale is commonly used for N. A linear stress scale is most frequently used, although logarithmic scales are sometimes used for both parameters. A common convention is to represent specimens unbroken at the completion of a test by an arrow extending beyond the test point. The fatigue limit is the value, which may be statistically determined, of the stress below which a material may endure an infinite number of stress cycles without failures. A fatigue limit is often arbitrarily defined at a specific long life, e.g., 10^7 cycles. The fatigue strength is used to specify a stress value from a *S–N* curve at a particular life of interest.

7.2 FATIGUE TESTING

Several ASTM standards, such as E466,[4] E467,[5] and E468,[6] address stress-based fatigue testing for metals. In general, uniaxial experiments, such as tension–tension fatigue tests and three-point and four-point bend tests, are conducted on bulk-metallic glasses (BMGs).

Initially, rotating-beam fatigue-testing machines were used for fatigue experiments. This type of instrument allows fatigue tests to be run in torsion, combined bending and torsion, or biaxial bending under a constant-amplitude loading at a constant frequency of cycling.[1] Later, closed-loop servohydraulic-testing machines were developed to test materials and components in cyclic loads to either characterize materials or simulate long-term operations. Specimens may be subjected to the constant-amplitude cycling with controlled loads, strains, or deflections, and the amplitude, mean, and cyclic frequency of the chosen variables can be selected. The frequency is usually less than 100 Hz. However, special high-frequency resonant vibration-testing devices can run at a frequency of up to 1 kHz.[7]

For tension–tension fatigue testing of bulk-metallic glasses, button-head fatigue specimens with a small radial sharp notch, as shown in Fig. 7.2a, may be used.[8,9] The stress-range value reflects the stress-concentration factor (K_t) of 1.55 at the notched section for this kind of specimens.[8,9] This factor of 1.55 may be overestimated.[10,11] However, recent finite-element-method simulations were in good agreement with this value.[12] The button-head fatigue specimens with a taper notch may also be used, as shown in Fig. 7.2b. To avoid surface effects, the samples are typically polished to at least 1,200 grit. The gripping system and the specimen geometry are modified versions of those employed for testing ceramics and other brittle materials.[8,9] Peter et al.[8] showed that these modified gripping systems largely eliminated the effect of crack initiation at the localized points of contact by increasing the area of the contact between the BMG sample and the grip.

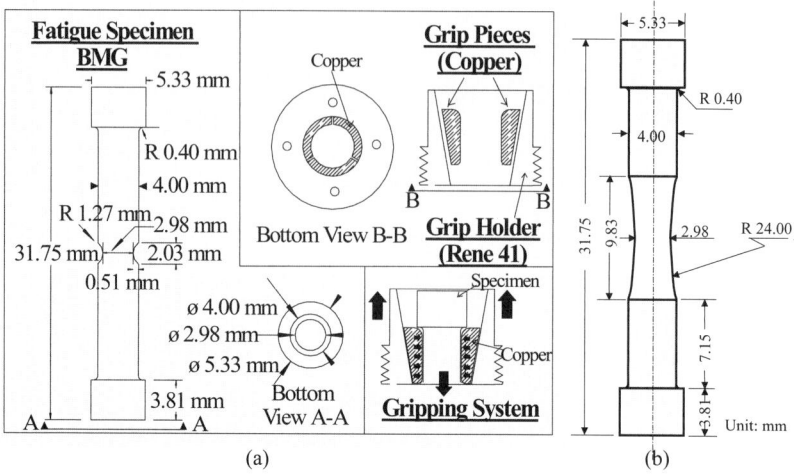

Fig. 7.2. Specimen (**a**) with a sharp notch, (**b**) with a taper notch for fatigue testing (reprinted from Intermetallics, W. H. Peter et al.[8,9] with permission from Elsevier)

Rectangular beams are often prepared for three- and four-point bending tests.[13–15] Stresses are calculated on the tensile surface within the outer span from

$$\sigma = \frac{3P(S_2 - S_1)}{2bh^2}, \qquad (7.1)$$

where P is the applied load, b is the specimen thickness, h is the specimen height, S_1 is the inner span, and S_2 is the outer span. Some irregular samples have also been used, e.g., the dumbbell-shaped plate with a 5-mm gage length whose cross section is 2 mm wide and 1 mm thick, and a plate, single-edge notched with a semicircle.[16]

7.3 HIGH-CYCLE FATIGUE

7.3.1 Stress–life behavior

Fatigue results are usually presented as *S–N* curves to reveal the most fundamental information in estimating the lifetime under different cyclic stresses. *S–N* curves vary widely for different classes of materials, and are affected by several factors. Any processing that changes the static mechanical properties or microstructure is also likely to affect the *S–N* curve.

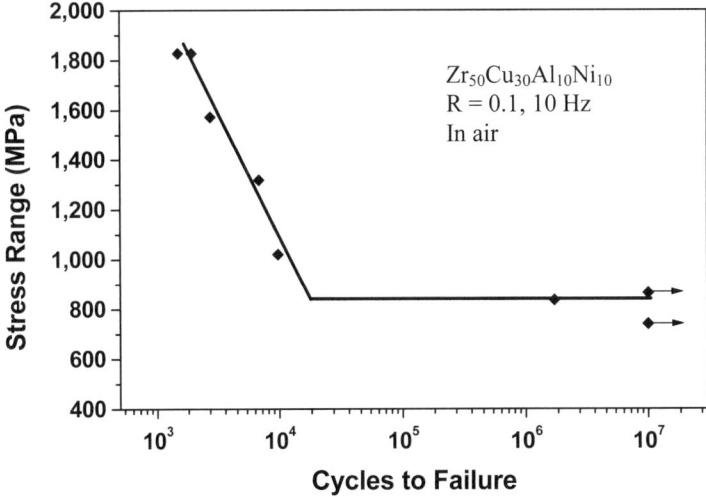

Fig. 7.3. The typical *S–N* curve of BMGs (reprinted from Intermetallics, G. Y. Wang et al.[17] with permission from Elsevier)

Additional factors of importance include the mean stress, material quality, specimen geometry, chemical environment, temperature, cyclic frequency, and residual stress. A typical *S–N* curve of a $Zr_{50}Cu_{30}Al_{10}Ni_{10}$ BMG that was tested in air with $R = 0.1$ and a frequency of 10 Hz is shown in Fig. 7.3. In general, the stress range for BMGs decreases significantly in the 10^3–10^4 cycle fatigue-life range.[17]

7.3.2 Effect of environment

The environment can have a dramatic influence on the fatigue properties. For example, Peter et al.[8,9] found that turning on an ionization gauge affected the fatigue properties of $Zr_{52.5}Cu_{17.9}Al_{10}Ni_{14.6}Ti_5$. The hot tungsten filament used in the gauge dissociated the residual water vapor in the vacuum system into the atomic hydrogen and oxygen and decreased the fatigue life.[8,9] When the ionization gauge was turned off, no distinct difference of the fatigue life in air and vacuum was found.[18] To ascertain the environmental effect on the fatigue lifetime of BMGs, testing in dry hydrogen, dry oxygen, and vacuum should be performed to determine the influence of hydrogen embrittlement and surface oxidation.

7.3.3 Effect of frequency

In general, a reduction of the fatigue life or number of cycles to failure occurs at lower frequencies due to the increase in the time for residual gas to adsorb on the exposed fatigue-crack surface. However, results from an LM002 composite of a $Zr_{47}Ti_{12.9}Nb_{2.8}Ni_{9.6}Cu_{11}Be_{16.7}$ BMG matrix with a dendritic Zr–Ti–Nb (β) phase seem to show the opposite trend in which lower frequencies yield longer lifetimes.[19] The lifetime at 0.1 and 0.01 Hz is more scattered than that at 10 Hz, as shown in Fig. 7.4.[20] The scattered data could be due to the lifetime at a lower frequency being more sensitive to defects, as cracks in BMGs generally initiate from inclusions or porosities.[21] Further research is required to ascertain if the effect of the heat produced from fatigue on lifetimes of BMGs at different frequencies is dominant[22,23] or the effect of the residual-gas adsorption to the exposed fatigue-crack surface on lifetimes of BMGs is significant.

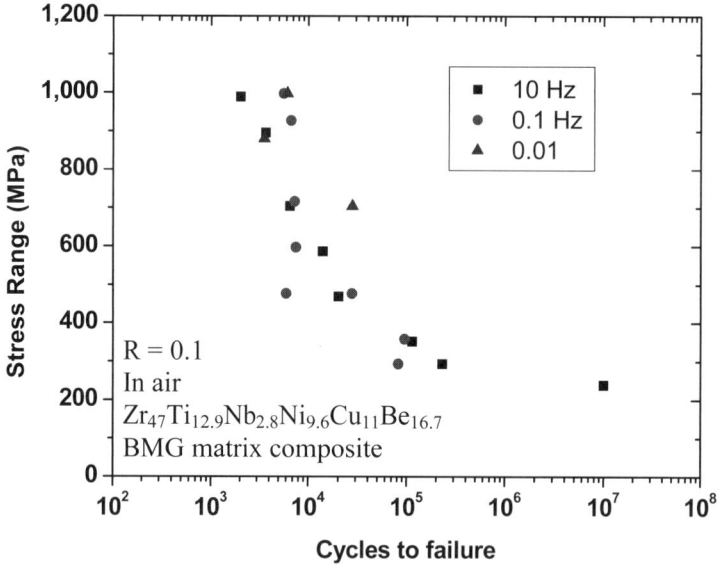

Fig. 7.4. The effect of frequency on the fatigue behavior of LM002 (Wang et al.[20])

7.3.4 Effect of temperature

Material properties are usually dependent on temperature. In metals, the tensile, yield strength, and modulus of elasticity generally decrease with increasing temperatures. The effect is associated with changes in the micro-structure due to phase transformations, grain growth, and dislocation re-structuring (softening).[21] Fatigue properties should also be affected by temperature. The combined action of the cyclic loading and test temperature should be expected to vary with different materials and temperatures. For most metals, an increase in the temperature reduces the fatigue resistance and generally causes an increase in the crack-propagation rate.[21] Research regarding the effect of temperature on the fatigue behavior of BMGs is limited. Hess and Dauskardt[24] studied the fatigue-crack-growth behavior at 100, 140, 180, and 220°C in a Zr–Ti–Cu–Ni–Be bulk-metallic glass. They found that the fatigue-crack-growth rate did not change significantly over this temperature range. More studies are required to understand the effect of the temperature on the fatigue behavior on BMGs.

7.3.5 Effect of mean stress

The mean stress has an important influence on *S–N* curves. For a given stress amplitude in crystalline metals, the tensile mean stress gives shorter fatigue lives than does the zero mean stress, and the compressive mean stress yields longer lives than does the zero mean stress. According to Hess, Menzel, and Dauskardt,[11] the compressive mean stress yields longer lives than does the zero mean stress. However, it is still unclear how the tensile mean stress affects the fatigue lives relative to the zero mean stress.

7.3.6 Effect of microstructure

Vitreloy 1, $Zr_{41.2}Cu_{12.5}Ni_{10}Ti_{13.8}Be_{22.5}$, exhibits a high strength of 1.9 GPa and an elastic-strain limit of 2% under the compressive or tensile loading.[25] This commercial alloy is being used to fabricate golf-club heads, tennis rackets, and cases of personal electronic devices, because of its high strength-to-stiffness ratio.[26] However, Vitreloy 1 fails catastrophically under uncons-trained conditions without significant macroscopic plasticity because of the formation of highly localized shear bands. This deformation behavior has prevented their wide application as engineering materials. To improve the mechanical properties of BMGs, bulk-metallic glass-matrix composites with ductile metals, refractory ceramic particles, and fibers have been developed.[27–30] The propagating shear band interacts with the particles, causing it to slow and deflect, thus delaying the failure and improving the toughness. Some bulk metallic glass-matrix composites have been reported with improved toughness, and tensile and compressive strains to failure.[28–30]

The high-cycle-fatigue behavior of a monolithic amorphous $Zr_{41.2}Ti_{13.8}Cu_{12.5}Ni_{10}Be_{22.5}$ glass (LM001) and its composite (LM002) has been investigated. The fatigue results of these samples in air are shown in Fig. 7.5. The fatigue lives of the monolithic glass are generally longer than those of the composite. The fatigue-endurance limits (σ_e), based on the applied stress ranges, for these samples tested in air were ~567 and 239 MPa, respectively. The fatigue ratios (σ_e divided by the ultimate tensile strengths of 1,850 and 1,500 MPa) were 0.31 and 0.16, respectively. Fatigue results reveal that the fatigue lives and fatigue-endurance limits of the composites are shorter than those of monolithic glasses.[19]

These results are plotted with similar data from Gilbert et al.[13,31] and Flores et al.[14] in Fig. 7.5. Gilbert et al.[13,31] used beam samples in four-point-bending fatigue tests on Vitreloy 1 with the same composition as LM001. Flores et al.[14] also used beam samples in four-point-bending-fatigue experiments on a BMG composite with the same composition as LM002. The fatigue-endurance limit of the LM001 is much greater than that of the Vitreloy 1 alloy studied by Gilbert et al.[13,31] However, the fatigue lifetimes are similar at higher stress level ($\sigma_{max} > 690$ MPa). The possible reason for this difference could be due to specimen geometries and testing procedures.[32] The results of the LM002 composite are consistent with those of Flores et al.[14], though they conducted four-point-bend-fatigue experiments on rectangular beams at 25 Hz, whereas the LM002 tests were performed on round, tapered samples in uniaxial tension at 10 Hz. Thus, it is possible that the test volume has little influence because the BMG composite has improved ductility and fracture toughness. The impact toughness of the ductile BMG composite improved by a factor of 2.5 and the average tensile strains to failure improved by a factor of 2.7, compared to Vitreloy 1.[30] The introduction of the crystalline phases in monolithic-amorphous materials appears to be an effective means of increasing the ductility and toughness. An interesting trend in Fig. 7.5 is that the ductility of BMGs increased, but the fatigue lifetime decreased with the introduction of the crystalline phase.

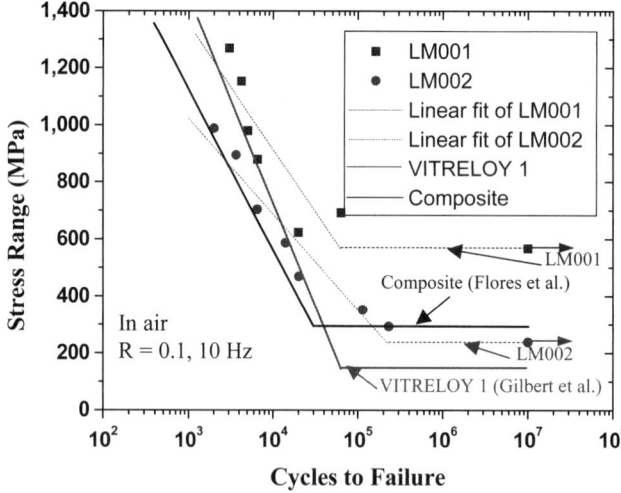

Fig. 7.5. The comparison of stress-range/fatigue-life results. LM001 is a $Zr_{41.2}Cu_{12.5}Ni_{10}Ti_{13.8}Be_{22.5}$ BMG, and LM002 is a $Zr_{47}Ti_{12.9}Nb_{2.8}Ni_{9.6}Cu_{11}Be_{16.7}$ BMG matrix composite (reprinted from Intermetallics, G. Y. Wang et al.[19] with permission from Elsevier)

7.3.7 Effect of composition

The fatigue results for $Zr_{50}Cu_{40}Al_{10}$, $Zr_{50}Cu_{30}Al_{10}Ni_{10}$, $Zr_{41.2}Ti_{13.8}Cu_{12.5}Ni_{10}Be_{22.5}$ (Batches 59 and 94), $Zr_{50}Cu_{37}Al_{10}Pd_3$, and $Zr_{52.5}Cu_{17.9}Al_{10}Ni_{14.6}Ti_5$ samples tested in air with $R = 0.1$ and a frequency of 10 Hz are shown in Fig. 7.6.[8,18,32,33] Batch 59 contains less oxygen than Batch 94. The value of the stress range for these BMGs in the fatigue life range of 10^3–10^4 cycles decreased significantly. $Zr_{50}Cu_{37}Al_{10}Pd_3$ and $Zr_{52.5}Cu_{17.9}Al_{10}Ni_{14.6}Ti_5$ show comparable fatigue lives at higher stress levels ($\geq 1{,}000$ MPa). The value of the stress range for $Zr_{50}Cu_{37}Al_{10}Pd_3$ and $Zr_{52.5}Cu_{17.9}Al_{10}Ni_{14.6}Ti_5$ in the fatigue life range of 10^3–10^4 cycles decreased more slowly than those of other BMGs with the increase of cycles to failure. The fatigue life of $Zr_{50}Cu_{37}Al_{10}Pd_3$ seems to be shorter than the other BMGs at high stress levels ($\geq 1{,}300$ MPa). The fatigue lives of $Zr_{41.2}Ti_{13.8}Cu_{12.5}Ni_{10}Be_{22.5}$ are generally shorter than the other BMGs. The fatigue life of $Zr_{41.2}Ti_{13.8}Cu_{12.5}Ni_{10}Be_{22.5}$ (Batch 94) is shorter than Batch 59. The fatigue lives of $Zr_{50}Cu_{40}Al_{10}$ and $Zr_{50}Cu_{30}Al_{10}Ni_{10}$ are comparable.

The values of σ_e for the $Zr_{41.2}Ti_{13.8}Cu_{12.5}Ni_{10}Be_{22.5}$ (Batch 94), $Zr_{41.2}Ti_{13.8}Cu_{12.5}Ni_{10}Be_{22.5}$ (Batch 59), $Zr_{50}Cu_{40}Al_{10}$, $Zr_{50}Cu_{30}Al_{10}Ni_{10}$, $Zr_{52.5}Cu_{17.9}Al_{10}Ni_{14.6}Ti_5$, and $Zr_{50}Cu_{37}Al_{10}Pd_3$ samples subjected to tension–tension loading were approximately 615, 703, 752, 865, 907, and 983 MPa,

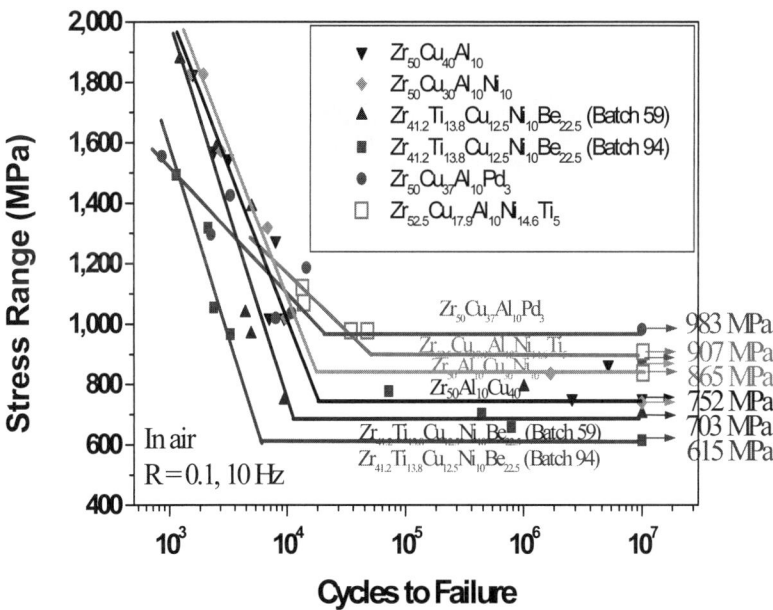

Fig. 7.6. Stress-range/fatigue-life data of $Zr_{50}Cu_{40}Al_{10}$, $Zr_{50}Cu_{30}Al_{10}Ni_{10}$, $Zr_{50}Cu_{37}Al_{10}Pd_3$, $Zr_{41.2}Ti_{13.8}Cu_{12.5}Ni_{10}Be_{22.5}$, and $Zr_{52.5}Al_{10}Ti_5Cu_{17.5}Ni_{14.5}$ specimens[8,18,32,33]

respectively. $Zr_{50}Cu_{37}Al_{10}Pd_3$ exhibited the highest fatigue-endurance limit among these BMGs. $Zr_{52.5}Cu_{17.9}Al_{10}Ni_{14.6}Ti_5$ has also a better fatigue resistance than $Zr_{41.2}Ti_{13.8}Cu_{12.5}Ni_{10}Be_{22.5}$, $Zr_{50}Cu_{40}Al_{10}$, and $Zr_{50}Cu_{30}Al_{10}Ni_{10}$. $Zr_{41.2}Ti_{13.8}Cu_{12.5}Ni_{10}Be_{22.5}$ exhibits a lower fatigue-endurance limit than other BMGs. The high-oxygen-containing Batch 94 has the lowest fatigue resistance.

The fatigue ratios (σ_e divided by the ultimate tensile strength) for $Zr_{41.2}Ti_{13.8}Cu_{12.5}Ni_{10}Be_{22.5}$ (Batch 94), $Zr_{41.2}Ti_{13.8}Cu_{12.5}Ni_{10}Be_{22.5}$ (Batch 59), $Zr_{50}Cu_{40}Al_{10}$, $Zr_{50}Cu_{30}Al_{10}Ni_{10}$, $Zr_{52.5}Cu_{17.9}Al_{10}Ni_{14.6}Ti_5$, and $Zr_{50}Cu_{37}Al_{10}Pd_3$ were 0.33, 0.38, 0.41, 0.46, 0.53, and 0.52, based on the ultimate tensile strengths of 1,850, 1,850, 1,821, 1,900, 1,700, and 1,899 MPa, respectively. The fatigue ratios of these BMGs show the same trend as the fatigue-endurance limits of these BMGs. However, $Zr_{52.5}Al_{10}Ti_5Cu_{17.9}Ni_{14.6}$ has the highest fatigue ratio because of its lowest ultimate tensile strength. Based on these results, it appears that the chemical compositions of Zr-based BMGs can affect the fatigue lives and endurance limits. In fact, the composition effect is because BMGs with different compositions have different free volumes. Yokoyama et al.[34] found that the fatigue limits of Zr–Cu–Al–Pd BMGs exhibited a good linear relationship with the volume change, which is probably corresponding to the excessive free-volume. Thus, by changing the chemical compositions of BMGs, it may be possible to develop BMGs with good fatigue resistance, provided that the influence of the chemical composition on the fatigue behavior is understood.

7.3.8 Effect of surface finish

Fatigue cracks generally start at the free surface of a material. Therefore, the surface conditions are important for the fatigue-crack nucleation and primarily affect the crack-initiation period of the fatigue life. The average surface deviation or commonly referred to as the surface roughness, R_a, can be measured from laser profilometry. In general, coarser grit finishes resulted in higher R_a values, and finer grit finishes yielded lower R_a values. An S–N plot of various surface finishes on BMG-11[35] and Vitreloy 1[13] is shown in Fig. 7.7. Obviously, the surface finish or average surface roughness has a profound impact on the observed fatigue-endurance limit. The fatigue-endurance limit was decreased from ~900 to 400 MPa when the average surface roughness increased from 0.16 μm (1,200 grit) to 0.67 μm (180 grit). However, the lifetime for a given surface finish at higher stress ranges was not affected. In fact, the 400 grit-finish appear to experience longer lifetimes in high-stress ranges in general than the 1,200/P4,000 grit-finish. BMGs seem to be highly sensitive to the surface finish with regard to fatigue-endurance limits.[35]

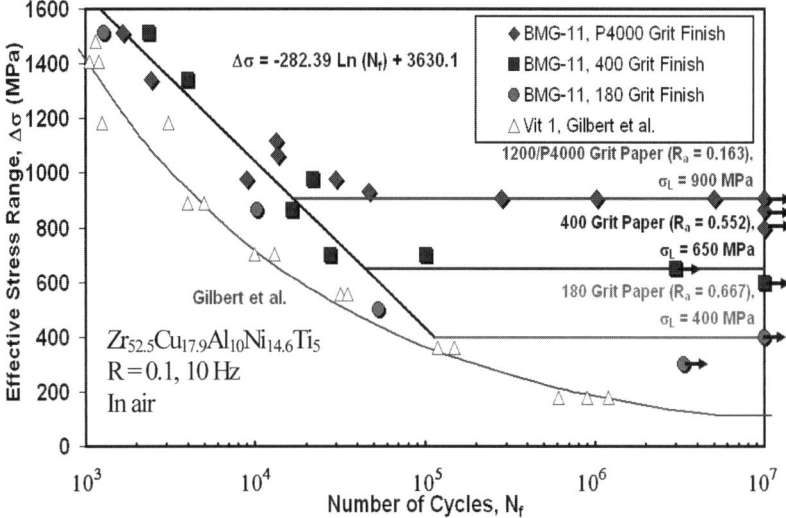

Fig. 7.7. Comparison of the stress-range/fatigue life data of notched BMG-11 specimens with various finishes (Peter[35])

7.3.9 Effect of test method

Fatigue experiments under cyclic tension and cyclic bending are not that different.[21] In both cases, the critical stress of an unnotched specimen is in cyclic tension in the surface layer of the material. The stress gradient perpendicular to the material surface is different for tension and bending, but the more important stress gradients occur along the material surface.[21] However, the fatigue behavior of BMGs seems to be very sensitive to defects. The test volumes under three-point bending, four-point bending, and tension–tension loading differ greatly. Freels et al.[36] studied the fatigue behavior of Cu-based BMGs under three-point and four-point bend loading and found that the fatigue lifetime under three-point bend loading was higher than that under four-point bend loading, as shown in Fig. 7.8. The greater test volume in the four-point bending condition could contain more critical defects, stress raisers, and free volumes, which will enhance the possibilities for the shear-band formation, crack initiation, and, therefore, decrease the fatigue-endurance limit.

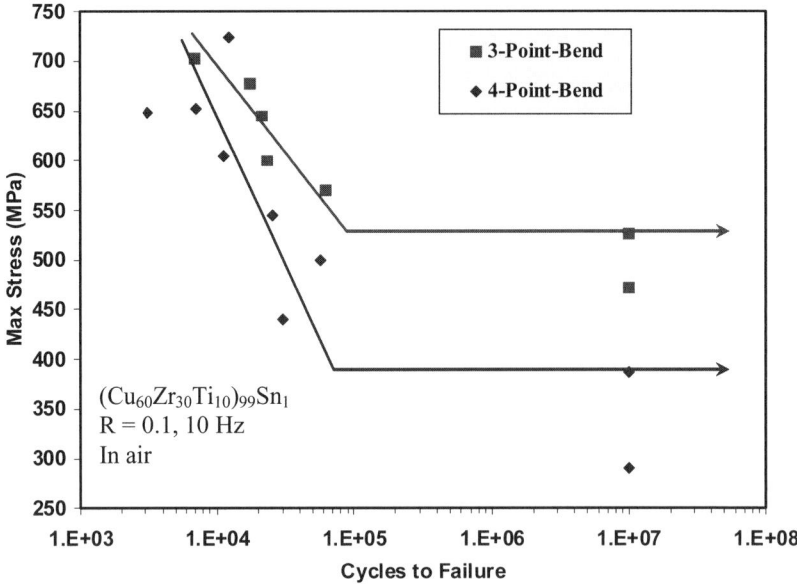

Fig. 7.8. Maximum applied stress/fatigue life data of $(Cu_{60}Zr_{30}Ti_{10})_{99}Sn_1$ BMG for both three-point and four-point bending loading conditions in air with $R = 0.1$ and a frequency of 10 Hz (reprinted from J. Mater. Res., M. Freels et al.[36] with permission from Materials Research Society)

7.3.10 Comparison with other materials

The yield strengths, ultimate tensile strengths, fatigue-endurance limits, and fatigue ratios of some crystalline materials, and Zr-based, Cu-based, and Fe-based BMGs and composites are listed in Tables 7.1 and 7.2, respectively. The fatigue ratio is defined as the ratio of the fatigue-endurance limit, based on the stress range, to the ultimate tensile strength. In Table 7.1, most fatigue-endurance limits of crystalline materials were obtained at $R = 0.1$ and some results were at $R = -1$. For the comparison with the fatigue-endurance limits of the BMGs at $R = 0.1$ in Table 7.2, the fatigue-endurance limits of crystalline materials at $R = -1$ can be converted into those at $R = 0.1$ using the Smith–Watson–Topper equation:[43]

$$\sigma_{ar} = \sigma_{max} \sqrt{\frac{1 - R}{2}}, \qquad (7.2)$$

where σ_{ar} is the stress variable, σ_{max} is the maximum stress, and R is the stress ratio. The following equation can be derived from (7.2)

$$\Delta\sigma_{(R)} = \sigma_{a(-1)}\sqrt{2(1-R)}, \tag{7.3}$$

where $\Delta\sigma_{(R)}$ is the stress range at a stress ratio of R, and $\sigma_{a(-1)}$ is the stress amplitude at $R = -1$.

Table 7.1. Fatigue-endurance limits and fatigue ratios based on the stress range of crystalline alloys

Material	Yield strength (MPa)	Tensile strength (GPa)	R-ratio	Fatigue-endurance limit (MPa)	Fatigue ratio[a]
300 M Steel[31,37,38]	1,670	2.0	0.1	800	0.400
D6AC Steel quenched and tempered at 260°C[37,38]	1,720	2.0	-1	926 (690)[b]	0.463
Ti–6Al–4V[39]	938	1.027	0.1	576	0.561
Hastelloy C2000 (Ni-based)	400	0.786	0.1	382	0.486
2090-T81 Al–Li alloy[39]	483	0.517	0.1	250	0.484
Silicon Brass[40]	205	0.460	-1	228 (170)[b]	0.496
Manganese Bronze[40]	195	0.490	-1	212 (158)[b]	0.433
Zirconium, grade 702[39]	310	0.430	–	145	0.337

[a]Fatigue ratio = fatigue-endurance limit/tensile strength.
[b]The fatigue-endurance limit based on the stress range at $R = 0.1$ is converted from the fatigue-endurance limit in the parenthesis at $R = -1$ using (7.3).

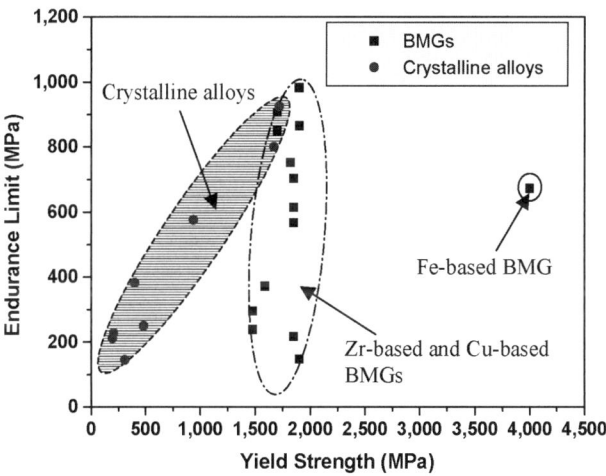

Fig. 7.9. Fatigue-endurance limit vs. yield-strength data of BMGs and crystalline materials.[13,14,16,31,35,37–42]

Table 7.2. Fatigue-endurance limits and fatigue ratios based on the stress range of BMGs

Material	Geometry (mm)	Frequency (Hz)	R-ratio	Testing	Fatigure-endurance limit (MPa)	Tensile strength (GPa)	Fatigue ratio
$Zr_{47}Cu_{11}Ni_{9.6}Ti_{12.9}Nb_{2.8}Be_{16.7}$ composites[14]	$3 \times 3 \times 30$	25	0.1	Four-point bend	~296	1.48	0.2
$Zr_{41.2}Cu_{12.5}Ni_{10}Ti_{13.8}Be_{22.5}$, Vitreloy 1[13]	$3 \times 3 \times 50$	25	0.1	Four-point bend	~152	1.9	0.08
$Zr_{55}Cu_{30}Al_{10}Ni_5$[16]	$1 \times 2 \times 5$	0.13	0.5	Tension	–	1.77	–
$Zr_{59}Cu_{20}Al_{10}Ni_8Ti_3$[41]	$6 \times 3 \times 1.5$	1	0.1	Tension	–	1.58	–
$Zr_{52.5}Cu_{17.9}Al_{10}Ni_{14.6}Ti_5$[41]	$6 \times 3 \times 1.5$	1	0.1	Tension	–	1.66	–
$Zr_{52.5}Cu_{17.9}Al_{10}Ni_{14.6}Ti_5$[35]	$\varnothing 2.98$	10	0.1	Tension (notch)	907	1.7	0.53
$Zr_{52.5}Cu_{17.9}Al_{10}Ni_{14.6}Ti_5$[42]	$3.5 \times 3.5 \times 30$	10	0.1	Four-point bend	850	1.7	0.50
$Zr_{50}Cu_{40}Al_{10}$	$\varnothing 2.98$	10	0.1	Tension (notch)	752	1.82	0.41
$Zr_{50}Cu_{30}Al_{10}Ni_{10}$	$\varnothing 2.98$	10	0.1	Tension (notch)	865	1.9	0.45
$Zr_{50}Cu_{37}Al_{10}Pd_3$	$\varnothing 2.98$	10	0.1	Tension (notch)	983	1.899	0.52
$Zr_{41.2}Cu_{12.5}Ni_{10}Ti_{13.8}Be_{22.5}$, LM001, Batch 59	$\varnothing 2.98$	10	0.1	Tension (notch)	703	1.85	0.38
$Zr_{41.2}Cu_{12.5}Ni_{10}Ti_{13.8}Be_{22.5}$, LM001, Batch 94	$\varnothing 2.98$	10	0.1	Tension (notch)	615	1.85	0.33
$Zr_{41.2}Cu_{12.5}Ni_{10}Ti_{13.8}Be_{22.5}$, LM001	$\varnothing 2.98$	10	0.1	Tension (taper)	567	1.85	0.31
$Zr_{56.2}Cu_{6.9}Ni_{5.6}Ti_{13.8}Nb_{5.0}Be_{12.5}$ LM002, composite	$\varnothing 2.98$	10	0.1	Tension (notch)	239	1.48	0.16
$Fe_{48}Cr_{15}Mo_{14}Er_2C_{15}B_6$	$3 \times 3 \times 25$	10	0.1	Four-point bend	673	~4.0	0.15
$(Cu_{60}Zr_{30}Ti_{10})_{99}Sn_1$	$3 \times 3 \times 25$	10	0.1	Four-point bend	350	–	–
$Cu_{47.5}Zr_{47.5}Al_5$	$3 \times 3 \times 25$	10	0.1	Four-point bend	218	1.85[a]	0.12
$Cu_{47.5}Zr_{38}Hf_{9.5}Al_5$ composite	$3 \times 3 \times 25$	10	0.1	Four-point bend	372	1.59[a]	0.23

[a]Compressive strength.

The fatigue-endurance limit of crystalline materials, Table 7.1, generally increases by increasing the tensile strength. However, such trend is not clear in the BMGs, Table 7.2 and Fig. 7.9. Some Fe-based and Cu-based BMGs show high tensile strengths, but they have low fatigue ratios (e.g., 0.12 and 0.15). The fatigue ratios of crystalline materials are usually between 0.3 and 0.5.

The fatigue behavior of BMGs, crystalline materials, and BMG composites are compared in Fig. 7.10. At higher stress levels, Zr-based BMGs and composites show comparable fatigue lifetimes. Fatigue experiments of Zr-based BMGs exhibit comparable fatigue-endurance limits ranging from about 700 to 900 MPa (Table 7.2 and Fig. 7.10). The comparison of the composite, $Zr_{41.2}Cu_{12.5}Ni_{10}Ti_{13.8}Be_{22.5}$ and Vitreloy 1, composition effects, and testing procedures were discussed in previous sections.

Zirconium-based BMGs exhibit higher fatigue-endurance limits than Vitreloy 1,[13] and Cu-based BMGs for bending-fatigue experiments. Moreover, the fatigue-endurance limits of Zr-based BMGs are comparable with or greater than those of the Fe-based BMG, high-strength alloys, such as 300 M steel and D6AC steel, and a titanium alloy. However, Vitreloy 1 and Cu-based BMGs bending-fatigue experiments reveal low fatigue-endurance limits of approximately 150–350 MPa, which are comparable with those of Hastelloy C2000, the Al–Li alloy, and zirconium.

Copper-based BMGs and composites have higher fatigue-endurance limits than silicon brass and manganese bronze. The $Cu_{47.5}Zr_{38}Hf_{9.5}Al_5$ composite and $(Cu_{60}Zr_{30}Ti_{10})_{99}Sn_1$ have comparable fatigue-endurance limits, which are greater than that of $Cu_{47.5}Zr_{47.5}Al_5$. Hence, the chemical composition can affect the fatigue-endurance limit in Cu-based BMGs and composites, as found in Zr-based BMGs and composites. Therefore, BMGs and composites with good fatigue resistances can be developed, provided that the influence of chemical composition on fatigue behavior is understood.

At higher stress range/tensile strength ratios, Zr-based BMGs and composites exhibit comparable fatigue lives. The fatigue ratios of Zr-based BMGs were found to be comparable with those of high-strength alloys, such as iron, titanium, nickel, aluminum alloys, and zirconium. The Zr-based and Cu-based BMG composites and Fe-based BMGs have comparable fatigue ratios. The results presented in Tables 7.1 and 7.2 reveal that Cu-based BMGs show lower fatigue ratios than the Cu-based BMG composite. This trend is different from that of the Zr-based BMGs and composite, which indicates that Zr-based BMGs have greater fatigue ratios than Zr-based BMG composites.

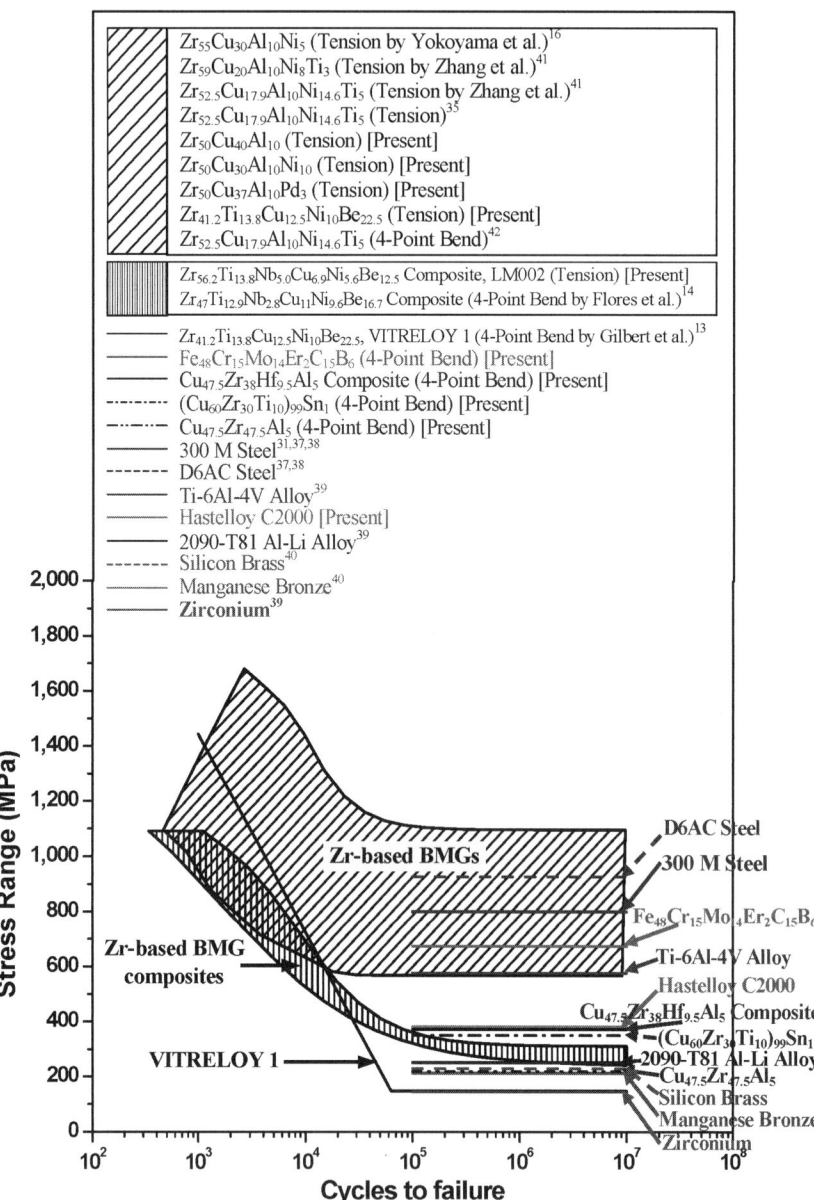

Fig. 7.10. Stress range vs. fatigue-life data of BMGs and crystalline alloys.[13,14,16,31,35,37–42]

7.4 FATIGUE-CRACK GROWTH

7.4.1 Fatigue-crack-growth behavior

Crack growth can be caused by cyclic loading and is called fatigue-crack growth. The rate of fatigue-crack growth is controlled by the stress-intensity factor, K. For a given material and test conditions, the crack-growth behavior can be described by the relationship between the cyclic-crack-growth rate, da/dN, and the stress-intensity-factor range, ΔK. In the intermediate region of crack growth, there is often a straight line on the log–log plot, which shows that da/dN is related to ΔK by the Paris power-law relationship[1]

$$\frac{da}{dN} = C(\Delta K)^m, \tag{7.4}$$

where ΔK is the stress-intensity-factor range defined as $\Delta K = K_{max} - K_{min}$, K_{min} and K_{max} are the applied minimum and maximum stress intensities, respectively, and C and m are experimentally measured scaling constants depending on the material microstructure and environmental conditions. At low growth rates, the curve generally becomes steep and appears to approach a vertical asymptote denoted ΔK_{th}, which is called the fatigue threshold. This quantity is interpreted as a lower limiting value of ΔK below which crack growth does not ordinarily occur or grow at an extremely slow rate of $\sim 10^{-10}$ m per cycle. At high growth rates, the curve may again become steep. This trend is due to the rapid unstable crack growth just prior to the final failure of the test specimen.

Compared to research on the monotonic fracture of BMGs, few studies have been performed on the fatigue-crack-propagation behavior of BMGs. Gilbert et al. performed fatigue-crack-growth-rate tests on a $Zr_{41.2}Ti_{13.8}Cu_{12.5}Ni_{10}Be_{22.5}$ BMG, and compared the results to high-strength polycrystalline materials, such as a 300-M ultrahigh-strength steel, and 2090-T81 aluminum–lithium alloy.[25,31] The fatigue-crack-growth-rate tests were conducted in a controlled room-air environment (22°C, 45% relative humidity) on 7-mm thick, 38-mm wide compact-tension [C(T)] specimens at a frequency of 25 Hz, and load ratios (R), the minimum load divided by the maximum load, of 0.1–0.5.[25,31]

The fatigue-crack-growth rates, da/dN, as a function of ΔK are plotted in Fig. 7.11. Results for a 300-M high-strength steel and a 2090-T81 aluminum alloys are compared with a Zr-based BMG. The value of ΔK_{th} was determined by decreasing the stress-intensity-factor range to a point where crack-growth rates were below 10^{-10} m per cycle. After determining ΔK_{th}, specimens were cycled under increasing ΔK conditions with the same K gradient, up to growth rates of 10^{-7} m per cycle. The fatigue-crack-growth rates of BMGs are similar to those of the high-strength steel and the aluminum alloy. When a regression fit to a simple Paris power-law equation (7.4), it was found that the crack-growth scaling constants were $m = 2.7–4.9$. The exponent, m, of ductile metallic alloys, usually lies between 2 and 4 over this regime of growth rates. In contrast, m is typically 20 or higher in brittle materials such as alumina. The fatigue-crack-growth threshold for the metallic glass, ΔK_{th}, is 1–3 MPa m$^{1/2}$, which is also comparable to many aluminum and steel alloys.[25,31] Hess et al.,[11] Flores et al.,[14] and Zhang et al.,[44] also conducted fatigue-crack-growth-rate experiments on BMGs. However, they found that the values of m in BMGs are typically lower, ~1.4–1.7. Gilbert et al. thought that the scatter in the crack-growth data of BMGs could be attributed to compressive residual stresses in the outer layers.[25,31] The BMGs demonstrated many of the same fatigue-crack-growth characteristics as the high-strength iron and aluminum alloys. Data followed a vertical trend with large increases in crack-growth rates accompanied by negligible increases in the stress-intensity range in the Paris power-law regions. This trend is the same as most crystalline metals. Cracking, meanwhile, was unstable in the crystallized alloy, and a catastrophic failure occurred immediately after loading.[25,31] These types of results illustrate the brittleness.

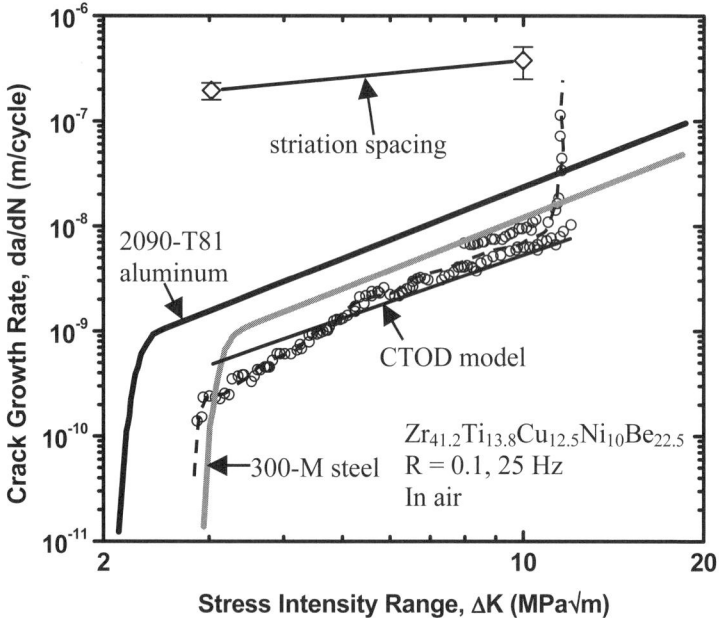

Fig. 7.11. Results in the form of growth rates, da/dN, plotted as a function of ΔK are compared with the behavior in a high-strength steel (300-M) and an age-hardened aluminum alloy (2090-T81) (reprinted from Metall. Mater. Trans. A, C. J. Gilbert et al.[31] with permission)

The fatigue life of BMGs is almost completely determined by the fatigue-crack-growth stage.[11] Therefore, understanding the mechanisms and behavior of fatigue-crack growth of BMGs is important for improving the fatigue lives of BMGs and BMG-matrix composites.[11] The fracture surfaces of BMGs after fatigue-crack-growth-rate experiments exhibited a quite distinct morphology from those in crystalline metals. The fracture-surface roughness increased markedly with increasing crack-growth rate. Fracture surfaces change from a mirror-like surface in the near-threshold fatigue-crack-growth region to a rough morphology exhibiting ridge-like features in the fast-fracture region.[25,31] However, a closer examination of the fatigue-fracture surfaces, especially in the higher growth-rate region, exhibits a clear feature of classic fatigue striations parallel to the crack front, representing the cycle-by-cycle advance of the crack.[25,31]

The mechanism of striation formation in BMGs is still unclear. However, in ductile crystalline metals, the steady-state fatigue-crack-growth behavior has been proposed to be due to the cyclic crack-tip blunting and resharpening. The distance over which blunting causes a deviation in near-tip stress fields is proportional to the crack-tip-opening displacement (CTOD), δ. If fatigue-crack-growth rates are dependent upon the blunting distance, then they may be shown to scale with the range of the crack-tip-opening displacement, $\Delta\delta$[10]:

$$\frac{da}{dN} \propto \Delta\delta \propto \frac{\Delta K^2}{\sigma_0 E'}, \tag{7.5}$$

where σ_0 is the cyclic flow stress, and $E' = E$ for the plane-stress condition and $E/(1 - v^2)$ for the plane-strain condition, E is Young's Modulus, and v is Poisson's ratio. This model for the striation formation was first suggested to apply to BMGs. The range of the crack-tip-opening displacement, $\Delta\delta$, using simple continuum-mechanics arguments is given by[25,31]

$$\Delta\delta = \beta \frac{\Delta K^2}{\sigma_0 E'}, \tag{7.6}$$

where β is a scaling constant (0.01–0.1 for the mode-I crack growth), which is a function of the degree of the slip reversibility and elastic–plastic properties of the material. The CTOD model (7.6) was used to successfully predict the fatigue-crack-growth behavior with a Paris power-law exponent, m, of 2.7 when $\beta = 0.01$ (Fig. 7.11), though the values of m are possibly lower, e.g., 1.4–1.7. One possible explanation for this trend may be related to the structural relaxation or damage of the near-tip material. This model strongly suggests a mechanism for the crack advance involving repetitive blunting and resharpening, i.e., a mechanism similar to that commonly observed in crystalline metals.[25,31]

While alternating blunting and resharpening with each cycle suggests that a striation is formed on a single loading cycle, the striation spacing is often larger than da/dN. This difference between the striation spacing and growth rate reveals that it is necessary for an accumulation of damages prior to the crack advance, similar to the growth-band formation in many polymers.[11] The examination of crack-growth surfaces reveals that striations do not extend over the width of the specimen. In fact, they are broken up in many places along the crack front. Therefore, the whole crack front does not extend consistently with a single-loading cycle, and the nonuniform extension of the crack front forms during the steady-state fatigue crack growth.[11] This trend could be the reason why the striation spacing is larger than da/dN.

7.4.2   Temperature effects

Hess et al. studied the temperature effects on the fatigue-crack-growth behavior of BMGs.[24] They used CT fatigue specimens of $Zr_{41.25}Ti_{13.75}Cu_{12.5}Ni_{10}Be_{22.5}$ (Vitreloy 1) with a width, W, of 38.1 mm and a thickness, B, of 3.1 mm. These measurements of fatigue-crack-growth rates were in accordance with the ASTM-E647 standard at elevated temperatures.[24] The side faces of the specimens were ground to ~100 μm to remove tempering stresses. To facilitate the optical monitoring of the crack length, the specimens were mechanically polished to a mirror finish on one side face. Specimens were fatigue precracked at room temperature prior to testing. Experiments were conducted in a temperature-controlled environmental chamber with a sinusoidal loading waveform of 20 Hz and a constant load ratio, R ($=K_{min}/K_{max}$) of 0.1. The crack length, a, was continuously monitored, using compliance techniques via back-face strain gages, and confirmed with an optical microscope on a high-resolution translation stage. They defined crack-closure levels as the point of the initial contact of mating fracture surfaces during the unloading cycle, which were monitored throughout testing using a back-face strain gage. The resulting effective-stress-intensity range, ΔK_{eff}, actually experienced by the crack tip, was defined by[24]

$$\Delta K_{eff} = K_{max} - K_{cl}, \qquad (7.7)$$

where K_{cl} is the crack-closure stress intensity factor determined from the initial deviation from the linearity of the unloading load vs. back-face strain data.[45–47] The fatigue-crack-growth threshold, ΔK_{th}, was operationally defined as the applied ΔK for the da/dN value approaching 10^{-10} m per cycle.

At temperatures of 100, 140, 180, and 220°C, fatigue crack-growth rates were measured and plotted as a function of the applied ΔK in Fig. 7.12.[24]

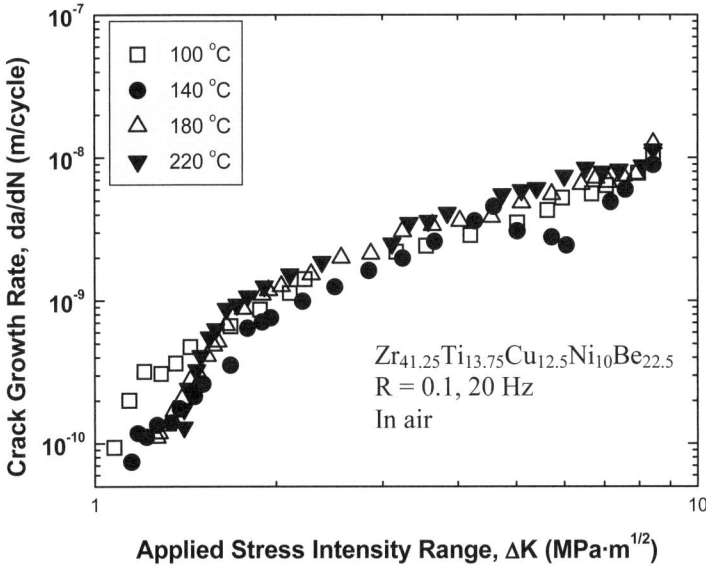

Fig. 7.12. Mode-I fatigue-crack-growth data for Vitreloy 1 tested at various temperatures (reprinted from Acta Mater., P. A. Hess et al.[24] with permission from Elsevier)

A distinct mid-growth-rate regime was apparent together with decreased growth rates in the near-threshold region. The mid-range growth rates were fitted to a Paris power-law relationship (7.4). C and m are material scaling constants reported in Table 7.3. Values for m were 1.4–1.55. A comparison of the fatigue data revealed that mid-range crack-growth rates were not significantly affected by the testing temperature. However, the fatigue-threshold values, ΔK_{th}, were found to increase with the testing temperature as shown in Fig. 7.13 and also summarized in Table 7.3.[24] Values of $\Delta K_{th,eff}$, which represent the fatigue threshold when the stress-intensity range is corrected for effects of crack closure, are also shown in Fig. 7.13. $\Delta K_{th,eff}$ exhibited the same increase with the temperature as the ΔK_{th} values, suggesting that crack-closure effects removed cannot account for the increase in ΔK_{th} with the temperature observed.[24]

Table 7.3. Fatigue results for Vitreloy 1 (Hess et al.[24])

Test temperature (°C)	C (m per cycle) ($\times 10^{-10}$ MPa m$^{1/2}$)	m	ΔK_{th} (MPa m$^{1/2}$)
25	6.6	1.4	1.25
100	7.1	1.4	1.07
140	5.9	1.4	1.14
180	7.4	1.45	1.24
220	6.8	1.55	1.40

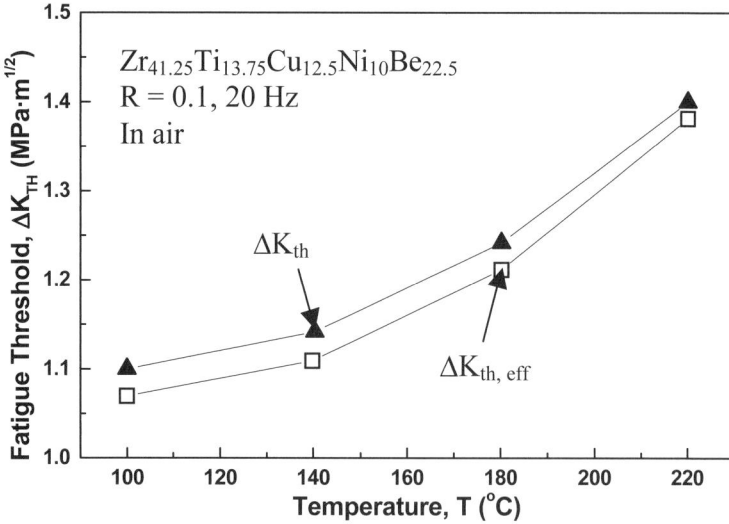

Fig. 7.13. Observed fatigue threshold at various testing temperatures. The *upper curve* is the fatigue threshold with respect to the applied stress-intensity range, while the *lower curve* is the effective fatigue threshold, corrected for closure effects (reprinted from Acta Mater., P. A. Hess et al.[24] with permission from Elsevier)

7.5   CHARACTERIZATION

7.5.1   Fractured surface

Following the development of *S–N* curves, detailed SEM analyses of the fracture surfaces indicate a distinct transition from the stable fatigue-crack propagation to overload fracture. In general, the whole fatigue-fracture surface comprises four main regions: the fatigue-crack initiation, crack propagation, final fast fracture, and apparent melting areas.[17] The fatigue-fracture surfaces of the $Zr_{50}Cu_{40}Al_{10}$ specimens tested in air are shown in Fig. 7.14. The fracture surface is basically perpendicular to the loading direction. The crack-initiation site on the notched surface was found, Fig. 7.14a, and a fatigue crack originates and propagates toward the inside of the specimen. The propagation region is of a thumb-nail shape, and it exhibits a striation-type fracture, Fig. 7.14b. The final fast-fracture region was rough and occupied most of the whole fracture surface, mark and vein pattern can be observed in the melting region at a high magnification in the SEM, Fig. 7.14d. The same fracture morphology was found in $Zr_{50}Cu_{30}Al_{10}Ni_{10}$ and $Zr_{50}Cu_{37}Al_{10}Pd_3$. Furthermore, similar crack-propagation, fast-fracture, and apparent melting morphologies were observed in $Zr_{41.2}Ti_{13.8}Cu_{12.5}Ni_{10}Be_{22.5}$ (LM001). However, the crack was found to initiate

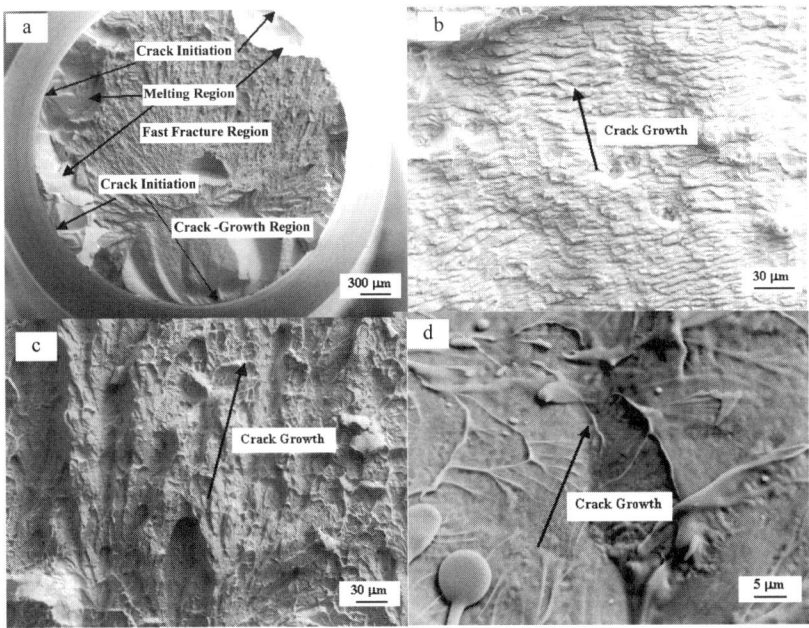

Fig. 7.14. (**a**) Overall fatigue fractography of the $Zr_{50}Cu_{40}Al_{10}$ specimen tested at $\sigma_{max} =$ 1,152 MPa; (**b**) fatigue-crack-growth region; (**c**) final-fast-fracture region; and (**d**) local-melting phenomena (reprinted from Intermetallics, G. Y. Wang et al.[33] with permission from Elsevier)

from casting defects, such as porosities or oxide inclusions, in LM001.[32] There is a distinct boundary between the crack-propagation and fast-fracture regions, which reveals that the fatigue and tensile fracture are probably controlled by different fracture mechanisms. In other words, the striation-type fracture mode was observed in the fatigue-crack-growth region, whereas the dimpled fracture was found in the fast-fracture region, Fig. 7.14b and c. Multiple crack-initiation sites were observed in Fig. 7.14a.

7.5.2 In situ technique

Thermography detection can be conducted, using a state-of-the-art Indigo Phoenix thermographic-infrared (IR) imaging system with a 320×256 pixels focal-plane-array InSb detector that is sensitive to a thermal radiation wave-length of 3–5 μm.[48–53] The temperature sensitivity is 0.015°C at 23°C, and the spatial resolution can be as small as 5.4 μm. The system has a maximum data-acquisition speed of 120 Hz at a full frame of 320×256 pixels and 38,400 Hz at 16×16 pixels. During fatigue testing, a thin submicron graphite coating was applied on the specimen gage-length section to decrease the surface-heat reflection. The IR camera was used at a speed of 300 Hz with 128×128 pixels.

During fatigue testing, an IR-camera system was employed to monitor the fatigue fracture at higher stress levels ($\sigma_{max} \geq 1{,}200$ MPa). Sparking phenomena were found when the $Zr_{50}Cu_{30}Al_{10}Ni_{10}$ specimen was cyclically loaded in air.[17,32] However, the same phenomena were not found with the $Zr_{50}Cu_{40}Al_{10}$, LM001, and $Zr_{50}Cu_{37}Al_{10}Pd_3$ specimens in air. The moment of the fatigue fracture of the $Zr_{50}Cu_{30}Al_{10}Ni_{10}$ specimen is shown in Fig. 7.15a. Sparking phenomena were clearly observed. Though the sparking phenomena were not detected for $Zr_{50}Cu_{40}Al_{10}$, LM001, and $Zr_{50}Cu_{37}Al_{10}Pd_3$, the fracture section was very bright at the moment of the specimen fracture, Fig. 7.15b, which means that the temperature of the fracture section was very high. Considering that the elastic energy of the BMG specimen was released in this final moment, the specimen temperature can instantly increase to more than 900°C,[54] which is sufficient to melt the BMG. The instant melting phenomenon at the fracture moment is also evident in the solidified droplet-like structure that is observed in the SEM on the fracture surface of the specimen, as shown in Fig. 7.14d.

Based on the results from the above experiments, especially those from the analyses of SEM, a mechanistic understanding of fatigue-crack initiation and propagation in BMGs will be described below.

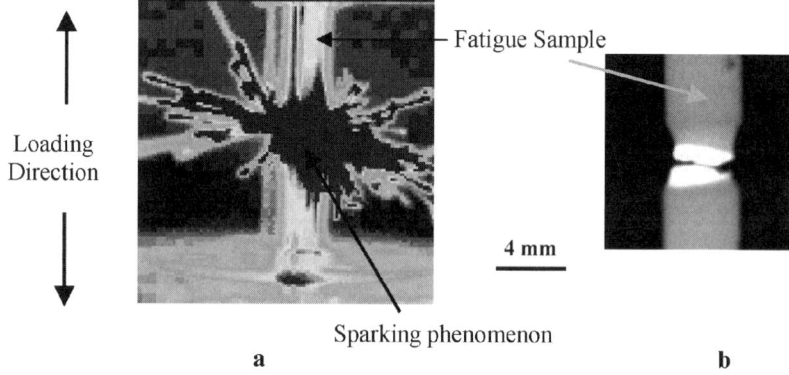

Fig. 7.15. The moment of fracturing (**a**) $Zr_{50}Cu_{30}Al_{10}Ni_{10}$ and (**b**) $Zr_{41.2}Ti_{13.8}Cu_{12.5}Ni_{10}Be_{22.5}$ specimens fatigue tested at room temperature in air (taken by the IR camera system) (reprinted from Intermetallics, G. Y. Wang et al.[17,32] with permission from Elsevier)

7.6 MECHANISM AND MODELING

7.6.1 Crack initiation

Fatigue-damage mechanisms of crystalline materials have been well studied and understood. Many theories, such as the sliding-off mechanism of the crack-extension process proposed by Laird and Smith,[55] can be used. In general, slip bands (SBs), twinning, deformation bands (DBs), and grain boundaries (GBs) are the preferential sites for the nucleation of fatigue cracks in crystalline alloys. However, since BMGs are amorphous, they have no crystal defects, twin boundaries, grain boundaries, or dislocations. Their fatigue crack-initiation and growth mechanisms could be different from the crystalline materials.

The basic fatigue-damage mechanism of BMGs is not known yet. Nevertheless, the deformation mechanism of metallic glasses is usually attributed to the presence of shear bands and plastic flow.[56,57] For the BMGs that have distinct casting defects, the crack could easily initiate at these defects. Some proposed fatigue crack-initiation mechanisms[58] have been suggested for BMGs that have no casting defects. Because there are a large number of free volumes in these BMGs, small shear bands will form at some local sites due to the movement of free volumes under the resolved normal and shear stresses during fatigue testing under tensile loading. Cameron and Dauskardt used a four-component amorphous Lennard-Jones solid with atoms of several different types to simulate the amorphous metals. They found that free-volume levels were increased and localized during the deformation. Cyclic loading in both shear and tension could result in increased free volumes of the system with each deformation cycle. This simulation could explain the rapid initiation of fatigue damage and/or shear-band formation in BMGs during fatigue.[59] Donovan and Stobbs[56] suggest that the edges of the deformation bands show greater atomic spacings, and voids appear to be formed close to the surface of BMGs. The viscosity in the shear bands could decrease due to the increase in the free-volume under cyclic loading. Because of the cyclic tension–tension loading, the gradual weakening, dilation, tearing, and the final opening of the shear band will result in the formation of a fatigue microcrack. Therefore, the nucleation of fatigue cracks can be attributed to the weak nature within shear bands.

Some shear bands were produced on the outer surface of the amorphous alloys, and this deformation feature is quite different from the slip bands in crystalline materials, which have a strict crystallographic plane. During cyclic deformation, some of the shear bands result in the formation of shear-off steps. Moreover, the stress concentration at the shear-off steps could contribute to the formation of fatigue microcracks.[17] Thus, the nucleation of fatigue cracks could be associated with the shear-off steps and shear bands due to the weakness of the shear bands and shear-off steps and the stress concentration at the shear-off steps.[17] Microcracks could form from casting defects,[32] nanostructures, shear bands, and shear-off steps. The shear bands could form due to the movement of free volumes and the formation. The propagation of shear bands is important for the initial stage of the fatigue-damage process. It is important to further study and theoretically model the fatigue mechanisms in BMGs. Moreover, the elimination of casting defects and the means to retard the shear-band initiation and shear-off step formation could be effective in increasing the fatigue resistance of BMGs.

7.6.2 Crack propagation

According to the observation of the fracture surface by SEM, a fatigue-crack-propagation mechanism was proposed to explain the crack-growth behavior. During the course of cyclic deformation of BMGs, at the crack tip several shear bands may be present. However, the cyclic stress might be only favorable for the formation of a fatigue microcrack along one shear-band direction, especially the shear bands near the crack tip because of the stress concentration produced at the crack tip. Under cyclic tension–tension loading, when loading increases gradually, the microcracks form along shear bands and propagate quickly. When loading gradually decreases, the micro-crack propagates slowly. Because elastic and plastic deformation probably form near the new crack tip, the crack propagation becomes difficult. When loading increases gradually again, new microcracks form along a new shear-band direction. The crack propagates along a new direction. A fractured sample was sectioned through the crack-propagation region, and the cross section of the crack-growth region was studied. The SEM observation of the cross section showed that each striation had a shear plane with an angle to the crack-propagation direction, which is ~10–20°.[60] A high-magnification picture from the crack-propagation region suggested that each striation is associated with the growth of the crack during some loading cycles. The final fracture surface shows a clear vein-like structure, which is identical to the features of the tensile fracture surfaces for most BMGs.[54] Along the propagation path of the fatigue crack, however, there is no vein-like structure, as shown in Fig. 7.14b. This trend demonstrates that the melting phenomenon of BMGs did not occur at the tip of the fatigue crack during the crack-propagation stage. This trend means that the released elastic energy due to crack propagation is too low to locally melt the metallic glass. This result is consistent with other observations.[11,13,31] In the melted region, the vein-like structure and droplets appear, as shown in Fig. 7.14d.

7.7 FRACTURE TOUGHNESS

From the theory of fracture mechanics, a stress-intensity factor, K, can be defined to characterize the severity of the crack situation, as affected by the crack size, stress, and geometry. A given material can resist a crack without the brittle fracture occurring as long as this K value is below a critical value,

K_c, which is a property of the material, called the fracture toughness. Values of K_c vary widely for different materials and are affected by the temperature, loading rate, and the thickness of the specimen. The three basic modes of the crack-surface displacement are: (I) an opening mode, (II) a sliding mode, and (III) a tearing mode. Most cracking problems of engineering interest involve primarily mode I and are due to tension stresses. K can be related to the applied stress and the crack length by an equation of the form[1]

$$K = F\sigma\sqrt{\pi a}, \tag{7.8}$$

where F is a dimensionless function that depends on the geometry and the loading configuration, and usually also on the ratio of the crack length to another geometric dimension, such as the member width or half-width. In general, fracture-toughness experiments are performed in accordance with ASTM standard E399.[61]

Gilbert et al. performed fracture-toughness experiments in a controlled room-air environment (22°C, 45% relative humidity) on 7-mm thick and 4-mm thick, 38-mm wide C(T) specimens of Vitreloy 1, machined from bulk plates.[25,31] Crack initiation was facilitated using a half-chevron-shaped starter notch prior to fracture-toughness testing. The samples were fatigue precracked for several millimeters beyond this notch. Thereafter, crack lengths were continuously monitored using unloading elastic-compliance measurements with a strain gauge attached to the back face of the specimen, and also checked periodically with a traveling microscope. Optical and compliance measurements of crack lengths were always found to be within 2% in these experiments. Fracture toughness, K_{Ic}, values were determined by monotonically loading the fatigue precracked specimens to failure. Procedures were in general accordance with ASTM standard E399.[61] As a reference, partially and fully crystalline samples with the same composition were also used. However, fatigue cracking was unstable in the partially and fully crystalline structures due to their extreme brittleness. Therefore, toughness values were obtained with Vickers indentation methods, with measurements averaged from at least five indents under an indentation load of 49 N. Results of the fracture-toughness testing showed that the fracture toughness of the amorphous alloy was remarkable, 30–68 MPa m$^{1/2}$. Nevertheless, the thermal exposure resulting in a partial or full crystallization led to approximately a 50-fold reduction in K_{Ic} values to 1.21 and 1.04 MPa m$^{1/2}$, respectively, Fig. 7.16.[25,31] The fracture toughness of BMGs is comparable to that of a typical polycrystalline aluminum or high-strength steel alloy. However, the fracture toughness after the partial or full crystallization is comparable to that of silica glass or very brittle ceramics. In fact, both the 7-mm and the 4-mm thick CT samples exceeded the plane-strain thickness requirement in ASTM E399 ($B > 2.5(K_{Ic}/\sigma_Y)^2 \sim 2$ mm, where B is the thickness and σ_Y is the yield stress of a material).[61] These fracture-toughness values do not strictly conform to ASTM E399. Therefore, the values are referred to as K_Q rather than K_{Ic}. A significant variability was observed in the fracture-toughness data. Sources of this variability may be associated with residual stresses at the surface of castings, compositional variation (particularly oxygen), crack branching and ligament bridging, and sensitivity to the loading rate.[25,31] The highest measured value was 68 MPa m$^{1/2}$ (measured with a 7-mm-thick sample), and the lowest was 30 MPa m$^{1/2}$ (measured with a 4-mm-thick sample). The thermal exposure, resulting in a partial or full crystallization, led to a dramatic reduction in fracture toughness to ~1 MPa m$^{1/2}$. Although fracture toughness of the amorphous microstructure is comparable to that of a typical polycrystalline aluminum or steel alloy, the fracture toughness of a partially or fully crystallized alloy is comparable to that of a silica glass.[25,31]

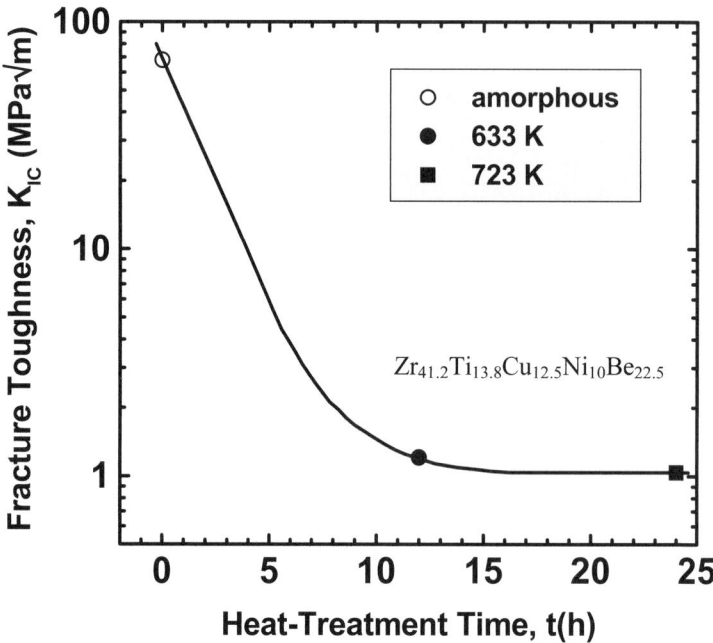

Fig. 7.16. Fracture toughness plotted as a function of annealing time for specimens heat treated at both 633 and 723 K in vacuum (reprinted from Metall. Mater. Trans. A, C. J. Gilbert et al.[31] with permission)

In general, the micromechanisms controlling the fracture toughness of amorphous alloys are poorly understood. The fundamental differences in both the atomic structure and observed deformation behavior (e.g., the extreme slip instability in tension, near-theoretical strength, and distinctive overload fracture-surface morphologies) make it clear that such mechanisms are quite distinct from the tensile fracture in crystalline metals, ceramics, or oxide glasses. However, the vein morphology observed on fracture surfaces after the fracture-toughness testing has been suggested to be a variant of the Taylor instability.[31] This instability is associated with the tendency of a fluid meniscus (under a positive pressure gradient) propagating in the direction of its convex curvature to break up into a series of fingers, which penetrate into the fluid meniscus. When this process dominates, the critical fracture event is associated with the onset of this instability, governed by the surface tension of the fluid and the applied pressure gradient. The notion that the material near the crack tip is softened, possibly by the adiabatic heating or a strain-softening phenomenon, is supported by the fracture-surface appearance.[25,31] A model for the fracture toughness of BMGs, based on the resistance of a blunt crack to this instability, gives an expression for K_{Ic} in terms of the surface tension, Γ, and Young's modulus, E:[31]

$$K_{\mathrm{Ic}} = 24\pi^3 \sqrt{3} \frac{\beta \Gamma E}{\alpha}, \qquad (7.9)$$

where β is a scaling constant dependent on the work-hardening behavior and $\alpha \sim 2.7$. Using values for $E = 95$ GPa and $\Gamma = 1$ J m^{-2}, a K_{Ic} value of ~ 13 MPa m$^{1/2}$ was obtained from (7.9). It is speculated that higher measured values of K_{Ic} than predicted values may be associated with a strain-rate effect, residual stresses, and/or with extensive crack branching and ligament bridging. Such mechanisms have been successfully applied in promoting toughness. Indeed, fracture toughnesses of 18 MPa m$^{1/2}$ have been measured when no crack branching is observed.[31]

Lewandowski, Wang, and Greer[62] suggested that a distinct scatter in the fracture toughness of BMGs, when nominally identical materials are tested using standardized test techniques, is because the metallic glasses under comparison have a wide range of Young's moduli, E. Thus, it is better not to quantify their mechanical behavior in terms of the fracture toughness but rather in terms of the energy of fracture, G, which is the energy required to create two new fracture surfaces. For ideally brittle materials, G is 2γ, where γ is the surface energy per unit area. Under the plane strain, $G = K^2/E(1 - v^2)$.

Lewandowski, Wang, and Greer.[62] also found that there is a distinct relationship between the fracture energy (G) and the elastic modulus ratio, shear modulus/bulk modulus [$\mu/B = 3(1 - 2v)/2(1 + v)$]. The fracture energy decreases with increasing μ/B. When μ/B is greater than 0.41–0.43, metallic glasses are generally extremely brittle. Thus, higher values of v give higher fracture energies, the transition between brittle and tough regimes where there is a large increase in G beyond the oxide glass values being for $v_{\mathrm{crit}} = 0.31$–0.32, Fig. 7.17. Schroers and Johnson[63] also found that the small

μ/B causes the tip of a shear band to extend rather than initiate a crack because the low shear modulus allows for the shear collapse before the extensional instability of the crack formation can occur. Thus, a large Poisson's ratio might be the indicator of the ductile characteristic of a BMG. They have demonstrated a link between the plasticity and high values for v for a particular platinum-rich glass.

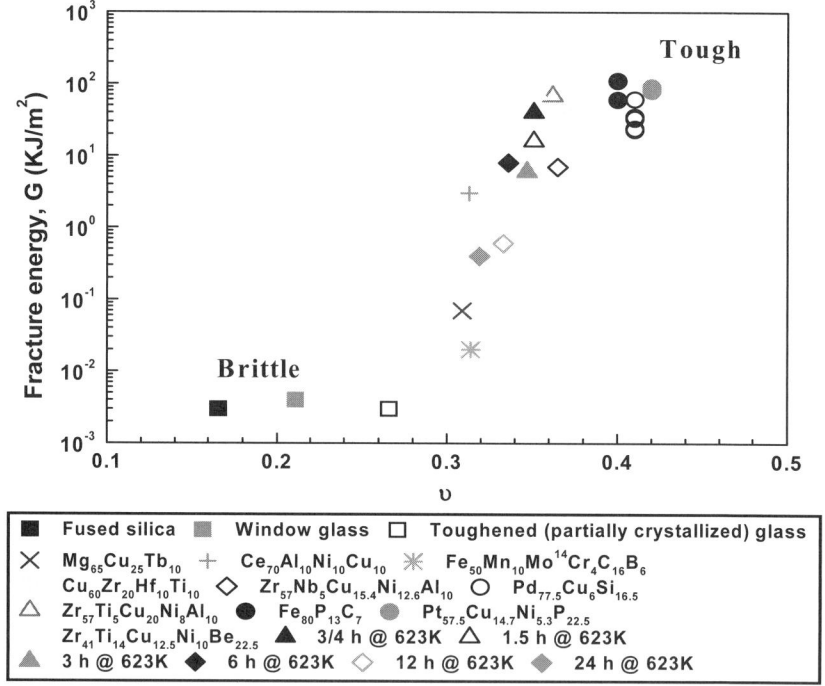

Fig. 7.17. The correlation between the fracture energy G and Poisson's ratios v for all the collected data on metallic glasses (as-cast and annealed) as well as for oxide glasses. The divide between the tough and brittle regimes is in the range $v_{crit} = 0.31–0.32$ (reprinted from Phil. Mag. Lett., J. J. Lewandowski et al.[62] with permission from Taylor & Francis)

7.8 UNRESOLVED ISSUES

The high strength, high hardness, and other unique properties of BMGs are a strong driving force for structural applications. Their fatigue behavior is important for engineering applications. However, little attention has been paid to the fatigue study, especially the fatigue behavior in controlled environments. The fatigue behavior of BMGs is poorly understood. As crystal characteristics, such as dislocations and grain boundaries, are not present in

BMGs, what is the real nature of the deformation mechanisms in BMGs? The characteristic for the formation of shear bands is still unknown during the cyclic deformation of metallic glasses. When will shear bands form? Another interesting question is how the fatigue crack initiates and propagates in metallic glasses because there are no grain boundaries in amorphous materials? Therefore, additional fatigue, fatigue-crack-growth-rate, and fracture-toughness studies are necessary to answer the above questions. The effects of many different factors, including the environment, frequency, temperature, mean stress, microstructure, composition, surface finish, and test method on the fatigue behavior of BMGs must be understood. Understanding the fatigue and fracture mechanisms of BMGs is of critical importance for the alloy design and practical application. In addition, theoretical models to predict fatigue lives of BMGs need to be developed so that engineers can use BMGs to design new products based on these theories.

7.9 CONCLUSIONS

1. The fatigue experiments have been successfully performed on BMGs and composites, showing that many factors including the mean stress, material quality, geometry, environment, temperature, cyclic frequency, and residual stress will affect the fatigue behavior.
2. The fatigue-crack-propagation behavior of BMGs is probably controlled by the stress-intensity-factor range. BMGs showed many of the same fatigue-crack characteristics as the high-strength steel and aluminum. Steady-state-fatigue-crack growth in BMGs has been proposed to be due to the cyclic crack-tip blunting and resharpening behavior. Models for the striation formation, from a continuum-mechanics standpoint, indicate that growth rates should scale with the range of the crack-tip-opening displacement. The temperature has no large effect on the fatigue-crack-propagation behavior of BMGs.
3. No indications, such as the vein-like structures or droplets, exist in the fatigue-crack-propagation region, which demonstrates that the released elastic energy due to the crack growth is too low to melt the metallic glass locally.
4. The vein patterns and droplets with a melted appearance were observed in the apparent melting region, which could be consistent with the phenomenon observed by the IR image. The highest temperature occurred at the moment of fracture.
5. The fatigue-endurance limits of Zr-based BMGs are generally comparable with those of high-strength structural alloys.

6. The fracture toughness of the amorphous alloy was comparable to that of a typical polycrystalline aluminum or high-strength steel alloy. However, the thermal exposure resulting in a partial or full crystallization led to approximately a 50-fold reduction in K_{Ic} values, which are comparable to those of silica glasses or very brittle ceramics.

ACKNOWLEDGMENTS

We acknowledge the financial support of the National Science Foundation: (1) the Division of the Design, Manufacture, and Industrial Innovation Program, under grant no. DMI-9724476, (2) the Combined Research-Curriculum Development (CRCD) Programs, under EEC-9527527 and EEC-0203415, (3) the Integrative Graduate Education and Research Training (IGERT) Program, under DGE-9987548, (4) the International Materials Institutes (IMI) Program, under DMR-0231320, and (5) the Major Research Instrumentation (MRI) Program, under DMR-0421219, to the University of Tennessee, Knoxville, with Dr. D. Durham, Ms. M. Poats, Dr. C. J. Van Hartesveldt, Dr. D. Dutta, Dr. W. Jennings, Dr. L. Goldberg, Dr. C. Huber, and Dr. C. R. Bouldin as Program Directors, respectively.

REFERENCES

1. N. E. Dowling, *Mechanical Behavior of Materials* (Prentice-Hall, New Jersey, 1999).
2. J. Y. Mann, *Fatigue of Materials* (Melbourne University Press, Australia, 1967).
3. V. V. Bolotin, *Mechanics of Fatigue* (CRC, New York, 1999).
4. *E466, Standard Practice for Conducting Constant Amplitude Axial Fatigue Tests of Metallic Materials, Annual Book of ASTM Standards, Metals Test Methods and Analytical Procedures* (Vol. 03.01, ASTM, Philadelphia, PA 1995).
5. *E467, Standard Practice for Verification of Constant Amplitude Dynamic Loads on Displacements in an Axial Load Fatigue Testing System, Annual Book of ASTM Standards*, Metals Test Methods and Analytical Procedures (Vol. 03.01, ASTM, Philadelphia, PA 1995).
6. *E468, Standard Practice for Presentation of Constant Amplitude Fatigue Test Results for Metallic Materials, Annual Book of ASTM Standards, Metals Test Methods and Analytical Procedures* (Vol. 03.01, ASTM, Philadelphia, PA 1995).
7. H. Tian, D. Fielden, M. J. Kirkham, and P. K. Liaw, Control of noise and specimen temperature during 1 kHz fatigue experiments, *J. Test. Eval.* **34**(2), 92–97 (2006).
8. W. H. Peter, P. K. Liaw, R. A. Buchanan, C. T. Liu, C. R. Brooks, J. A. Horton, Jr., C. A. Carmichael, Jr., and J. L. Wright, Fatigue behavior of $Zr_{52.5}Al_{10}Ti_5Cu_{17.9}$ $Ni_{14.6}$ bulk metallic glass, *Intermetallics* **10**, 1125–1129 (2002).
9. W. H. Peter, R. A. Buchanan, C. T. Liu, and P. K. Liaw, The fatigue behavior of a zirconium-based bulk metallic glass in vacuum and air, *J. Non-Cryst. Solids* **317**, 187–192 (2003).

10. B. C. Menzel and R H. Dauskardt, Stress–life fatigue behavior of a Zr-based bulk metallic glass, *Acta Mater.* **54**, 935–943 (2006).

11. P. A. Hess, B. C. Menzel, and R. H. Dauskardt, Fatigue damage in bulk metallic glass. II. Experiments, *Scripta Mater.* **54**, 355-361 (2006).

12. G. Y. Wang, P. K. Liaw, and M. Denda, unpublished results.

13. C. J. Gilbert, J. M. Lippmann, and R. O. Ritchie, Fatigue of a Zr–Ti–Cu–Ni–Be bulk amorphous metal: Stress/life and crack-growth behavior, *Scripta Mater.* **38**, 537–542 (1998).

14. K. M. Flores, W. L. Johnson, and R. H. Dauskardt, Fracture and fatigue behavior of a Zr–Ti–Nb ductile phase reinforced bulk metallic glass matrix composite, *Scripta Mater.* **49**, 1181–1187 (2003).

15. D. C. Qiao, P. K. Liaw, C. Fan, Y. H. Lin, G. Y. Wang, H. Choo, and R. A. Buchanan, Fatigue and fracture behavior of $(Zr_{58}Ni_{13.6}Cu_{18}Al_{10.4})_{99}Nb_1$ bulk-amorphous alloy, *Intermetallics* **14**, 1043–1050 (2006).

16. Y. Yokoyama, N. Nishiyama, K. Fukaura, and H. Sunada, Fatigue properties and microstructures of $Zr_{55}Cu_{30}Al_{10}Ni_5$ bulk glassy alloys, *Mater. Trans. JIM* **41**, 675–680 (2000).

17. G. Y. Wang, P. K. Liaw, W. H. Peter, B. Yang, Y. Yokoyama, M. L. Benson, B. A. Green, M. J. Kirkham, S. A. White, T. A. Saleh, R. L. McDaniels, R. V. Steward, R. A. Buchanan, C. T. Liu, and C. R. Brooks, Fatigue behavior of bulk-metallic glasses, *Intermetallics* **12**, 885–892 (2004).

18. G. Y. Wang, P. K. Liaw, Y. Yokoyama, W. H. Peter, B. Yang, M. Freels, R. A. Buchanan, C. T. Liu, and C. R. Brooks, Influence of air and vacuum environment on fatigue behavior of Zr-based bulk metallic glasses, *J. Alloys Compd*, **434–435**, 68–70 (2007).

19. G. Y. Wang, P. K. Liaw, A. Peker, M. Freels, W. H. Peter, R. A. Buchanan, and C. R. Brooks, Comparison of fatigue behavior of a bulk metallic glass and its composite, *Intermetallics* **14**, 1091–1097 (2006).

20. G. Y. Wang, P. K. Liaw, A. Peker, Y. Yokoyama, M. Freels, W. Peter, R. Buchanan, and C. Brooks, The effect of frequency on fatigue behavior of bulk metallic glass and composites, presented at The TMS Annual Meeting, San Antonio, TX, (2006), unpublished.

21. J. Schijve, *Fatigue of Structures and Materials* (Kluwer, Boston, 2001).

22. H. Tian, P. K. Liaw, H. Wang, D. Fielden, J. P. Strizak, L. K. Mansur, and J. R. DiStefano, Influence of mercury environment on the fatigue behavior of spallation neutron source (SNS) target container materials, *Mater. Sci. Eng. A* **314**(1–2), 140–149 (2001).

23. H. Tian, P. K. Liaw, D. E. Fielden, L. Jiang, B. Yang, C. R. Brooks, M. D. Brotherton, H. Wang, J. P. Strizak, and L. K. Mansur, Effects of frequency on fatigue behavior of type 316 low-carbon, nitrogen-added stainless steel in air and mercury for the spallation neutron source, *Metall. Mater. Trans. A* **37**(1), 163–173 (2006).

24. P. A. Hess and R. H. Dauskardt, Mechanisms of elevated temperature fatigue crack growth in Zr–Ti–Cu–Ni–Be bulk metallic glass, *Acta Mater.* **52**, 3525–3533 (2004).

25. C. J. Gilbert, R. O. Ritchie, and W. L. Johnson, Fracture toughness and fatigue-crack propagation in a Zr–Ti–Ni–Cu–Be bulk metallic glass, *Appl. Phys. Lett.* **71**, 476–478 (1997).

26. http://www.liquidmetal.com/

27. H. Choi-Yim, R. Busch, U. Kosster, and W. L. Johnson. Synthesis and characterization of particulate reinforced $Zr_{57}Nb_5Al_{10}Cu_{15.4}Ni_{12.6}$ bulk metallic glass composites, *Acta Mater.* **47**, 2455–2462 (1999).

28. R. D. Conner, H. Choi-Yim, and W. L. Johnson, Mechanical properties of $Zr_{57}Nb_5Al_{10}Cu_{15.4}Ni_{12.6}$ metallic glass matrix particulate composites, *J. Mater. Res.* **14**, 3292–3297 (1999).

29. C. C. Hays, C. P. Kim, and W. L. Johnson, Microstructure controlled shear band pattern formation and enhanced plasticity of bulk metallic glasses containing in situ formed ductile phase dendrite dispersions, *Phys. Rev. Lett.* **84**, 2901–2904 (2000).

30. F. Szuecs, C. P. Kim, and W. L. Johnson, Mechanical properties of $Zr_{56.2}Ti_{13.8}Nb_{5.0}Cu_{6.9}Ni_{5.6}Be_{12.5}$ ductile phase reinforced bulk metallic glass composite, *Acta Mater.* **49**, 1507–1513 (2001).

31. C. J. Gilbert, V. Schroeder, and R. O. Ritchie, Mechanisms for fracture and fatigue-crack propagation in a bulk metallic glass, *Metall. Mater. Trans. A* **30**, 1739–1753 (1999).

32. G. Y. Wang, P. K. Liaw, A. Peker, B. Yang, M. L. Benson, W. Yuan, W. H. Peter, L. Huang, M. Freels, R. A. Buchanan, C. T. Liu, and C. R. Brooks, Fatigue behavior of Zr–Ti–Ni–Cu–Be bulk-metallic glasses, *Intermetallics* **13**, 429–435 (2005).

33. G. Y. Wang, P. K. Liaw, W. H. Peter, B. Yang, M. Freels, Y. Yokoyama, M. L. Benson, B. A. Green, T. A. Saleh, R. L. McDaniels, R. V. Steward, R. A. Buchanan, C. T. Liu, and C. R. Brooks, Fatigue behavior and fracture morphology of $Zr_{50}Al_{10}Cu_{40}$ and $Zr_{50}Al_{10}Cu_{30}Ni_{10}$ bulk-metallic glasses, *Intermetallics*, **12**, 1219–1227 (2004).

34. Y. Yokoyama, P. K. Liaw, M. Nishijima, K. Hiraga, R. A. Buchanan, and A. Inoue, Fatigue-strength enhancement of cast $Zr_{50}Cu_{40}Al_{10}$ glassy alloys, *Mater. Trans. JIM* **47**, 1286–1293 (2006).

35. W. H. Peter, *Fatigue Behavior of A Zirconium-Based Bulk Metallic Glass*, Dissertation (2005).

36. M. Freels, P. K. Liaw, G. Y. Wang, Q. S. Zhang, and Z. Q. Hu, Stress–life fatigue behavior and fracture-surface morphology of a Cu-based bulk-metallic glass, *J. Mater. Res.* **22**, 374–381 (2007).

37. R. Hertzberg, *Deformation and Fracture Mechanics of Engineering Materials*, 3rd ed. (Wiley, New York, 1989).

38. *Structural Alloys Handbook* (Mechanical Properties Data Center, Traverse City, MI, 1977).

39. *ASM Handbook, Properties and Selections: Nonferrous Alloys and Special Purpose Materials* (Vol. 2, ASM, Metals Park, OH, 1990).

40. *Metals Handbook*, 9th ed. (Vol. 2, American Society for Metals, Metals Park, OH, 1979).

41. Z. F. Zhang, J. Eckert, and L. Schultz, Fatigue and fracture behavior of bulk metallic glass, *Metall. Mater. Trans. A* **35**, 3489–3498 (2004).

42. M. L. Morrison, R. A. Buchanan, P. K. Liaw, B. A. Green, G. Y. Wang, C. T. Liu, and J. A. Horton, "Four-point bending fatigue behavior of the Zr-based Vitreloy 105 bulk metallic glass." *Mater. Sci. Eng. A* **467**(1-2), 190–197 (2007).

43. K. N. Smith, P. Watson, and T. H. Topper, A stress–strain function for the fatigue of metals, *J. Mater.* **5**, 767–778 (1970).

44. H. Zhang, Z. G. Wang, K. Q. Qiu, Q. S. Zang, and H. F. Zhang, Cyclic deformation and fatigue crack propagation of a Zr-based bulk amorphous metal, *Mater. Sci. Eng. A* **356**, 173–180 (2003).

45. P. K. Liaw T. R. Leax, and W. A. Logsdon, Near-threshold fatigue crack-growth behavior in metals, *Acta Metall.* **31**, 1581–1587 (1983).

46. P. K. Liaw, A. Saxena, V. P. Swaminathan, and T. T. Shih, Effects of load ratio and temperature on the near-threshold fatigue crack-propagation behavior in a CrMoV steel, *Metall. Trans. A* **14**(8), 1631–1640 (1983).

47. L. J. Chen, P. K. Liaw, R. L. McDaniels, and D. L. Klarstrom, The low-cycle fatigue and fatigue-crack-growth behavior of HAYNES (R) HR-120 alloy, *Metall. Mater. Trans. A* **34**(7), 1451–1460 (2003).

48. L. Jiang, H. Wang, P. K. Liaw, C. R. Brooks, and D. L. Klarstrom, Characterization of the temperature evolution during high-cycle fatigue of the ULTIMET superalloy: Experiment and theoretical modeling, *Metall. Mater. Trans. A* **32**(9), 2279–2296 (2001).

49. H. Wang, L. Jiang, P. K. Liaw, C. R. Brooks, and D. L. Klarstrom, Infrared temperature mapping of ULTIMET alloy during high-cycle fatigue tests, *Metall. Mater. Trans. A* **31**, 1307–1310 (2000).

50. P. K. Liaw, H. Wang, L. Jiang, B. Yang, J. Y. Huang, R. C. Kuo, and J. G. Huang, Thermographic detection of fatigue damage of pressure vessel steels at 1,000 Hz and 20 Hz, *Scripta Mater.* **42**, 389–395 (2000).

51. B. Yang, P. K. Liaw, G. Wang, W. H. Peter, R. A. Buchanan, Y. Yokoyama, J. Y. Huang, R. C. Kuo, J. G. Huang, D. E. Fielden, and D. L. Klarstrom, Thermal-imaging technologies for detecting damage during high-cycle fatigue, *Metall. Mater. Trans. A* **35**(1), 15–23 (2004).

52. L. Jiang, H. Wang, P. K. Liaw, C. R. Brooks, L. Chen, and D. L. Klarstrom, Temperature evolution and life prediction in fatigue of superalloys, *Metall. Mater. Trans. A* **35**(3), 839–848 (2004).

53. B. Yang, P. K. Liaw, G. Wang, M. Morrison, C. T. Liu, R. A. Buchanan, and Y. Yokoyama, In-situ thermographic observation of mechanical damage in bulk-metallic glasses during fatigue and tensile experiments, *Intermetallics* **12**(10–11): 1265–1274 (2004).

54. C. T. Liu, L. Heatherly, D. S. Easton, C. A. Carmichael, J. H. Schneibel, C. H. Chen, J. L. Wright, M. H. Yoo, J. A. Horton, and A. Inoue. Test environment and mechanical properties of Zr-base bulk amorphous alloys, *Metall. Mater. Trans. A* **29**, 1811–1820 (1998).

55. C. Laird, *Fatigue Crack Propagation* (ASTM, Philadelphia, PA. 1967).

56. P. E. Donovan and W. M. Stobbs, The structure of shear bands in metallic glasses, *Acta Metall.* 29, 1419–1436 (1981).

57. P. S. Steif, F. Spaepen, and J. W. Hutchinson, Strain localization in amorphous metals, *Acta Metall.* **30**, 447–455 (1982).

58. Z. F. Zhang, J. Eckert, and L. Schultz, Tensile and fatigue fracture mechanisms of a Zr-based bulk metallic glass, *J. Mater. Res.* **18**(2), 456–465 (2003).

59. K. K. Cameron and R. H. Dauskardt, Fatigue damage in bulk metallic glass. I. Simulation, *Scripta Mater.* **54**, 349–353 (2006).

60. G. Y. Wang, P. K. Liaw, A. Peker, Y. Yokoyama, W. H. Peter, B. Yang, M. L. Benson, W. Yuan, L. Huang, M. Freels, R. A. Buchanan, C. T. Liu, and C. R. Brooks, Fatigue behavior of Zr-based bulk metallic glasses (BMG), The TMS Annual Meeting, Charlotte, North Carolina (2004).

61. *E399, Standard Test Method for Plane-Strain Fracture Toughness of Metallic Materials, Annual Book of ASTM Standards, Metals Test Methods and Analytical Procedures* (Vol. 03.01, ASTM, Philadelphia, PA 1995).

62. J. J. Lewandowski, W. H. Wang, and A. L. Greer, Intrinsic plasticity or brittleness of metallic glasses, *Philos. Mag. Lett.* **85**, 77–87 (2005).

63. J. Schroers, W. L. Johnson, Ductile bulk metallic glass, *Phys. Rev. Lett.* **93**, 255506 (2004).

Chapter 8

CORROSION BEHAVIOR

Brandice A. Green, Peter K. Liaw, and Raymond A. Buchanan[†]

Department of Materials Science and Engineering, University of Tennessee, Knoxville, TN 37996, USA

8.1 INTRODUCTION

There has been a growing interest in the corrosion behavior of bulk metallic glasses (BMGs). The enhanced glass formability of these amorphous alloys has made the fabrication of bulk size components easier and thereby improving their prospects for engineering applications.[1] The corrosion behavior of BMGs is particularly pertinent when considering biomedical applications and is also relevant for other applications, e.g., watches and electronic casings. The corrosion resistance is also a critical factor for the consideration of their use in hostile or chemical environments.

The corrosion properties of an amorphous alloy are expected to be superior to those of its crystalline counterpart due to its chemical homogeneity and lack of microstructure. Amorphous alloys lack grain boundaries, dislocations, and other defects that are commonly the culprits behind the localized corrosion observed in crystalline alloys. The rapid cooling rates required to produce amorphous alloys are believed to promote chemical homogeneity since there is no time for appreciable solid-state diffusion.[2] The short time available for significant diffusion suggests that amorphous alloys should lack second phases, precipitates, and segregates. However, this assertion has been shown not to always be true for BMGs.[3–6] It should be noted that the second phases (crystalline inclusions) observed in some BMGs are a result of heterogeneous nucleation often caused by impurities in the melt.[4,7,8]

[†]*deceased*

The persistent quest for BMG systems with increased glass-forming ability (i.e., decreased crystalline inclusions) has provided the field with an influx of new BMG compositions. The importance of corrosion resistance in achieving some of the applications for BMGs mentioned previously has led to the extensive chronicling of the electrochemical behaviors of various BMG compositions. The majority of the electrochemical reports prioritize presenting the corrosion resistance of the new compositions and emphasize less the corrosion mechanisms of amorphous alloys. For instance, explanations of pitting mechanisms in ideally structurally and chemically homogeneous materials are not as available. This chapter summarizes the electrochemical reports available on common BMG systems (Cu-, Fe-, Ni-, and Zr-based) and presents general and localized corrosion mechanisms of BMGs that are available in the literature. Before a discussion of the corrosion behaviors of BMGs, a summary of basic aqueous corrosion concepts is presented. The objective of this summary is to present to the reader basic concepts related to the aqueous corrosion behavior of metals in an attempt to elucidate the electrochemical results presented later in the chapter. A complete discussion of aqueous corrosion is beyond the scope of this chapter. A more comprehensive treatment of this subject area may be found elsewhere.[9–11]

8.2   BASICS OF AQUEOUS CORROSION

Corrosion is broadly defined as the deterioration of materials due to reactions with their environment.[9] More specifically, corrosion is considered to be a surface phenomenon whereby mass is transferred from the material to the environment by chemical, physical, or electrochemical transport processes. Electrochemical corrosion involves the release of ions into the environment and the movement of electrons through the material. One of the most important examples of electrochemical corrosion is the corrosion of a metal in an aqueous solution as shown in Fig. 8.1. The driving force for the current flow I_{corr} from the anodic site to the cathodic site is the potential difference in the solution ($E_X - E_M$). The corrosion current I_{corr} is a relevant parameter because it is proportional to the corrosion penetration rate (CPR) of the metal at the open-circuit condition (i.e., external current = 0; potential = E_{corr}).

The current density is also of interest because it is directly related to the interfacial potential. When the current density has a linear relationship to the interfacial potential, it is said to exhibit Tafel behavior. However, many metals have potentials and current densities that have a nonlinear relationship called *active–passive behavior*. These alloys may be susceptible to localized corrosion, which can significantly limit their applications.

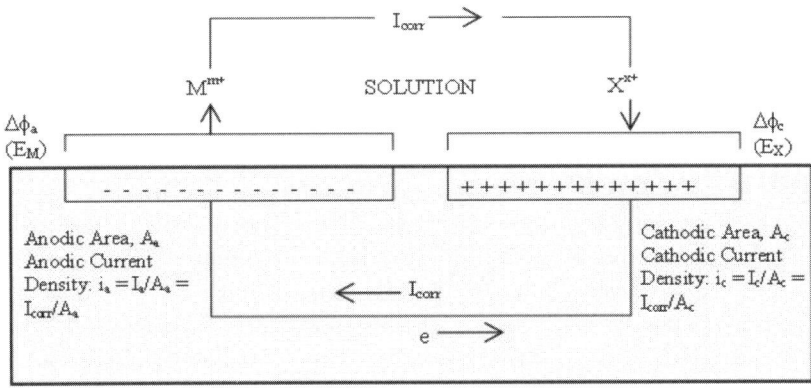

Fig. 8.1. The electrochemical corrosion circuit.[9] When the metal (M) undergoes oxidation, metal ions (M^{m+}) are forced into solution, and m electrons (me) travel through the metal from the anodic area to the cathodic area. X^{x+} represents a dissolved species that supports reduction. $\Delta\phi_a$ and $\Delta\phi_c$ represent the difference in electrical potential over the anodic and cathodic areas, respectively (reprinted with permission of ASM International®. All rights reserved. www.asminternational.org)

To achieve a comprehensive picture of a metal's electrochemical behavior, the corrosion behavior of the material is assessed at E_{corr} and other potentials. One way this approach can be accomplished is by using a potentiostat to vary the potential of the material while recording the corresponding currents. The process is designated as anodic polarization when the potential is above E_{corr}, and cathodic polarization when the potential is below E_{corr}.

A typical polarization curve of a material that has undergone localized corrosion, specifically pitting corrosion, is illustrated in Fig. 8.2. E_{corr} corresponds to the potential at which the current density approaches zero. The region where the current density remains approximately $10\ \text{mA m}^{-2}$ (the passive current density) as the potential is increased is called the *passive region*. This region corresponds to the stabilization of a passive film. Materials that exhibit a high resistance to localized corrosion form an adherent, nonporous film that protects the sample surface. The potential at which the current density suddenly increases is designated as the pitting potential E_{pit}. The pitting potential is associated with the breakdown of the passive film. The continued local breakdown of the film exposes the surface to rapid active dissolution.[9] The corrosion current density i_{corr} can be obtained from polarization experiments by the extrapolation of the anodic and cathodic portions of the polarization curves as shown in Fig. 8.2.

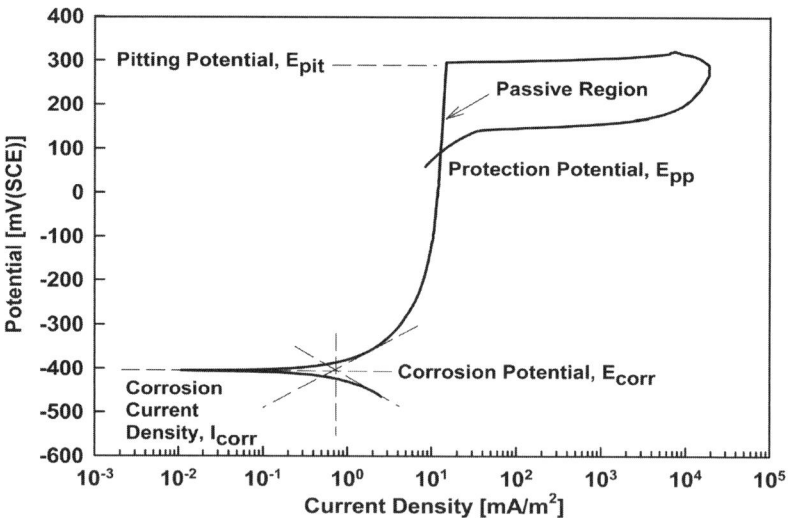

Fig. 8.2. Representative polarization curve of a material that is susceptible to pitting corrosion[12] (reprinted from *Intermetallics*, W. H. Peter, R. A. Buchanan, C. T. Liu, P. K. Liaw, M. L. Morrison, J. A. Horton, C. A. Carmichael Jr., and J. L. Wright, Localized corrosion behavior of a zirconium-based bulk metallic glass relative to its crystalline state, **10**, 1157–1162, Copyright (2002), with permission from Elsevier)

Additional information about a material's ability to repassivate (i.e., reform the passive film) can be obtained by performing cyclic-anodic-polarization experiments. These tests are similar to polarization experiments except that the potential scan is reversed (indicated by the loop in Fig. 8.2) at a specified current density. Peter et al.[12] outlined several key corrosion parameters that can be obtained from cyclic-anodic-polarization scans. The repassivation potential or protection potential E_{pp} is the potential at which the current density returns to the passive value on the reverse scan. Between E_{pit} and E_{pp}, pits are initiating and propagating.

High positive values of the pitting overpotential ($E_{pit} - E_{corr}$, η_{pit}) and the protection overpotential ($E_{pp} - E_{corr}$, η_{pp})[13] reflect a strong resistance to pitting at E_{corr} including at surface flaws and after incubation times.[12] Though cyclic-anodic-polarization experiments provide additional information, polarization and immersion experiments have been the most common methods to assess the corrosion behavior of both metallic glasses and bulk metallic glasses.

8.2.1 Pit-initiation mechanisms

As mentioned previously, pitting corrosion is a result of the local breakdown of the passive film. This loss of film integrity can be attributed to preexisting conditions on the passive film, such as areas over defects and impurities, or chemical/physical damage of the film. Pit-initiation mechanisms aim to describe the breakdown of the passive film, i.e., explain the local failure of the passive film. The phenomena surrounding pit initiation are complicated and still not fully understood even for crystalline alloys. Nevertheless, there are three main categories of pit-initiation mechanisms (1) penetration, (2) film breaking, and (3) thinning.[14–17]

The penetration mechanism, first introduced by Evans,[18] involves the migration of chloride ions or other aggressive anions from the film/solution interface to the metal/film interface due to the electric field within the passive film. In this model, chloride ions are not believed to destroy the film. They promote the local dissolution at the metal/film interface and eventually result in the formation of a pit.

Rather than assuming a static passive film as in penetration models, the film-breaking mechanism utilized the concept of a dynamic film that continuously undergoes breakdown and repair events due to its formation under mechanical stresses. The metal is exposed to the electrolyte during breakdown events via pores and flaws. In a nonaggressive solution, the film is able to heal. Hoar[19] proposed, however, that the presence of chloride ions could decrease the pH within a pore and inhibit repassivation, which enables the local dissolution of the metal. In this mechanism, a stable pit is established due to breakdown only when the conditions to pit growth are favorable.[14]

The thinning mechanism assumes a local reduction in the passive film due to the adsorption of aggressive anions that form surface complexes.[16] These soluble complexes are transferred to the electrolyte at a higher rate than other regions on the film. At a given anode potential, these thinner areas experience a greater electric field that locally ruptures the film and forms a pit. Recently, Szklarska-Smialowska[17] proposed a mechanism of pit nucleation that is caused by the electrical breakdown of the passive film.

8.3 CORROSION OF BULK METALLIC GLASSES

The commercial fabrication of BMGs by companies such as Liquidmetal® Technologies gives credence to a potential market for commercial applications of BMGs. As a result, increased research into the various properties of BMGs has occurred. More recently, the corrosion behaviors of BMGs have been of great interest. Since new compositions are constantly being developed, much of the corrosion investigations concentrate on reporting the corrosion behavior of new compositions. Some of the more common BMG compositions for which corrosion investigations have been reported are Cu-,[20-28] Fe-,[29-33] Ni-,[34-46] and Zr-based[12,13,47,48-76] systems. To a lesser extent, the electrochemical behaviors of Ca-,[77] Mg-,[78-80] and Ti-based[81] BMG systems have also been examined. The corrosion behaviors of Cu-, Fe-, Ni-, and Zr-based BMGs will be discussed with a brief treatment of the corrosion behavior of some of the less common BMG systems.

8.3.1 Cu-based BMGs

The ability to fabricate Cu-based BMGs has been a relatively new occurrence. Lin and Johnson[82] successfully fabricated Cu–Zr–Ti–Ni glassy alloys with thicknesses of at least 4 mm. Inoue and coworkers,[83] subsequently, developed Cu–Zr–Ti amorphous alloys containing at least 50 at.% Cu with some compositions having critical diameters between 4 and 5 mm. The Cu–Zr–Ti[84] and Cu–Hf–Ti[83] BMGs earned attention because of their impressive mechanical properties (compressive fracture strengths of 2,060–2,150 MPa). In 2002, Inoue and Zhang[85] reported another Cu-based BMG system, Cu–Zr–Al. The remarkable strengths of the Cu-based BMGs have further perpetuated the concept of their use as engineering materials. Thus, the chemical stability and corrosion resistance of Cu-systems have become increasingly of interest. Unfortunately, the corrosion resistance of most Cu-based bulk amorphous alloys has not been as impressive as their mechanical properties. Nevertheless, modifications to the composition have been shown to improve the corrosion behavior.[20-24,26-28]

8.3.1.1 Effects of composition

Small additions of Nb[20,22,23,26,28] have been shown to be helpful in increasing the corrosion resistance of Cu-based BMGs. Other elements such as Cr,[21,27] Ta,[20,28] and Mo[20,21,24,28] have also proved to be effective in improving the electrochemical behavior. The electrochemical behaviors of selected Cu-based BMGs from the literature are presented in Fig. 8.3 and Table 8.1. Asami et al.[20] studied the effect of small additions of Nb, Mo, and Ta to a $Cu_{60}Zr_{30}Ti_{10}$ (at.%, from henceforth all compositions will be expressed in at.%) BMG in 1 M HCl, 1 M HNO_3, 1 M NaOH, and 0.5 M NaCl solutions. The observation of $Cu_{59.4}Zr_{29.7}Ti_{9.9}Nb_1$, $Cu_{59.4}Zr_{29.7}Ti_{9.9}Mo_1$, and $Cu_{59.4}Zr_{29.7}Ti_{9.9}Ta_1$ demonstrated that Nb was the most effective in decreasing the corrosion rate in all of the solutions. Tantalum was believed not to be as effective as the Nb because the actual concentration of Ta, as determined

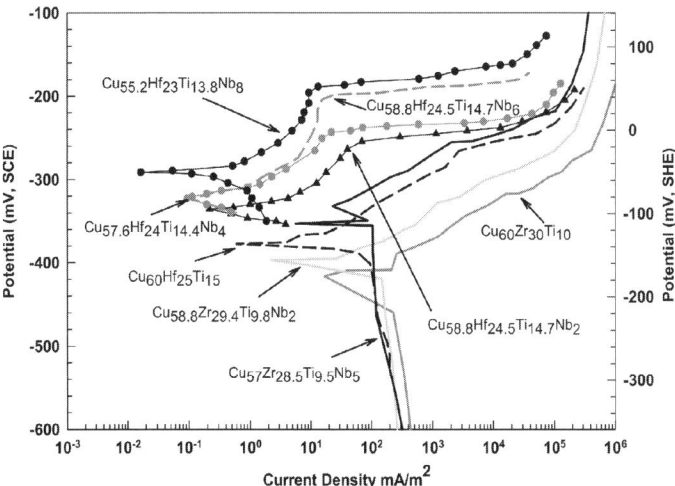

Fig. 8.3. Polarization curves of Cu-based BMGs in 0.5 M NaCl open to air at 298 K from selected reports in the literature[22,23]

Table 8.1. Summary of corrosion parameters for selected Cu-based BMGs from literature reports

Authors	Material (at.%)	T (K)	Electrolyte	η_{pit} (mV)	CPR (μm year^{-1})
Qin et al.[26]	$Cu_{55}Zr_{40}Al_5$	298	0.5 M NaCl	-435^a	200
	$Cu_{50}Zr_{45}Al_5$	298	0.5 M NaCl	-435^a	120
	$Cu_{55}Zr_{40}Al_5Nb_5$	298	0.5 M NaCl	-335^a	13
Asami et al.[20]	$Cu_{60}Zr_{30}Ti_{10}$	298	0.5 M NaCl	–	290
	$Cu_{59.4}Zr_{29.7}Ti_{9.9}Nb_1$	298	0.5 M NaCl	–	120
	$Cu_{59.4}Zr_{29.7}Ti_{9.9}Mo_1$	298	0.5 M NaCl	–	140
	$Cu_{59.4}Zr_{29.7}Ti_{9.9}Ta_1$	298	0.5 M NaCl	–	200
Qin et al.[23]	$Cu_{60}Hf_{25}Ti_{15}$	298	0.5 M NaCl	A	100
	$Cu_{58.8}Hf_{24.5}Ti_{14.7}Nb_2$	298	0.5 M NaCl	70	6
	$Cu_{57.6}Hf_{24}Ti_{14.4}Nb_4$	298	0.5 M NaCl	80	<1
	$Cu_{55.2}Hf_{23}Ti_{13.8}Nb_8$	298	0.5 M NaCl	100	<1
Asami et al.[20]	$Cu_{60}Zr_{30}Ti_{10}$	298	1 M HCl	–	660
	$Cu_{59.4}Zr_{29.7}Ti_{9.9}Nb_1$	298	1 M HCl	–	350
	$Cu_{59.4}Zr_{29.7}Ti_{9.9}Mo_1$	298	1 M HCl	–	360
	$Cu_{59.4}Zr_{29.7}Ti_{9.9}Ta_1$	298	1 M HCl	–	410
Qin et al.[23]	$Cu_{60}Hf_{25}Ti_{15}$	298	1 M HCl	A	340
	$Cu_{58.8}Hf_{24.5}Ti_{14.7}Nb_2$	298	1 M HCl	A	170
	$Cu_{57.6}Hf_{24}Ti_{14.4}Nb_4$	298	1 M HCl	A	76
	$Cu_{55.2}Hf_{23}Ti_{13.8}Nb_8$	298	1 M HCl	50	<1

A active dissolution.

All CPRs were determined from weight-loss measurements.

$^a E_{pit}$ values in mV, SCE.

by electron-microprobe analyses, was only 0.2 at.%. The corrosion rate also decreased in all solutions by increasing the Nb content (up to 5 at.%). It is clearly shown in Fig. 8.3 that Nb additions result in more positive values of E_{corr}, increased nobility, which suggests better corrosion resistance. This trend supports the observation of lower i_{corr} values (graphs shifted to the left) and greater η_{pit} values.

Liu and Liu[21] investigated the electrochemical behavior of $Cu_{47}Zr_{11}Ti_{34}Ni_8$ and $(Cu_{47}Zr_{11}Ti_{34}Ni_8)_{99.5}M_{0.5}$ (M = Cr, Mo, and W) BMGs in 0.5 M H_2SO_4 and 1 M NaOH aqueous solutions. Additions of Cr, Mo, and W led to extended passive regions, lower passive current densities, and lower corrosion rates. The alloy with a Mo addition, however, showed the most improved corrosion resistance in both solutions. All the BMGs with additions had passive films enriched in ZrO_2 and TiO_2 but depleted in Cu- oxides, which is less chemically stable and dense than both ZrO_2 and TiO_2.[24] The Mo addition was believed to be more effective in improving the corrosion resistance because its lower ionization energy, when compared to Cr and W, led to a faster film formation.[21,24]

8.3.1.2 Effects of test environment

Less attention has been given to the dependence of the corrosion behavior of Cu-based BMGs on the environment. Investigations addressing environment effects focus on the importance of the experimental solution. As is expected, a given composition will prove more corrosion resistant in a less aggressive electrolyte.[20,22,23,25,26] The dependence of the corrosion parameters of Cu-based BMGs on the electrolyte is illustrated in Table 8.1. Tam and coworkers[25] found – in their corrosion investigation of Cu–Zr–Al–Nb BMGs in 1 M HCl, 0.5 M NaCl, and 0.5 M H_2SO_4 – that regardless of the composition, the corrosion rate was higher in the more aggressive solutions (1 M HCl and 0.5 M NaCl) and lower in the less aggressive solution (0.5 M H_2SO_4). In both solutions containing chloride ions, the BMGs exhibited active behavior demonstrating the deleterious effects of chloride ions. Active–passive behavior was observed, however, in the H_2SO_4 solution. Similarly, Qin et al.[26] found that Cu–Zr–Al BMGs and a Cu–Zr–Al–Nb BMG had high corrosion rates and active polarization behavior in 0.5 M NaCl and lower corrosion rates with active–passive polarization behavior in 0.5 M H_2SO_4.

8.3.2 Fe-based BMGs

The early investigations of Fe amorphous alloys performed by Naka et al.[86–90] concentrated on the corrosion behavior of Fe-based alloys with P and C in acidic solutions containing chloride ions. The corrosion resistance (both generalized and localized) of the amorphous Fe-based alloys was due to the protective oxide films that form with appropriate additions of C and P.[91] Ferrous compositions were limited mainly to amorphous ribbon samples due to the low glass-forming ability and thermal stability. In the 1990s, new bulk forming ferrous compositions were introduced.[92–94] Some of these systems, which were alloyed with at least one metalloid and usually other metals, reported diameters up to 6 mm and notable mechanical properties. $Fe_{77}Ga_3P_{9.5}C_4B_4Si_{2.5}$ exhibited a Young's modulus of 182 GPa and compression yield and fracture strengths of 2,980 and 3,160 MPa, respectively.[95] The unique properties of these alloys coupled with their inexpensive cost have led to interest in their use as new structural materials. The development of Fe-based BMGs has caused researchers to revisit the topic of their electrochemical behavior, this time, focusing on the corrosion properties of bulk amorphous samples.

8.3.2.1 Effects of composition

With Fe-based amorphous ribbons, metalloids such as P and B were shown to drastically increase the corrosion resistance in aggressive solutions. The information obtained about the corrosion behavior of glassy ferrous alloys has been used as a guide to develop corrosion-resistant Fe-based BMGs. Pang and coworkers[29] investigated the corrosion behavior of bulk $Fe_{50-X}Cr_{16}Mo_{16}C_{18}B_X$ alloys ($X = 4$, 6, and 8 at.%) in HCl solutions. Both the corrosion rate and the passive current density decreased as the amount of B increased. Another investigation by the group[31] showed that the $Fe_{43}Cr_{16}Mo_{16}C_{10}B_5P_{10}$ BMG had a lower passive current density than that of $Fe_{43}Cr_{16}Mo_{16}C_{15}B_{10}$ BMG in a 1 M HCl solution. In a 6 M HCl solution upon anodic polarization, the current steeply increases for the $Fe_{43}Cr_{16}Mo_{16}C_{15}B_{10}$ BMG. However, $Fe_{43}Cr_{16}Mo_{16}C_{10}B_5P_{10}$ exhibited a larger passive range with a passive current density of 100 mA m^{-2}. The decrease in the passive current density and expansion of the passive region as a function of B and P additions to Fe-based BMGs tested in a 6 M HCl solution is shown in Fig. 8.4.

The effects of metal additions on the electrochemical behavior of Fe-based BMGs have also been examined. Pang et al.[30] studied the corrosion behavior of the $Fe_{75-X-Y}Cr_XMo_YC_{15}B_{10}$ BMG in a 1 M HCl solution. The anodic polarization of $Fe_{60}Mo_{15}C_{15}B_{10}$ resulted in general corrosion. However, upon polarization, the $Fe_{60-X}Cr_XMo_{15}C_{15}B_{10}$ ($X = 7.5–30$ at.%) alloys exhibited low passive currents until transpassive dissolution of Cr. Thus, Cr was effective in increasing the corrosion resistance with the corrosion rate of $Fe_{30}Cr_{30}Mo_{15}C_{15}B_{10}$ estimated as being 1 μm year^{-1}. Increasing Mo to 22.5 at.% was shown to be detrimental to the corrosion resistance.

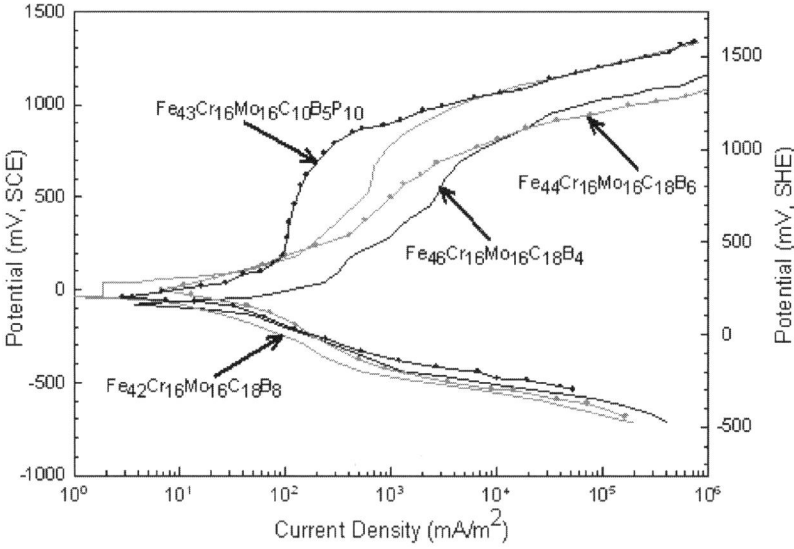

Fig. 8.4. Polarization curves of Fe-based BMGs in 6 M HCl open to air at 298 K from selected reports[29,32]

Table 8.2. Summary of corrosion behavior for selected Fe-based BMGs from literature reports

Authors	Material (at.%)	T (K)	Electrolyte	η_{pit} (mV)	CPR ($\mu m\ year^{-1}$)
Qiao et al.[96]	$Fe_{61}Y_2Zr_8Co_6Al_1Mo_7B_{15}$	295	0.6 M NaCl	137	3[a]
Qiao et al.[97]	$Fe_{48}Cr_{15}Mo_{14}Er_2C_{15}B_6$	295	0.6 M NaCl	P, Pit	3[a]
	$Fe_{48}Cr_{15}Mo_{14}Er_2C_{15}B_6$	295	1 M HCl	P, Pit	40[a]
Pang et al.[30]	$Fe_{30}Cr_{30}Mo_{15}C_{15}B_{10}$	295	1 M HCl	P, N	1[a]
Pang et al.[29]	$Fe_{46}Cr_{16}Mo_{16}C_{18}B_4$	298	1 M HCl	P, N	23
	$Fe_{44}Cr_{16}Mo_{16}C_{18}B_6$	298	1 M HCl	P, N	8
	$Fe_{42}Cr_{16}Mo_{16}C_{18}B_8$	298	1 M HCl	P, N	5
	$Fe_{46}Cr_{16}Mo_{16}C_{18}B_4$	298	6 M HCl	P, N	37
	$Fe_{44}Cr_{16}Mo_{16}C_{18}B_6$	298	6 M HCl	P, N	10
	$Fe_{42}Cr_{16}Mo_{16}C_{18}B_8$	298	6 M HCl	P, N	7
	$Fe_{46}Cr_{16}Mo_{16}C_{18}B_4$	298	12 M HCl	A, N	71
	$Fe_{44}Cr_{16}Mo_{16}C_{18}B_6$	298	12 M HCl	A, N	47
	$Fe_{42}Cr_{16}Mo_{16}C_{18}B_8$	298	12 M HCl	A, N	25

A active dissolution, *P* passivated, *Pit* pitting upon anodic polarization, *N* no pitting upon anodic polarization.

[a]CPR values determined using Faraday's Law rather than weight-loss measurements.

8.3.2.2 Effects of test environment

The environmental factors addressed concerning the corrosion behavior of Fe-based BMGs have been the effect of the solution on the corrosion behavior. Intuitively, the corrosion resistance decreases as the solution aggressiveness increases. In the case of Fe-based BMG corrosion studies, the concentration of the solution is what is generally altered. Pang et al.[29] performed immersion experiments on the $Fe_{46}Cr_{16}Mo_{16}C_{18}B_4$ BMG in 1, 6, and 12 M HCl electrolytes. As expected, the corrosion rate increased as the HCl concentration increased. The alloy experienced pitting on the surface after 168 h of immersion in the 12 M HCl solution at room temperature. However, in 1 and 6 M HCl solutions $Fe_{46}Cr_{16}Mo_{16}C_{18}B_4$ was spontaneously passivated with no visible pits on the surface. Asami et al.[31] found that the $Fe_{43}Cr_{16}Mo_{16}C_{10}B_5P_{10}$ BMG was spontaneously stable in a 1 M HCl solution. The same alloy was described as not having a stable passive region with its current increasing drastically when anodically polarized in a 12 M HCl solution. This trend of degrading corrosion properties due to more aggresive solutions is demonstrated in Table 8.2.

8.3.3 Ni-based BMGs

In general, Ni-based compositions are quite resistant to general and localized corrosion. This being the case, it is common to assess the corrosion behaviors of these alloys at elevated temperatures in extremely aggressive solutions such as 12 M HCl. A significant number of the electrochemical investigations of Ni-based amorphous alloys have been performed on ribbons[38–42,98,99] due to the difficulty of producing completely amorphous bulk samples (i.e., >1.5 mm). The elemental constituents that have typically been used to ensure good corrosion resistance have been either additions of metalloids such as P[38,40–42] or additions of metals such as Ta[38,39,42] and Nb.[35,38] More recently, there have been efforts to decrease the additions of P because of its correlation with a loss of ductility.[45] Overall, the tactical approach that proves to be more effective in increasing the corrosion resistance is dependent on whether or not the solution is strongly oxidizing (e.g., 9 M HNO_3).

8.3.3.1 Effects of composition

There has been a general consensus that additions of certain metals to amorphous Ni-based alloys tend to result in increased corrosion resistance. Shimamura et al.[38] conducted an extensive study on the effect of P and certain metals on the corrosion behavior of Ni-based amorphous ribbons. The electrochemical behavior of Ni amorphous alloys containing Ti, Ta, Zr, Nb, and/or P was investigated in boiling 9 M HNO_3 solutions with and without Cr^{6+} ions and in a boiling 6 M HCl solution. Ta additions were found to be the most effective at decreasing corrosion rates. Additions of critical amounts of Ta resulted in undetectable corrosion rates ($<10^{-3}$ mm year^{-1}). For instance, in a boiling 9 N HNO$_3$ solution after 168 h, the estimated corrosion rate of Ni$_{60}$Ti$_{40}$ was close to 1 mm year^{-1}. After adding 30 at.% of Ta (Ni$_{60}$Ti$_{10}$Ta$_{30}$), however, the altered alloy was described as being immune to corrosion and even maintained its luster after immersion for 168 h.

More recent studies have shown that other metals are also beneficial in producing more corrosion-resistant Ni-based amorphous alloys. In a study performed by Habazaki and coworkers,[36] polarization experiments were conducted on Ni$_{75-x}$Cr$_x$Ta$_5$P$_{16}$B$_4$ ($X = 5$, 10, and 15 at.%) BMGs in a 6 M HCl solution that was open to air at 303 K. The passive current density was shown to decrease as Cr was increased with Ni$_{60}$Cr$_{15}$Ta$_5$P$_{16}$B$_4$ having a passive current density of approximately 10^2 mA m^{-2}. Nevertheless, not all metal additions improve the corrosion resistance of Ni-based alloys. Pang et al.[44] investigated the anodic polarization behavior of Ni$_{60-x}$Co$_x$Nb$_{20}$Ti$_{10}$Zr$_{10}$ ($X = 0$, 5, and 20 at.%) in a 6 M HCl solution. The Co additions did not significantly alter the polarization behavior; however, all three compositions were spontaneously passivated in the solution and experienced no pitting upon anodic polarization. The passive current densities for the alloys were approximately 10^2 mA m^{-2}. The compositional effects just discussed are presented in Fig. 8.5. The beneficial effects of other elements, such as P and Ta, on increasing the passivating abilities of some Ni-based amorphous alloys are shown in Table 8.3.

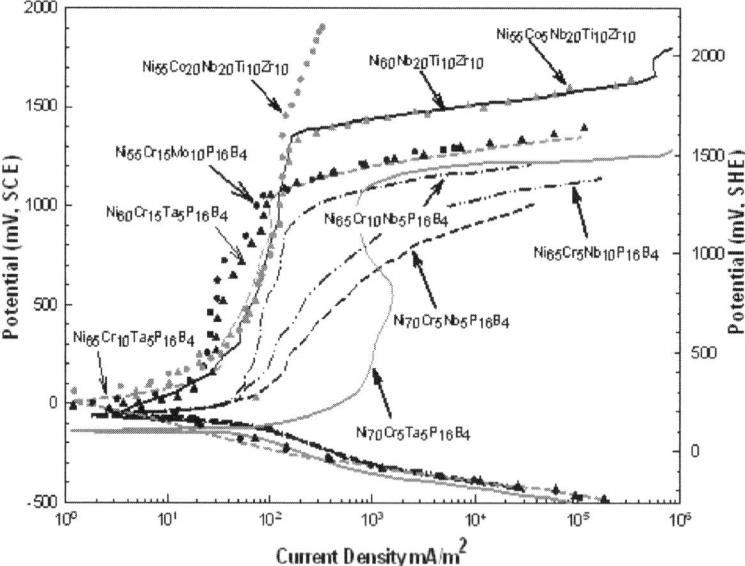

Fig. 8.5. Polarization curves of various Ni-based BMGs in 6 M HCl open to air at either 298 or 303 K from selected literature reports[35,36,44]

There is some ambiguity concerning what alloy additions play a significant role in increasing the corrosion resistance of Ni-based amorphous systems. Attempts to deconvolute the effects of alloying on corrosion behavior are made by considering the effects that the additions have on the composition and the stability of the passive film recognizing that the film can be influenced by the environment.

8.3.3.2 Effects of test environment

As stated previously, the investigation of the electrochemical behavior for most Ni-based amorphous alloys requires the usage of aggressive solutions, due to their strong resistance to dissolution in less aggressive media. The corrosion parameters of selected Ni-based compositions in 6 and 12 M HCl solutions are presented in Table 8.3. The 6 M HCl electrolyte is the most common testing electrolyte for amorphous Ni-based alloys. Katagiri et al.[34] observed that the passive current densities of the $Ni_{67}Cr_5Ta_5Mo_3P_{16}B_4$ BMG were almost identical in 6 and 12 M HCl solutions at 303 K. The authors did observe a difference in the transpassivation behavior in the two electrolytes with the milder damage to the specimen in the 6 M HCl solution. Since 6 and 12 M HCl solutions are both quite aggressive, it is possible that there would have been a more pronounced difference between the passive current densities if 1 and 6 M HCl solutions were compared. Literature reports have suggested that a passive film enriched with Ta is the reason for the high corrosion resistance of amorphous Ni-based alloys with significant amounts of Ta in aggressive solutions.[38,39,41,42] Shimamura and coworkers[38] asserted that amorphous Ni–Ta alloys required more than 35 at.% Ta in a boiling 6 M HCl solution to form a tantalum oxyhydroxide ($TaO_2[OH]$) passive film. Adding small amounts of P to Ni–Ta alloys was shown to significantly reduce corrosion rates. The corrosion rate of $Ni_{70}Ta_{30}$ in a boiling 6 M HCl solution was more than 10,000 times greater than that of $Ni_{68}Ta_{30}P_2$ alloy in the same solution. The authors believed that P promoted a passive film made exclusively of $TaO_2(OH)$ film by accelerating selective dissolution of elements unnecessary for the passive film formation.[38] In solutions with a high oxidizing power, they found that additions of P to Ni–Ta alloys were not needed to aid in the formation of the passive film.

8.3.4 Zr-based BMGs

The majority of electrochemical investigations of Zr-based BMGs have been either based on the Zr–Ni–Cu–Al alloys developed by Inoue et al.[100] or Zr–Ti–Ni–Cu–Be[101] and Zr–Ti–Ni–Cu[102] alloys, which have been developed by Peker, Lin, and Johnson. These Zr-based families are attractive due to their good glass-forming ability and impressive mechanical properties. Both the Zr–Ni–Cu–Al and the Zr–Ti–Ni–Cu–Be alloys have reported low cooling rates less than 10 K s^{-1} and diameters greater than 10 mm.[1,100,101] Some of the unique mechanical properties exhibited by Zr-based BMGs are high strength,[8,103,104] high elastic strain,[103,105] and low Young's modulus.[8,106]

Table 8.3. Corrosion parameters from literature reports of selected Ni-based amorphous alloys in various chloride solutions

Authors	Material (at.%)	T (K)	Sample size (mm)	Electrolyte	η_{pit} (mV)	CPR (μm year^{-1})
Habazaki et al.[35]	$Ni_{70}Cr_5Nb_5P_{16}B_4$	303	1	6 M HCl	P, N	–
	$Ni_{65}Cr_{10}Nb_5P_{16}B_4$	303	1	6 M HCl	P, N	<1
	$Ni_{65}Cr_5Nb_{10}P_{16}B_4$	303	1	6 M HCl	P, N	–
Habazaki et al.[36]	$Ni_{70}Cr_5Ta_5P_{16}B_4$	303	1	6 M HCl	P	–
	$Ni_{65}Cr_{10}Ta_5P_{16}B_4$	303	1	6 M HCl	P	<1
	$Ni_{60}Cr_{15}Ta_5P_{16}B_4$	303	1	6 M HCl	P	–
	$Ni_{55}Cr_{15}Mo_{10}P_{16}B_4$	303	1	6 M HCl	P	<1
Pang et al.[44]	$Ni_{60}Nb_{20}Ti_{10}Zr_{10}$	298	1	6 M HCl	P, N	<1
	$Ni_{55}Co_5Nb_{20}Ti_{10}Zr_{10}$	298	1.5	6 M HCl	P, N	<1
	$Ni_{40}Co_{20}Nb_{20}Ti_{10}Zr_{10}$	298	1	6 M HCl	P, N	<1
Lee et al.[40]	$Ni_{70}Ta_{30}$	303	Ribbon	12 M HCl	A, N	550
	$Ni_{69}Ta_{30}P_1$	303	Ribbon	12 M HCl	A–P, N	70
	$Ni_{67}Ta_{30}P_3$	303	Ribbon	12 M HCl	P, N	3.5
	$Ni_{60}Ta_{30}P_{10}$	303	Ribbon	12 M HCl	P, N	1.5
Lee et al.[41]	$Ni_{85}Ta_{10}P_5$	303	Ribbon	12 M HCl	A–P	–
	$Ni_{80}Ta_{10}P_{10}$	303	Ribbon	12 M HCl	A–P	–
	$Ni_{75}Ta_{10}P_{15}$	303	Ribbon	12 M HCl	P	–
	$Ni_{70}Ta_{10}P_{20}$	303	Ribbon	12 M HCl	P	–
Katagiri et al.[34]	$Ni_{67}Cr_5Ta_5Mo_3P_{16}B_4$	303	Ribbon	12 M HCl	P, N	–
	$Ni_{67}Cr_5Ta_5Mo_3P_{16}B_4$	303	1	12 M HCl	P, N	–
Kawashima et al.[43]	$Ni_{55}Nb_{35}Ta_5P_5$	303	1	12 M HCl	A–P, N	600
	$Ni_{55}Nb_{25}Ta_{15}P_5$	303	1	12 M HCl	A–P, N	–
	$Ni_{55}Nb_{10}Ta_{30}P_5$	303	1	12 M HCl	P, N	<0.7
	$Ni_{55}Ta_{40}P_5$	303	1	12 M HCl	P, N	–

A–P observed active–passive transition, *P* passivation, *N* no pitting.
All CPRs were determined with weight-loss measurements.

8.3.4.1 *Effects of composition*

Additions of noble elements (e.g., Nb, Pd, Ti, and Ta) have been utilized as a method to develop compositions with increased glass-forming ability and/or increased resistance to general and localized corrosion.[49,50,58–60,63] Raju and coworkers[63] investigated the corrosion behavior of Zr–Cu–Al–Ni–X (X = Nb or Ti up to 5 at.%) BMGs in weakly alkaline sulfate and chloride electrolytes (0.1 M Na_2SO_4 and 0.01 M NaCl, respectively). In the sulfate solution, the increase in Nb content resulted in a slight decrease in E_{corr} and an increase in the passive current density. A similar trend was observed for Ti additions. However, the increase in Nb and Ti did lead to increased values of η_{pit} and η_{pp} in the NaCl solution.

In a similar study, Asami et al.[49] investigated the electrochemical behavior of $Zr_{60-X}Nb_XAl_{10}Ni_{10}Cu_{20}$ (X = 0, 5, 10, 15, and 20 at.%) BMGs in 0.5 M H_2SO_4 and 1 M HCl solutions. A substitution of 20 at.% of Nb for Zr decreased the CPRs in the 1 M HCl solution from 100 to 1 μm year^{-1}. An increase in Nb also led to an increase in the E_{pit}. The η_{pit}, however, did not always increase as the Nb content was raised due to the fact that increases in Nb sometimes led to higher values of E_{corr}. The polarization curves of selected Zr-based BMGs in NaCl electrolytes (0.5 or 0.6 M) are plotted in Fig. 8.6. The effectiveness of Nb additions in raising E_{pit} for a Zr–Al–Co

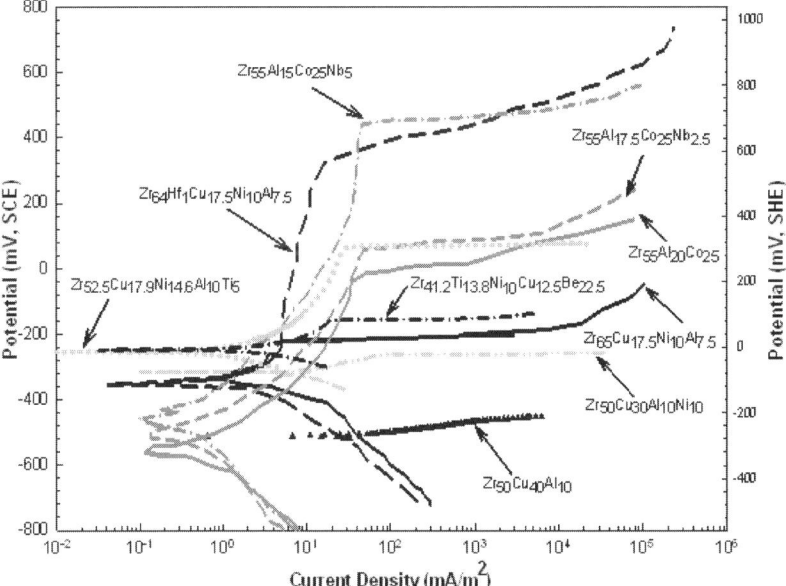

Fig. 8.6. Polarization curves of various Zr-based BMGs in either a 0.5 or a 0.6 M NaCl electrolyte from selected literature reports[12,52,55,60,107]

BMG is illustrated in Fig. 8.6.[60] Positive influences of Hf on the passivating ability of a Zr–Cu–Ni–Al BMG[52] are also shown in Fig. 8.6. Nevertheless, additions of noble metals do not always guarantee better corrosion resistance. Qin et al.[58] investigated bulk amorphous alloys with compositions of $Zr_{55}Al_{10}Cu_{30}Ni_{5-X}Pd_X$ ($X = 0$, 1, 3, and 5 at.%) in 0.6 M NaCl and found that additions of Pd decreased η_{pit}.

8.3.4.2 Effects of test environment

Though the composition of Zr-based BMGs does affect the corrosion properties, it is evident from Table 8.4 that the environment is also a significant factor in the corrosion behavior. Table 8.4 provides corrosion parameters for several Zr-based BMGs in various electrolytes. The insidious effect of the chloride-ion concentration on the localized corrosion resistance of Zr-based bulk amorphous alloys is clear from the table. Gebert et al.[74] found that $Zr_{55}Cu_{30}Al_{10}Ni_5$ was immune to localized corrosion in a 0.1 M Na_2SO_4 solution and a 0.1 M NaOH solution for the entire potential ranges tested (−1,000 to 2,000 mV, SCE). However, a susceptibility to pitting corrosion was observed during anodic polarization experiments at chloride concentrations as low as 10^{-3} M. The pitting potential was shown to decrease with increasing chloride concentrations. However, this trend was subdued with anodic preformation of a passive film. Similarly, Mudali et al.[56] found that increasing the concentration of NaCl from 0.01 to 0.2 M in a 0.5 M H_2SO_4 electrolyte significantly decreased the E_{pit} of the $Zr_{59}Ti_3Cu_{20}Al_{10}Ni_8$ BMG. Other investigations have shown a decrease in the resistance to localized corrosion with an increase in the chloride-ion concentration.[71,74] Subsequent corrosion investigations have also illustrated that many Zr-based systems are more resistant to localized corrosion in solutions such as H_2SO_4, Na_2SO_4, and NaOH[12,49,52–54,56,60,61,67–69] when compared to solutions that contain chloride ions.

Other factors, such as temperature, may also affect the electrochemical properties of Zr-based BMGs. Gebert and coworkers[67] investigated the effect of temperature on the corrosion behavior of the $Zr_{55}Cu_{30}Al_{10}Ni_5$ BMG. Anodic polarization was performed at 298, 423, and 523 K in a 0.001 M NaCl electrolyte on $Zr_{55}Cu_{30}Al_{10}Ni_5$ specimens that were prepassivated and specimens that had no treatment. For both the nontreated and prepassivated BMG samples, E_{pit} decreased as the temperature increased. Thus, the $Zr_{55}Cu_{30}Al_{10}Ni_5$ BMG exhibited an increased tendency of pitting in the chloride solution as the temperature was increased.

Table 8.4. Reported corrosion parameters of Zr-based BMGs in various solutions

Authors	Material (at.%)	T (K)	Electrolyte	η_{pit} (mV)	η_{pp} (mV)	CPR (μm year^{-1})
			Chloride solutions			
Gebert et al.[74]	$Zr_{55}Cu_{30}Al_{10}Ni_5$	295	10^{-3} M NaCl	Pit	–	–
			10^{-2} M NaCl	Pit	–	–
			10^{-1} M NaCl	Pit	–	–
Gebert et al.[67]	$Zr_{55}Cu_{30}Al_{10}Ni_5$ (as-prepared)	298	10^{-3} M NaCl	450[a]	–	–
		423	10^{-3} M NaCl	50[a]	–	–
		523	10^{-3} M NaCl	-100[a]	–	–
	$Zr_{55}Cu_{30}Al_{10}Ni_5$ (prepassivated)	298	10^{-3} M NaCl	750[a]	–	–
		423	10^{-3} M NaCl	700[a]	–	–
		523	10^{-3} M NaCl	100[a]	–	–
Raju et al.[63]	$Zr_{55}Cu_{30}Al_{10}Ni_5$	298	10^{-2} M NaCl	–	–	–
	$Zr_{60}Cu_{20}Al_{10}Ni_5Nb_2$	298	10^{-2} M NaCl	415	40	–
	$Zr_{59}Cu_{20}Al_{10}Ni_8Nb_3$	298	10^{-2} M NaCl	625	125	–
	$Zr_{57}Cu_{15.4}Al_{10}Ni_{12.6}Nb_5$	298	10^{-2} M NaCl	750	430	–
	$Zr_{59}Cu_{20}Al_{10}Ni_8Ti_3$	298	10^{-2} M NaCl	625	50	–
	$Zr_{52.5}Cu_{17.9}Al_{10}Ni_{14.6}Ti_5$	298	10^{-2} M NaCl	500	165	–
Liu et al.[52]	$Zr_{65}Cu_{17.5}Ni_{10}Al_{7.5}$	298	0.5 M NaCl	125	–	–
	$Zr_{64}Hf_1Cu_{17.5}Ni_{10}Al_{7.5}$	298	0.5 M NaCl	690	–	–
Pang et al.[60]	$Zr_{55}Al_{20}Co_{25}$	298	0.5 M NaCl	500	–	<1[b]
	$Zr_{55}Al_{17.5}Co_{25}Nb_{2.5}$	298	0.5 M NaCl	540	–	<1[b]
	$Zr_{55}Al_{15}Co_{25}Nb_5$	298	0.5 M NaCl	860	–	<1[b]
Chieh et al.[61]	$Zr_{52.5}Cu_{17.9}Ni_{14.6}Al_{10}Ti_5$	295	0.6 M NaCl	340	–	1.5
He et al.[68]	$Zr_{52.5}Cu_{17.9}Ni_{14.6}Al_{10}Ti_5$	295	0.5 M NaCl	225	–	–
Peter et al.[12]	$Zr_{52.5}Cu_{17.9}Al_{10}Ni_{14.6}Ti_5$					
	600 grit finish	295	0.6 M NaCl	320	40	1.3
	0.5 μm finish	295	0.6 M NaCl	100	-20	9
Morrison et al.[55]	$Zr_{41.2}Ti_{13.8}Ni_{10}Cu_{12.5}Be_{22.5}$	295	0.6 M NaCl	97	26	2.5
Asami et al.[49]	$Zr_{60}Al_{10}Ni_{10}Cu_{20}$	298	1 M HCl	A	–	100[b]
	$Zr_{55}Nb_5Al_{10}Ni_{10}Cu_{20}$	298	1 M HCl	25	–	–
	$Zr_{50}Nb_{10}Al_{10}Ni_{10}Cu_{20}$	298	1 M HCl	100	–	–
	$Zr_{45}Nb_{15}Al_{10}Ni_{10}Cu_{20}$	298	1 M HCl	50	–	–
	$Zr_{40}Nb_{20}Al_{10}Ni_{10}Cu_{20}$	298	1 M HCl	75	–	1[b]
Liu et al.[52]	$Zr_{65}Cu_{17.5}Ni_{10}Al_{7.5}$	298	1 M HCl	A	–	–
	$Zr_{65}Hf_1Cu_{17.5}Ni_{10}Al_{7.5}$	298	1 M HCl	A	–	–
Chieh et al.[61]	$Zr_{52.5}Cu_{17.9}Ni_{14.6}Al_{10}Ti_5$	295	1 M HCl	–	–	20,500
Dhawan et al.[53]	$Zr_{46.75}Ti_{8.25}Cu_{7.5}Ni_{10}Be_{27.5}$	300	0.5 M HCl	A	–	410[c]
	$Zr_{65}Cu_{17.5}Ni_{10}Al_{7.5}$	300	0.5 M HCl	A	–	240[c]

(Continued)

Table 8.4. (*cont.*)

Authors	Material (at.%)	*T* (K)	Electrolyte	η_{pit} (mV)	η_{pp} (mV)	CPR (μm year^{-1})
		Biological solutions				
Qiu et al.[50]	$Zr_{65}Cu_{17.5}Ni_{10}Al_{7.5}$	310	ABF	600	–	–
	$Zr_{63}Nb_2Cu_{17.5}Ni_{10}Al_{7.5}$	310	ABF	975	–	–
	$Zr_{60}Nb_5Cu_{17.5}Ni_{10}Al_{7.5}$	310	ABF	1,430	–	–
Morrison et al.[13]	$Zr_{52.5}Cu_{17.9}Ni_{14.6}Al_{10}Ti_5$	310	PBS	474	225	0.8
Morrison et al.[55]	$Zr_{41.2}Ti_{13.8}Ni_{10}Cu_{12.5}Be_{22.5}$	310	PBS	410	184	1
Hiromoto et al.[62]	$Zr_{55}Al_{10}Ni_{10}Cu_{15}$ (after 1.8 ks)	310	Hanks	400	–	0.1^c
	$Zr_{55}Al_{10}Ni_{10}Cu_{15}$ (after 605 ks)	310	Hanks	730	–	0.03^c

A active dissolution, *Pit* pitting, *P* passivated.

$^a E_{pit}$ values in mV, SCE.

^{b}CPRs determined from weight-loss measurements.

^{c}Calculated values.

8.3.5 Other BMG systems

Though the majority of electrochemical investigations of BMGs are of Cu-, Fe-, Ni-, or Zr-based systems, there are other BMGs, such as Ca-,[77] Mg-,[78–80] and Ti-based,[81] whose corrosion properties are being examined. Ca- and Ti-based BMGs are of interest because of their potential applications as biomaterials. The Mg-based system is attractive for applications that require high-strength with light-weight materials.[108]

Morrison and coworkers[77] examined the corrosion behavior of three Ca-based BMGs ($Ca_{65}Mg_{15}Zn_{20}$, $Ca_{55}Mg_{18}Zn_{11}Cu_{16}$, and $Ca_{50}Mg_{20}Cu_{30}$) in a 0.05 M Na_2SO_4 electrolyte. The $Ca_{65}Mg_{15}Zn_{20}$ BMG underwent pitting at E_{corr} and had a CPR of 5,691 μm year^{-1}. However, $Ca_{50}Mg_{20}Cu_{30}$ and $Ca_{55}Mg_{18}Zn_{11}Cu_{16}$ were both slightly passivated at E_{corr} with CPRs of 1,503 and 311 μm year^{-1}, respectively. The electrochemical behavior of the $Ti_{43.3}Zr_{21.7}Ni_{7.5}Be_{27.5}$ BMG at 310 K in a phosphate buffered saline (PBS) solution was also examined by Morrison et al.[81] $Ti_{43.3}Zr_{21.7}Ni_{7.5}Be_{27.5}$ exhibited passive behavior at E_{corr} but a susceptibility to localized corrosion at increased potentials. The η_{pit} value was 589 mV. The material had a CPR of 2.9 μm year^{-1}. The author concluded that the resistance to localized corrosion in the PBS solution was equal to or better than that of a 316 L stainless steel in an identical solution.

Gebert and coworkers[78] studied the electrochemical behavior of $Mg_{65}Y_{10}Cu_{15}Ag_{10}$ and $Mg_{65}Y_{10}Cu_{25}$ BMGs in a borate buffer solution (pH

8.4). Their electrochemical behaviors were compared to those of Mg and the crystalline $Mg_{65}Y_{10}Cu_{25}$ alloy. The corrosion behaviors of $Mg_{65}Y_{10}Cu_{25}$ amorphous and crystalline alloys were comparable to each other and superior to that of Mg. $Mg_{65}Y_{10}Cu_{15}Ag_{10}$ exhibited superior electrochemical behavior to the other three alloys. However, the passive current density of $Mg_{65}Y_{10}Cu_{15}Ag_{10}$ was 1,400 mA m^{-2}.

8.3.6 Corrosion of BMGs in relation to crystalline alloys

There is still considerable interest in how the electrochemical properties of metallic glasses and BMGs compare to those of conventional crystalline alloys. Amorphous alloys are believed to exhibit corrosion resistance because of (1) their compositions, which are not constrained by solubility limits, can be alloyed with elements that promote passivation,[74] and (2) their lack of microstructural features, such as grain boundaries, dislocations, and precipitates, which commonly serve as sites for the local passive film breakdown. Comparisons of amorphous alloys with their crystalline counterparts have been performed to determine if structural disorder influences the corrosion behavior.[12,73,76,109–114] Schroeder et al.[73] found that the $Zr_{41.2}Ti_{13.8}Cu_{12.5}Ni_{10}Be_{22.5}$ BMG was only slightly more resistant to pitting corrosion than its crystalline counterpart in a 0.5 M NaCl solution. The BMG was assessed to be no more resistant to generalized corrosion in a 0.5 M $NaClO_4$ solution. Köster et al.[115] determined that there was no significant difference between the polarization behavior of the amorphous and nanocrystalline states of $Zr_{69.5}Cu_{12}Ni_{11}Al_{7.5}$ in 0.1 N NaOH. However, Peter and coworkers[12] found that the $Zr_{52.5}Cu_{17.9}Ni_{14.6}Al_{10}Ti_5$ BMG exhibited lower corrosion rates and greater resistance to pitting corrosion in a 0.6 M NaCl solution than the crystalline alloy of the same composition. Other investigations have compared amorphous alloys to crystalline alloys of similar composition. Work by Naka et al.[90] reported higher corrosion rates of an amorphous $Fe_{70}Cr_{10}B_{20}$ (ribbon) in a 1 M HCl solution at 303 K than a crystalline $Fe_{90}Cr_{10}$ alloy. However, the $Fe_{50}Cr_{30}B_{20}$ amorphous alloy experienced a lower corrosion rate in 1 M HCl than the $Fe_{70}Cr_{30}$ crystalline alloy. It has been cautioned that an accurate assessment of the role of structural disorder on the corrosion behavior can be made only when the corrosion properties of an amorphous alloy are compared to those of a single-phased crystalline alloy of the same composition.[116] When considering the corrosion properties, the effect of the structure must be addressed. However, the alloy composition probably has a larger influence on the electrochemical behavior than whether the structure is periodic or disordered.[116]

8.3.7 Summary

Though the corrosion behavior of Cu-, Fe-, Ni-, and Zr-based BMG systems has been discussed in some detail, a brief treatment summarizing the key aspects of each system is beneficial. Ni-based systems exhibit the highest resistance to general and localized corrosion. Even in extremely aggressive solutions, the Ni-based amorphous alloys demonstrate impressive passivity. The best compositions are spontaneously passive in aggressive solutions and experience extremely low corrosion rates. Many Fe-based BMG compositions have also been shown to be corrosion resistant supported by their assessment in aggressive solutions. Even so, in general, the Fe-based BMGs are not as resistant to corrosion as Ni-based BMGs. The passive current density of Fe-based BMGs in 6 M HCl is in the range of 100–1,000 mA m^{-2}, which is considerably higher than that of the Ni-based BMG alloys (25–100 mA m^{-2}) in the same solution. Also, the corrosion rates of Fe-based BMGs are usually higher than those of Ni-based BMGs.

Unlike Ni- and Fe-based systems, Zr- and Cu-based systems are not strongly protected from localized corrosion in aggressive solutions, particularly solutions containing chloride ions. Zr- and Cu-based BMGs are especially susceptible to pitting corrosion in NaCl and HCl solutions. However, Zr-based BMGs exhibit better corrosion properties than those of Cu-based BMGs. In 0.5 NaCl, Cu-based BMGs show little evidence of a passive region, meaning active dissolution occurs or pitting resistance is extremely low. Many of the Zr-based BMG compositions have pronounced passive regions in 0.5/0.6 M NaCl solutions with current densities of 5–40 mA m^{-2}. From observing the average corrosion resistance of Zr-based BMGs in 0.5/0.6 M NaCl solutions, it can be assumed that their corrosion resistance is not maintained in a 6 M HCl solution.

The addition of certain elements to BMG compositions has been successful in improving their aqueous corrosion behavior. In general, additions of metal elements with known passivating abilities (e.g., Cr, Mo, Nb, Pd, Ta, and Ti) increase the resistance to general corrosion, which is identified by a decrease in CPRs. These metal additions have also accomplished enhanced resistance to localized corrosion, as evident by lower passive current densities and increased pitting potentials. In particular, the benefits of Nb additions to Cu-, Ni-, and Zr-based BMGs have been clearly recorded. Fe-based BMGs seem to benefit from the addition of a nonmetal in conjunction with passivating metals and nonmetals for instance P and Cr. Altering BMG compositions, however, should be weighed with how the change could affect the glass-forming ability. Qin et al.[58] found that Pd additions to $Zr_{55}Cu_{30}Al_{10}Ni_{5-X}Pd_X$ BMGs made the alloys more vulnerable to pitting corrosion. High-resolution transmission electron microscopy revealed higher volume fractions of crystalline nuclei as the Pd content was raised. The relationship between crystalline inclusions and an increased vulnerability to localized attack will be addressed further in Sect. 8.4.2.

It is evident that many of the available studies concerning corrosion behavior of BMGs concentrate on reporting the electrochemical response of new compositions. Albeit this approach may not be optimal for understanding the governing electrochemical mechanisms of BMGs, the available literature does provide a practical way to speculate about the success of various BMG systems in certain applications. Ni- and Fe-based BMGs have been shown to exhibit high resistances to corrosion in even extremely aggressive environments. Thus, it is probably safe to assume that both these systems would maintain their corrosion resistance in a less aggressive solution such as 0.6 M NaCl, simulated seawater. The current Ni- and Fe-based compositions are more suited for coatings, due to their limited glass-forming abilities. However, since corrosion is a surface phenomenon, these compositions could still be beneficial in providing corrosion-resistant coatings for materials. Zr- and Cu-based BMGs, on the other hand, would probably not be as successful in sea applications owing to their mediocre performance in simulated seawater. These systems seem more suited for applications where they have little contact with aqueous media or are in contact with only mild solutions. For instance, the corrosion resistance of Zr-based BMGs has been suitable for sporting applications (e.g., golf clubs and racquets). Some compositions of Zr-based BMGs have exhibited good corrosion resistance in biological media such as PBS solutions. The success of crystalline Zr and Zr alloys in orthopedic applications has made some in the medical field more amenable to considering Zr-based BMGs with proven corrosion resistance in biological environments for orthopedic applications.

8.4 CORROSION MECHANISMS IN BMGS

8.4.1 General corrosion

The mechanism for general corrosion of amorphous alloys is analogous to corrosion in conventional crystalline alloys. That is, the oxidation reaction on the surface ($M \rightarrow M^{m+} + me$) produces a loss of material. The electrons that are lost, me, can be related to the corrosion rate by Faraday's Law. If the alloy exhibits active–passive behavior, a passive film may form. A passive film that is continuous, nonporous, and adherent to the surface can cause the corrosion rate to decrease significantly. The lower corrosion rates are because the controlling mechanism for ion migration through the film is slow solid-state diffusion. The corrosion rate may be decreased further by a film that is a poor conductor of electrons.

Due to the importance of passive films to corrosion rates, efforts have been made to comprehensively characterize passive films. As mentioned previously, one of the methods for improving corrosion properties is alloying with elements known for their passivating abilities. Nb, for example, was shown to be effective in lowering corrosion rates in Cu-based BMGs[20,26] and changing the behavior at E_{corr} from active to spontaneously passivating.[23] Additions of certain metals with known passivating abilities have been shown to lead to the enrichment of metal cations in the passive film. This enrichment, in turn, promotes the formation of protective oxide or oxyhydroxide films.[38] Metalloids, such as P, also enhance the passivating ability, but it is uncertain whether the improvements are due to synergistic effects with metal elements[38,42,91] or due solely to the metalloid.[117]

8.4.2 Pitting corrosion

Since BMGs are ideally homogenous materials, the susceptibility of some systems to pitting was at first unexpected. However, the pitting susceptibility of BMGs has been justified by the presence of crystalline inclusions.[52,58,74] It is possible that crystalline inclusions are not the only inhomogeneity that can lead to the localized corrosion of BMGs. In a review of electrochemical properties of metallic glasses, Archer et al.[116] suggested that the local corrosion attack could be due to physical and/or chemical inhomogeneities. Despite the uncertainties, there have been only a few reports dedicated to clarifying pitting mechanisms of BMGs. Gebert et al.[74] reported chloride-ion attack at the transition zone between a crystalline inclusion and the amorphous matrix of the $Zr_{55}Cu_{30}Al_{10}Ni_5$ BMG. The inclusion/amorphous transition zone was compared to a grain boundary in a polycrystalline material. Using this parallel, Gebert et al. suggested an adsorption mechanism for pit initiation where chloride ions are preferentially adsorbed at active regions of the passive film that are located over the transition zones. The localization of the corrosion was correlated with chemical composition changes associated with the transition zone or the corresponding overlying film.

Mudali et al.[56] suggested that the breakdown of the passive oxide on the $Zr_{59}Ti_3Cu_{20}Al_{10}Ni_8$ BMG in NaCl was a result of weak portions of the film that grow over crystalline inclusions or inhomogeneous chemical clusters in the amorphous alloy. The weaker regions of the passive film would be vulnerable to the adsorption and penetration of chloride ions. The mechanism also stated that an autocatalytic reaction with chloride ions leads to the local selective dissolution of zirconium and aluminum. More recently, Mudali and coworkers[118] concluded that decreased corrosion resistance due to the presence of crystalline inclusions was not a suitable mechanism if a stable ZrO_2 film was present. There have been reports of enrichment of copper at the metal surface/passive film interface in Zr-based amorphous alloys.[56,119] Mudali et al. suggested that this copper enrichment causes the enhanced migration of chloride ions, which leads to the production of cupric chloride. It is asserted that oxidizing cupric ions cause the local damage of the ZrO_2 passive film.

The present authors agree with the sentiment that the mere presence of crystalline inclusions does not provide an adequate explanation for pitting in BMGs. It seems likely that the susceptibility of amorphous materials to localized corrosion is dominated by chemical factors. Some elements are inherently more resistant to localized corrosion due to the nature of their passive film. Thus, there is a limit to how resistant a particular composition is to localized dissolution in a solution based predominately on its chemistry. Other factors, however, can make the composition more vulnerable to localized corrosion. Crystalline inclusions may indeed be one of these factors. These crystalline inclusions will result in less resistance to localized corrosion behavior if they adversely affect the passive film. For instance, a crystalline inclusion with a composition different from its glass matrix would probably have a more negative effect on pitting resistance than an inclusion with the same composition. This is due to the fact that chemical inhomogeneities of the metal surface affect the protectiveness of the passive film even more so than physical defects.[120]

8.5   CONCLUDING REMARKS

The investigation of the electrochemical behavior of BMGs is a relatively new area. Thus, much of the literature focuses on the important work of characterizing the corrosion behavior of new alloy compositions. This research has led to the discovery of certain elements that are quite effective in improving general and local corrosion resistance. Significantly less has been reported on mechanistic explanations of pitting in BMGs. This trend is understandable since pitting mechanisms in conventional materials are not completely understood. Though the available mechanistic theories are not completely established, they do explain the observed pitting phenomena of BMGs. Many of the reports stress the relationship between increased vulnerability of pitting corrosion and the presence of crystalline regions. However, more work is needed to explore the relationship between crystalline inclusions, changes in local chemistry, and protectiveness of the passive film. Future research utilizing in situ techniques (e.g., scanning electrochemical microscopy) in conjunction with chemical analysis tools may shed light on the issue. The suggested approach may provide a more detailed and conclusive explanation for the vulnerability of some amorphous alloys to pitting corrosion.

REFERENCES

1. W. L. Johnson, Bulk metallic glasses – A new engineering material, *Curr. Opin. Solid State Mater. Sci.* **1**(3), 383–386 (1996).
2. K. Hashimoto, in *Amorphous Metallic Alloys*, edited by F. E. Luborsky (Butterworths, London, 1983), pp. 471–486.
3. A. Gebert, J. Eckert, and L. Schultz, Effect of oxygen on phase formation and thermal stability of slowly cooled $Zr_{65}Al_{7.5}Cu_{17.5}Ni_{10}$ metallic glass, *Acta Mater.* **46**(15), 5475–5482 (1998).
4. C. T. Liu, M. F. Chisholm, and M. K. Miller, Oxygen impurity and microalloying effect in a Zr-based bulk metallic glass alloy, *Intermetallics* **10**(11–12), 1105–1112 (2002).
5. B. S. Murty, D. H. Ping, K. Hono, and A. Inoue, Direct evidence for oxygen stabilization of icosahedral phase during crystallization of $Zr_{65}Cu_{27.5}Al_{7.5}$ metallic glass, *Appl. Phys. Lett.* **76**(1), 55–57 (2000).
6. D. J. Sordelet, X. Yang, E. A. Rozhkova, M. F. Besser, and M. J. Dramer, Influence of oxygen content in phase selection during quenching of $Zr_{80}Pt_{20}$ melt spun ribbons, *Intermetallics* **12**(10–11), 1211–1217 (2004).
7. C. T. Liu and Z. P. Lu, Effect of minor alloying additions on glass formation in bulk metallic glasses, *Intermetallics* **13**, 415–418 (2005).
8. A. Inoue, Stabilization of metallic supercooled liquid and bulk amorphous alloys, *Acta Mater.* **48**(1), 279–306 (2000).
9. E. E. Stansbury and R. A. Buchanan, *Fundamentals of Electrochemical Corrosion* (ASM International, Materials Park, 2000).
10. D. A. Jones, *Principles and Prevention of Corrosion*, 2nd ed. (Prentice Hall, Upper Saddle River, 1995).
11. H. H. Uhlig and R. W. Revie, *Corrosion and Corrosion Control*, 3rd ed. (Wiley-Interscience, New York, 1985).
12. W. H. Peter, R. A. Buchanan, C. T. Liu, P. K. Liaw, M. L. Morrison, C. A. Carmichael, Jr., and J. L. Wright, Localized corrosion behavior of a zirconium-based bulk metallic glass relative to its crystalline state, *Intermetallics* **10**(11–12), 1157–1162 (2002).
13. M. L. Morrison, R. A. Buchanan, R. V. Leon, C. T. Liu, B. A. Green, P. K. Liaw, and J. A. Horton, The electrochemical evaluation of a Zr-based bulk metallic glass in a phosphate-buffered saline electrolyte, *J. Biomed. Mater. Res. A* **74**(3), 430–438 (2005).
14. G. S. Frankel, Pitting corrosion of metals: A review of the critical factors, *J. Electrochem. Soc.* **145**(6), 2186–2198 (1998).
15. H. Böhni, in *Uhlig's Corrosion Handbook*, edited by R. W. Revie (Wiley, New York, 2000), pp. 173–190.
16. H.-H. Strehblow, in *Corrosion Mechanisms in Theory and Practice*, edited by P. Marcus and J. Oudar (Dekker, New York, 1995).
17. Z. Szklarska-Smialowska, Mechanism of pit nucleation by electrical breakdown of the passive film, *Corros. Sci.* **44**(5), 1143–1149 (2002).
18. U. R. Evans, The passivity of metals. Part I. The isolation of the protective film, *J. Chem. Soc., Chem. Commun.* 1020–1040 (1927).
19. T. P. Hoar, The breakdown and repair of oxide films on iron, *Trans. Faraday Soc.* **45**, 683–693 (1949).
20. K. Asami, C.-L. Qin, T. Zhang, and A. Inoue, Effect of additional elements on the corrosion behavior of a Cu–Zr–Ti bulk metallic glass, *Mater. Sci. Eng. A* **375–377**, 235–239 (2004).

21. B. Liu and L. Liu, The effect of microalloying on thermal stability and corrosion resistance of Cu-based bulk metallic glasses, *Mater. Sci. Eng. A* **415**(1–2), 286–290 (2006).

22. C. L. Qin, K. Asami, T. Zhang, W. Zhang, and A. Inoue, Corrosion behavior of Cu–Zr–Ti–Nb bulk glassy alloys, *Mater. Trans.* **44**(4), 749–753 (2003).

23. C. L. Qin, W. Zhang, K. Asami, N. Ohtsu, and A. Inoue, Glass formation, corrosion behavior and mechanical properties of bulk glassy Cu–Hf–Ti–Nb alloys, *Acta Mater.* **53**(14), 3903–3911 (2005).

24. L. Liu and B. Liu, Influence of the micro-addition of Mo on glass forming ability and corrosion resistance of Cu-based bulk metallic glasses, *Electrochim. Acta* **51**(18), 3724–3730 (2006).

25. M. K. Tam, S. J. Pang, and C. H. Shek, Effects of niobium on thermal stability and corrosion behavior of glassy Cu–Zr–Al–Nb alloys, *J. Phys. Chem. Solids* **67**(4), 762–766 (2006).

26. C. L. Qin, W. Zhang, H. Kimura, K. Asami, and A. Inoue, New Cu–Zr–Al–Nb bulk glassy alloys with high corrosion resistance, *Mater. Trans.* **45**(6), 1958–1961 (2004).

27. B. Liu, L. Liu, M. Sun, C. L. Qiu, and Q. Chen, Influence of Cr micro-addition on the glass forming ability and corrosion resistance of Cu-based bulk metallic glasses, *Acta Metall. Sin.* **41**(7), 738–742 (2005).

28. C. L. Qin, K. Asami, T. Zhang, W. Zhang, and A. Inoue, Effects of additional elements on the glass formation and corrosion behavior of bulk glassy Cu–Hf–Ti alloys, *Mater. Trans.* **44**(5), 1042–1045 (2003).

29. S. J. Pang, T. Zhang, K. Asami, and A. Inoue, Bulk glassy Fe–Cr–Mo–C–B alloys with high corrosion resistance, *Corros. Sci.* **44**(8), 1847–1856 (2002).

30. S. J. Pang, T. Zhang, K. Asami, and A. Inoue, Formation of bulk glassy $Fe_{75-x-y}Cr_xMo_yC_{15}B_{10}$ alloys and their corrosion behavior, *J. Mater. Res.* **17**(3), 701–704 (2002).

31. K. Asami, S. J. Pang, T. Zhang, and A. Inoue, Preparation and corrosion resistance of Fe–Cr–Mo–C–B–P bulk glassy alloys, *J. Electrochem. Soc.* **149**(8), B366–B369 (2002).

32. S. J. Pang, T. Zhang, K. Asami, and A. Inoue, Synthesis of Fe–Cr–Mo–C–B–P bulk metallic glasses with high corrosion resistance, *Acta Mater.* **50**(3), 489–497 (2002).

33. J. Jayaraj, Y. C. Kim, K. B. Kim, H. K. Seok, and E. Fleury, Corrosion studies on Fe-based amorphous alloys in simulated PEM fuel cell environment, *Sci. Tech. Adv. Mater.* **6**(3–4), 282–289 (2005).

34. H. Katagiri, S. Meguro, M. Yamasaki, H. Habazaki, T. Sato, A. Kawashima, K. Asami, and K. Hashimoto, An attempt at preparation of corrosion-resistant bulk amorphous Ni–Cr–Ta–Mo–P–B alloys, *Corros. Sci.* **43**(1), 183–191 (2001).

35. H. Habazaki, H. Ukai, K. Izumiya, and K. Hashimoto, Corrosion behaviour of amorphous Ni–Cr–Nb–P–B bulk alloys in 6M HCl solution, *Mater. Sci. Eng. A* **318**(1–2), 77–86 (2001).

36. H. Habazaki, T. Sato, A. Kawashima, K. Asami, and K. Hashimoto, Preparation of corrosion-resistant amorphous Ni–Cr–P–B bulk alloys containing molybdenum and tantalum, *Mater. Sci. Eng. A* **304–306**, 696–700 (2001).

37. C. L. Qin, W. Zhang, H. Nakata, H. Kimura, K. Asami, and A. Inoue, Effect of tantalum on corrosion resistance of Ni–Nb(–Ta)–Ti–Zr glassy alloys at high temperature, *Mater. Trans.* **46**(4), 858–862 (2005).

38. K. Shimamura, A. Kawashima, K. Asami, and K. Hashimoto, Corrosion behavior of amorphous nickel-valve metal-alloys in boiling concentrated nitric and hydrochloric acids, *Sci. Rep. Res. Tohoku A* **33**(1), 196–210 (1986).

39. A. Mitsuhashi, K. Asami, A. Kawashima, and K. Hashimoto, The corrosion behavior of amorphous nickel base alloys in a hot concentrated phosphoric acid, *Corros. Sci.* **27**(9), 957–970 (1987).

40. H.-J. Lee, E. Akiyama, H. Habazaki, A. Kawashima, K. Asami, and K. Hashimoto, The effect of phosphorus addition on the corrosion behavior of amorphous Ni–30Ta–P alloys in 12 M HCl, *Corros. Sci.* **37**(2), 321–330 (1995).

41. H.-J. Lee, E. Akiyama, H. Habazaki, A. Kawashima, K. Asami, and K. Hashimoto, The effect of phosphorus on the passivation behavior of Ni–10Ta–P alloys in 12 M HCl, *Corros. Sci.* **37**(8), 1313–1324 (1995).

42. H.-J. Lee, E. Akiyama, H. Habazaki, A. Kawashima, K. Asami, and K. Hashimoto, The roles of tantalum and phosphorus in the corrosion behavior of Ni–Ta–P alloys in 12 M HCl, *Corros. Sci.* **39**(2), 321–332 (1997).

43. A. Kawashima, H. Habazaki, and K. Hashimoto, Highly corrosion-resistant Ni-based bulk amorphous alloys, *Mater. Sci. Eng. A* **304–306**, 753–757 (2001).

44. S. Pang, T. Zhang, K. Asami, and A. Inoue, Bulk glassy Ni(Co–)Nb–Ti–Zr alloys with high corrosion resistance and high strength, *Mater. Sci. Eng. A* **375–377**, 368–371 (2004).

45. S.-J. Pang, C.-H. Shek, T. Zhang, K. Asami, and A. Inoue, Corrosion behavior of glassy $Ni_{55}Co_5Nb_{20}Ti_{10}Zr_{10}$ alloy in 1N HCl solution studied by potentiostatic polarization and XPS, *Corros. Sci.* **48**(3), 625–633 (2006).

46. K. Hashimoto, 2002 W. R. Whitney Award Lecture: In pursuit of new corrosion-resistant alloys, *Corrosion* **58**(9), 715–722 (2002).

47. M. Mehmood, B.-P. Zhang, E. Akiyama, H. Habazaki, A. Kawashima, K. Asami, and K. Hashimoto, Experimental evidence for the critical size of heterogeneity areas for pitting corrosion of Cr–Zr alloys in 6 M HCl, *Corros. Sci.* **40**(1), 1–17 (1998).

48. A. Gebert, U. Kuehn, S. Baunack, N. Mattern, and L. Schultz, Pitting corrosion of zirconium-based bulk glass-matrix composites, *Mater. Sci. Eng. A* **415**(1–2), 242–249 (2006).

49. K. Asami, H. Habazaki, A. Inoue, and K. Hashimoto, Recent development of highly corrosion resistant bulk glassy alloys, *Mater. Sci. Forum* **502**, 225–230 (2005).

50. C. L. Qiu, L. Liu, M. Sun, and S. M. Zhang, The effect of Nb addition on mechanical properties, corrosion behavior, and metal-ion release of ZrAlCuNi bulk metallic glasses in artificial body fluid, *J. Biomed. Mater. Res. A* **75**(4), 950–956 (2005).

51. K. Mondal, B. S. Murty, and U. K. Chatterjee, Electrochemical behaviour of amorphous and nanoquasicrystalline Zr–Pd and Zr–Pt alloys in different environments, *Corros. Sci.* **47**(11), 2619–2635 (2005).

52. L. Liu, C. L. Qiu, H. Zou, and K. C. Chan, The effect of the microalloying of Hf on the corrosion behavior of ZrCuNiAl bulk metallic glass, *J. Alloys Compd.* **399**(1–2), 144–148 (2005).

53. A. Dhawan, S. Roychowdhury, P. K. De, and S. K. Sharma, Potentiodynamic polarization studies on bulk amorphous alloys and $Zr_{46.75}Ti_{8.25}Cu_{7.5}Ni_{10}Be_{27.5}$ and $Zr_{65}Cu_{17.5}Ni_{10}Al_{7.5}$, *J. Non-Cryst. Solids* **351**(10–11), 951–955 (2005).

54. F.-Q. Zu, Z.-H. Chen, J.-J. Tao, L.-J. Liu, J. Yu, and Y. Xi, Corrosion resistance of Zr–Al–Ni–Cu (Nb) bulk amorphous alloys, *Trans. Nonferr. Metal. Soc.* **14**(5), 961–965 (2004).

55. M. L. Morrison, R. A. Buchanan, A. Peker, W. H. Peter, J. A. Horton, and P. K. Liaw, Cyclic-anodic-polarization studies of a $Zr_{41.2}Ti_{13.8}Ni_{10}Cu_{12.5}Be_{22.5}$, *Intermetallics* **12**(10–11), 1177–1181 (2004).

56. U. K. Mudali, S. Baunack, J. Eckert, L. Schultz, and A. Gebert, Pitting corrosion of bulk glass-forming zirconium-based alloys, *J. Alloys Compd.* **377**, 290–297 (2004).

57. D. Zander and U. Köster, Corrosion of amorphous and nanocrystalline Zr-based alloys, *Mater. Sci. Eng. A* **375–377**, 53–59 (2004).

58. F. X. Qin, H. F. Zhang, Y. F. Deng, B. Z. Ding, and Z. Q. Hu, Corrosion resistance of Zr-based bulk amorphous alloys containing Pd, *J. Alloys Compd.* **375**(1–2), 318–323 (2004).

59. F. X. Qin, H. F. Zhang, P. Chen, F. F. Chen, D. C. Qiao, and Z. Q. Hu, Corrosion behavior of bulk amorphous $Zr_{55}Al_{10}Cu_{30}Ni_{5-x}Pd_x$ alloys, *Mater. Lett.* **58**(7–8), 1246–1250 (2004).

60. S. J. Pang, T. Zhang, K. Asami, and A. Inoue, Formation, corrosion behavior, and mechanical properties of bulk glassy Zr–Al–Co–Nb alloys, *J. Mater. Res.* **18**(7), 1652–1658 (2003).

61. T. C. Chieh, J. Chu, C. T. Liu, and J. K. Wu, Corrosion of $Zr_{52.2}Cu_{17.9}Ni_{14.6}Al_{10}Ti_5$ bulk metallic glasses in aqueous solutions, *Mater. Lett.* **57**(20), 3022–3025 (2003).

62. S. Hiromoto, A. P. Tsai, M. Sumita, and T. Hanawa, Polarization behavior of bulk Zr-base amorphous alloy immersed in cell culture medium, *Mater. Trans.* **43**(12), 3112–3117 (2002).

63. V. R. Raju, U. Kühn, U. Wolff, F. Schneider, J. Eckert, R. Reiche, and A. Gebert, Corrosion behaviour of Zr-based bulk glass-forming alloys containing Nb or Ti, *Mater. Lett.* **57**(1), 173–177 (2002).

64. S. Hiromoto, K. Asami, A. P. Tsai, M. Sumita, and T. Hanawa, Surface composition and anodic polarization behavior of zirconium-based amorphous alloys in a phosphate-buffered saline solution, *J. Electrochem. Soc.* **149**(4), B117–B122 (2002).

65. U. Wolff, A. Gebert, J. Eckert, and L. Schultz, Effect of surface pretreatment on the electrochemical activity of a glass-forming Zr–Ti–Al–Cu–Ni alloy, *J. Alloys Compd.* **346**(1–2), 222–229 (2002).

66. S. Hiromoto and T. Hanawa, Re-passivation current of amorphous $Zr_{65}Al_{7.5}Ni_{10}Cu_{17.5}$ alloy in a Hanks' balanced solution, *Electrochim. Acta* **47**(9), 1343–1349 (2002).

67. A. Gebert, K. Buchholz, A. M. El-Aziz, and J. Eckert, Hot water corrosion behaviour of Zr–Cu–Al–Ni bulk metallic glass, *Mater. Sci. Eng. A* **316**(1–2), 60–65 (2001).

68. G. He, Z. Bian, and G. L. Chen, Corrosion behavior of a Zr-base bulk glassy alloy and its crystallized counterparts, *Mater. Trans.* **42**(6), 1109–1111 (2001).

69. S. J. Pang, T. Zhang, H. Kimura, K. Asami, and A. Inoue, Corrosion behavior of Zr–(Nb–)Al–Ni–Cu glassy alloys, *Mater. Trans.* **41**(11), 1490–1494 (2000).

70. S. Hiromoto, A. P. Tsai, M. Sumita, and T. Hanawa, Effects of surface finishing and dissolved oxygen on the polarization behavior of $Zr_{65}Al_{7.5}Ni_{10}Cu_{17.5}$ amorphous alloy in phosphate buffered solution, *Corros. Sci.* **42**(12), 2167–2185 (2000).

71. S. Hiromoto, A. P. Tsai, M. Sumita, and T. Hanawa, Effect of chloride ion on the anodic polarization behavior of the $Zr_{65}Al_{7.5}Ni_{10}Cu_{17.5}$ amorphous alloy in phosphate buffered solution, *Corros. Sci.* **42**(9), 1651–1660 (2000).

72. A. Gebert, K. Buchholz, A. Leonhard, K. Mummert, J. Eckert, and L. Schultz, Investigations on the electrochemical behaviour of Zr-based bulk metallic glasses, *Mater. Sci. Eng. A* **267**(2), 294–300 (1999).

73. V. Schroeder, C. J. Gilbert, and R. O. Ritchie, Comparison of the corrosion behavior of a bulk amorphous metal, $Zr_{41.2}Ti_{13.8}Cu_{12.5}Ni_{10}Be_{22}$, with its crystallized form, *Scripta Mater.* **38**(10), 1481–1485 (1998).

74. A. Gebert, K. Mummert, J. Eckert, L. Schultz, and A. Inoue, Electrochemical investigations on the bulk glass forming $Zr_{55}Cu_{30}Al_{10}Ni_5$ alloy, *Mater. Corros.* **48**(5), 293–297 (1997).

75. K. Asami, Y. Murakami, H. M. Kimura, and M. Kikuchi, Characterization of amorphous Zr–Cu alloy surfaces by electron-probe microanalysis and auger-electron spectroscopy, *Sci. Rep. Res. Tohoku A* **41**(1), 77–81 (1995).

76. G. K. Dey, R. T. Savalia, S. K. Sharma, and S. K. Kulkarni, Corrosion studies on amorphous and crystalline $Zr_{67}Ni_{33}$, *Corros. Sci.* **29**(7), 823–831 (1989).

77. M. L. Morrison, R. A. Buchanan, O. N. Senkov, D. B. Miracle, and P. K. Liaw, Electrochemical behavior of Ca-based bulk metallic glasses, *Metall. Mater. Trans. A* **37**(4), 1239–1245 (2006).

78. A. Gebert, R. V. Subba Rao, U. Wolff, S. Baunack, J. Eckert, and L. Schultz, Corrosion behaviour of the $Mg_{65}Y_{10}Cu_{15}Ag_{10}$ bulk metallic glass, *Mater. Sci. Eng. A* **375–377**, 280–284 (2004).

79. A. Gebert, U. Wolff, A. John, J. Eckert, and L. Schultz, Stability of the bulk glass-forming $Mg_{65}Y_{10}Cu_{25}$ alloy in aqueous electrolytes, *Mater. Sci. Eng. A* **299**(1–2), 125–135 (2001).

80. A. Gebert, U. Wolff, A. John, and J. Eckert, Corrosion behaviour of $Mg_{65}Y_{10}Cu_{25}$ metallic glass, *Scripta Mater.* **43**(3), 279–283 (2000).

81. M. L. Morrison, R. A. Buchanan, A. Peker, P. K. Liaw, and J. A. Horton, Electrochemical behavior of a Ti-based bulk metallic glass, *J. Non-Cryst. Solids* **353** (22–23), 2115–2124 (2007).

82. X. H. Lin and W. L. Johnson, Formation of Ti–Zr–Cu–Ni bulk metallic glasses, *J. Appl. Phys.* **78**(11), 6514–6519 (1995).

83. A. Inoue, W. Zhang, T. Zhang, and K. Kurosaka, High-strength Cu-based bulk glassy alloys in Cu–Zr–Ti and Cu–Hf–Ti ternary systems, *Acta Mater.* **49**(14), 2645–2652 (2001).

84. A. Inoue, W. Zhang, T. Zhang, and K. Kurosaka, Thermal and mechanical properties of Cu-based Cu–Zr–Ti bulk glassy alloys, *Mater. Trans.* **42**(6), 1149–1151 (2001).

85. A. Inoue and W. Zhang, Formation, thermal stability and mechanical properties of Cu–Zr–Al bulk glassy alloys, *Mater. Trans.* **43**(11), 2921–2925 (2002).

86. M. Naka, K. Hashimoto, and T. Masumoto, Corrosion resistivity of amorphous Fe alloys containing Cr, *J. Jpn. Inst. Metals* **38**(9), 835–841 (1974).

87. M. Naka, K. Hashimoto, and T. Masumoto, High corrosion resistance of Cr-bearing amorphous Fe alloys in neutral and acidic solutions containing chloride, *Corrosion* **32**(4), 146–152 (1976).

88. M. Naka, K. Hashimoto, and T. Masumoto, Effect of metalloidal elements on corrosion resistance of amorphous iron-chromium alloys, *J. Non-Cryst. Solids* **28**(3), 403–413 (1978).

89. M. Naka, K. Hashimoto, and T. Masumoto, Effect of addition of chromium and molybdenum on the corrosion behavior of amorphous Fe-20B, Co-20B and Ni-20B alloys, *J. Non-Cryst. Solids* **34**(2), 257–266 (1979).

90. M. Naka, K. Hashimoto, A. Inoue, and T. Masumoto, Corrosion-resistant amorphous Fe–C alloys containing chromium and/or molybdenum, *J. Non-Cryst. Solids* **31**(3), 347–354 (1979).

91. T. Masumoto and K. Hashimoto, Chemical properties of amorphous metals, *Annu. Rev. Mater. Sci.* **8**, 215–233 (1978).

92. A. Inoue, T. Zhang, and A. Takeuchi, Bulk amorphous alloys with high mechanical strength and good soft magnetic properties in Fe–TM–B (TM=IV–VIII group transition metal) system, *Appl. Phys. Lett.* **71**(4), 464–466 (1997).

93. T. D. Shen and R. B. Schwarz, Bulk ferromagnetic glasses prepared by flux melting and water quenching, *Appl. Phys. Lett.* **75**(1), 49–51 (1999).

94. A. Inoue, Y. Shinohara, and J. S. Gook, Thermal and magnetic properties of bulk Fe-based glassy alloys prepared by copper mold casting, *Mater. Trans.* **36**(12), 1427–1433 (1995).

95. A. Inoue, B. L. Shen, A. R. Yavari, and A. L. Greer, Mechanical properties of Fe-based bulk glassy alloys in Fe–B–Si–Nb and Fe–Ga–P–C–B–Si systems, *J. Mater. Res.* **18**(6), 1487–1492 (2003).

96. D. C. Qiao, P. K. Liaw, C. T. Liu, M. Morrison, R. A. Buchanan, and C. R. Brooks, Unpublished.

97. D. C. Qiao, P. K. Liaw, S. J. Poon, and R. A. Buchanan, Unpublished.

98. B.-P. Zhang, A. Kawashima, H. Habazaki, K. Asami, and K. Hashimoto, The effect of structural heterogeneity on the pitting corrosion behavior of melt-spun amorphous Ni–Zr alloys, *Corros. Sci.* **39**(10–11), 2005–2018 (1997).

99. B.-P. Zhang, H. Habazaki, A. Kawashima, K. Asami, and K. Hashimoto, The corrosion behavior of amorphous Ni–Cr–P alloys in concentrated hydrofluoric acid, *Corros. Sci.* **33**(10), 1519–1528 (1992).

100. A. Inoue, T. Zhang, N. Nishiyama, K. Ohba, and T. Masumoto, Preparation of 16 mm diameter rod of amorphous $Zr_{65}Al_{7.5}Ni_{10}Cu_{17.5}$ alloy, *Mater. Trans.* **34**(12), 1234–1237 (1993).

101. A. Peker and W. L. Johnson, A highly processable metallic-glass – $Zr_{41.2}Ti_{13.8}Cu_{12.5}Ni_{10.0}Be_{22.5}$, *Appl. Phys. Lett.* **63**(17), 2342–2344 (1993).

102. X. H. Lin and W. L. Johnson, Formation of Ti–Zr–Cu–Ni bulk metallic glasses, *J. Appl. Phys.* **78**(11), 6514–6519 (1995).

103. H. A. Bruck, A. J. Rosakis, and W. L. Johnson, The dynamic compressive behavior of beryllium bearing bulk metallic glasses, *J. Mater. Res.* **11**(2), 503–511 (1996).

104. Z. Bian, G. He, and G. L. Chen, Microstructure and mechanical properties of as-cast $Zr_{52.5}Cu_{17.9}Ni_{14.6}Al_{10}Ti_5$ bulky glass alloy, *Scripta Mater.* **43**(11), 1003–1008 (2000).

105. J. Basu and S. Ranganathan, Bulk metallic glasses: A new class of engineering materials, *Sadhana-Acad. Proc. Eng. Sci.* **28**, 783–798 (2003).

106. A. Inoue, Bulk amorphous and nanocrystalline alloys with high functional properties, *Mater. Sci. Eng. A* **304–306**, 1–10 (2001).

107. B. A. Green, R. A. Buchanan, Y. Yokoyama, and P. K. Liaw, Unpublished.

108. A. Inoue and T. Masumoto, Mg-based amorphous alloys, *Mater. Sci. Eng. A* **173**(1–2), 1–8 (1993).

109. M. Naka, K. Hashimoto, and T. Masumoto, Corrosion behavior of amorphous and crystalline $Cu_{50}Ti_{50}$ and $Cu_{50}Zr_{50}$ alloys, *J. Non-Cryst. Solids* **30**(1), 29–36 (1978).

110. J. E. Sweitzer, G. J. Shiflet, and J. R. Scully, Localized corrosion of $Al_{90}Fe_5Gd_5$ and $Al_{87}Ni_{8.7}Y_{4.3}$ alloys in the amorphous, nanocrystalline and crystalline states: Resistance to micrometer-scale pit formation, *Electrochim. Acta* **48**(9), 1223–1234 (2003).

111. K. Buchholz, A. Gebert, K. Mummert, J. Eckert, and L. Schultz, Corrosion behaviour of bulk amorphous and crystalline $Zr_{55}Al_{10}Cu_{30}Ni_5$ alloys at ambient and elevated temperature, *Mater. Sci. Forum* **343–346**, 213–218 (2000).

112. H. J. Lee, E. Akiyama, H. Habazaki, A. Kawashima, K. Asami, and K. Hashimoto, The corrosion behavior of amorphous and crystalline Ni–10Ta–20P alloys in 12 M HCl, *Corros. Sci.* **38**(8), 1269–1279 (1996).

113. H. Bala and S. Szymura, Acid corrosion of amorphous and crystalline Cu–Zr alloys, *Appl. Surf. Sci.* **35**(1), 41–51 (1988).

114. H. Bala and S. Szymura, Acid corrosion behavior of amorphous and crystalline $Ti_{75}Ni_{20}Si_5$ alloys, *Thin Solid Films* **149**(2), 171–176 (1987).

115. U. Köster, D. Zander, Triwikantoro, A. Rüdiger, and L. Jastrow, Environmental properties of Zr-based metallic glasses and nanocrystalline alloys, *Scripta Mater.* **44**(8–9), 1649–1654 (2001).

116. M. D. Archer, C. C. Corke, and B. H. Harji, The electrochemical properties of metallic glasses, *Electrochim. Acta* **32**(1), 13–26 (1987).

117. B. Elsener and A. Rossi, XPS investigation of passive films on amorphous Fe–Cr alloys, *Electrochim. Acta* **37**(12), 2269–2276 (1992).

118. U. K. Mudali, U. Kuhn, J. Eckert, I. Schultz, and A. Gebert, Corrosion behaviour of zirconium based bulk metallic glasses, *Trans. Indian Inst. Metals* **59**(1), 123–138 (2006).

119. S. Baunack, U. K. Mudali, A. Gebert, and J. Eckert, Characterization of oxide layers on amorphous Zr-based alloys by Auger electron spectroscopy with sputter depth profiling, *Appl. Surf. Sci.* **252**(1), 162–166 (2005).

120. Z. Szklarska-Smialowska, *Pitting Corrosion of Metals* (NACE, Houston, 1986).

SUBJECT INDEX